Interactive Intermediate Algebra

Robert H. Alwin
Robert D. Hackworth

H&H Interactive Mathematics Series

H&H Publishing Company, Inc.
Clearwater, Florida

H&H Publishing Company, Inc.

1231 Kapp Drive
Clearwater, FL 34625
(813) 442-7760
(800) 366-4079
FAX (813) 442-2195
e-mail: info@HHPublishing.com
web site: www.HHPublishing.com

INTERACTIVE
INTERMEDIATE
ALGEBRA

Robert H. Alwin
Robert D. Hackworth

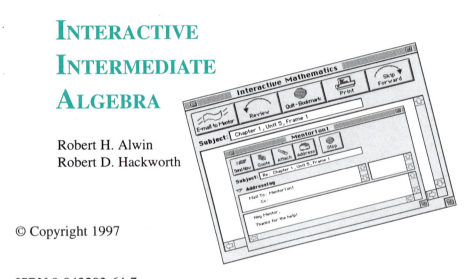

© Copyright 1997

ISBN 0-943202-64-7

Library of Congress Catalog Number 97-072390

All rights reserved. No part of this book may be reproduced or transmitted in any form or by any means, electronic or mechanical, including photocopy, recording, or any information storage or retrieval system, without written permission of the publisher.

This text in previous versions was originally published by Prentice-Hall, Inc.

Printing is the lowest number: 10 9 8 7 6 5 4 3 2 1

PREFACE

Interactive Intermediate Algebra is designed to aid the student in learning its material.

The text possesses strong teaching qualities that virtually assure the conscientious student of steady, confident progress. At the same time, the text's interactive processes assure the student of continuous opportunities to get help where and when it is needed. Solid mathematics, quality instruction, and personal assistance are the hallmark of this series of texts.

The text is designed in a format that constantly creates an interaction between the student, the mathematics to be learned, and an experienced teacher either in a classroom or in cyberspace. The majority of the book is a series of questions/answers that achieves a true dialogue with its student reader. Most of the time, the dialogue teaches the mathematics effectively, and students learn easily without benefit of outside assistance. Sometimes, however, the dialogue may be insufficient; it is in such cases where the student can find extra, personal assistance through conversation with a friend or teacher. These helpful conversations can occur in person, but also are available with a cyberspace instructor who is always as close as the nearest computer.

The quality of the mathematical content and instructional use of this text have been thoroughly documented in a wide range of situations and with great numbers of students. Specifically, there are no units or chapters where common difficulties can be expected. When a student has a problem understanding the material, the difficulties will be unique to that student. Something of importance was overlooked, misunderstood, etc. That is the reason for providing personal assistance in those instances and getting the student back into the flow of the book's dialogue. The personal assistance will both answer the questions of individual students and give guidance in the best ways to study mathematics.

Because of the design of this book, the student can expect other dramatic improvements in the study and learning of mathematics. One of those advantages is the student's ability to vary the pace of instruction to match their ability to learn. At times, a student needs to work slowly through the instruction and carefully ponder each new concept. At other times, the student grasps the material quickly and can accelerate their learning rapidly through the material. As a result, most students will move through this material more quickly than in a regular class situation.

Another great advantage of this book is the fact that students have an opportunity to learn to ask intelligent, specific questions about the mathematics in a context that encourages and does not threaten the student. Study skill deficiencies can be corrected within the context of these materials while maintaining traditional levels of achievement and rigor.

The content of this book is comparable to a second year of Algebra as it is normally taught in high school. Intermediate Algebra primarily involves an expansion and extension of the topics of an Elementary Algebra course. Most of the titles of the chapters of this book will be quite familiar to the student who has successfully completed an earlier Algebra course, but each topic will be presented at a higher level of development. For example, in an Elementary Algebra course the study of exponents is limited to integers without any direct relationship being shown between exponents and radical expressions. Here in Intermediate Algebra the study of exponents is extended to rational number exponents with much attention being devoted to exponents and their relationships to radical expressions.

The spiraling effect on learning is clearly evident in this book. After Chapter 1, each chapter in this book reviews some portion of the work of Elementary Algebra and then extends the concepts to new number systems or more sophisticated problems requiring more techniques for their solutions. In other words, the spiral effect begins each new topic at its level of prior knowledge and then lifts the knowledge of the topic to a new, higher level.

Chapter 1 reviews the vocabulary of Elementary Algebra while teaching the learner that words and their meanings will be of great importance to success in this course. The learner is instructed that the acquisition of vocabulary has been directly linked to expertise in the subject itself. That is, knowing the language of Intermediate Algebra is a necessary step to successfully learning the concepts and skills described by that language. A good vocabulary goes beyond the goals of effective communication between the learner and the instruction. A good vocabulary enables the learner to efficiently store new information, retrieve it more effectively, and, most importantly, assimilate it into his/her own knowledge base where it may become the source of abilities which are far beyond the scope of this course. Chapter 1 stresses the importance of the meanings of words in a mathematics course.

Each chapter ends with an Application Unit teaching word problem-solving processes. Unlike most text presentations of word problems, the student is taught a process for breaking down the information of a word problem and developing methods for extending these processes to the more difficult problems which will be encountered later.

The book contains many opportunities for the student to make wise decisions about the level of understanding achieved. Each chapter begins with an Objectives Test that illustrates, by example, all of the concepts and skills which are to be presented. Each problem is accompanied by a designation showing where it will be taught in the chapter. Each chapter is broken into units in a way that divides the content into topics of such length that it gives the student a way of organizing their own studying to make it most effective. Most units are designed to be completed in a single study session and, for best results, should be handled in that way. The unit ends with a Feedback which is an opportunity for the student to test the level of understanding achieved. It is strongly suggested that the Feedback be taken at the beginning of the next study session so that some time intervenes between the learning and assessment. The chapter ends with a Mastery Test that is similar to the Objectives Test. When a student can do all the problems correctly on a Mastery Test, it indicates a complete understanding of all the material in the chapter. Again, each problem is accompanied by a designation showing where it was taught in the chapter. Answers for all problems in the Tests and Feedbacks are included at the back of the book. Good learning can occur only when the student is aware of what is known and what still needs to be learned.

We are greatly indebted to thousands of teachers and hundreds of thousands of students who have successfully used our previous materials and given us the benefit of their own experiences and ideas. The strength of this new book is a result of that special kind of study of the work and challenges we face in teaching mathematics to students — some in desperate need for overcoming past experiences that were markedly unsuccessful and have interfered with motivations to try again.

Robert H. Alwin
Robert D. Hackworth

Directions to the Student

It is important to use this text correctly to achieve its full benefits. Its format encourages you to become a responsible participant in your own learning as you enter a dialogue with the book. The majority of the book involves giving you information, asking you a question about it, waiting for your answer, and then confirming or denying your understanding. You need to respond thoughtfully just as if you were in a conversation with a close friend about some matter of particular interest.

Each of these question-answer cycles is numbered and is called a frame. The first sentence of a frame will generally tell you what to look for or focus upon. Then you will be asked a question about the focus of the frame. Think about it and then write your answer. After you have written your answer, look in the right-hand column to see whether you completely understood the purpose of the frame. When you have completely understood, proceed to the next frame. When something is not completely understood, try to see what was not clearly communicated. If the misunderstanding continues, then seek personal assistance and resolve the matter before continuing.

Each frame contains something to be learned. Read the first sentence to find where the authors want you to focus your attention. Writing your answers to the question is important because it requires a commitment and that commitment is necessary if you are to learn. Checking your answer immediately is important. When the answer is correct, the learning is reinforced and, therefore, remembered. When the answer in incorrect, you are immediately alerted to some misunderstanding. This is not necessarily bad because many of your most important learning experiences will occur following some awareness of a problem. Treat these situations as good learning opportunities.

Each chapter begins with a set of problems that illustrate the objectives of the chapter. You are not expected to know how to do the problems on an Objectives Test. Each problem indicates the unit in which it will be learned. Use these problems to acquaint yourself with what is to be learned in the chapter, and do not allow yourself to be frightened by any new symbols or seemingly difficult situations. In the rare case that all the problems on an Objectives Test are truly easy, it is possible to consider jumping to the Chapter Mastery Test.

Each unit concludes with a Feedback exercise for you to test your understanding of the particular skills learned in the unit. In general, it is best not to take the Feedback immediately after finishing the frames of a unit. Give yourself some time before assessing your understanding; it will improve your progress.

Each chapter ends with a Mastery Test for you to assess your understanding of all the material in the chapter. Each problem is accompanied by a designation indicating where it is taught in the chapter. Use that information to restudy any portions of the chapter that the text indicates you have not mastered.

Throughout your study of this book, put your emphasis on understanding the material rather than memorizing without understanding. Regardless of your past experiences with learning mathematics, do not be tempted into memorizing. Those that memorize may remember for a week or a month, but then suffer a complete loss and must start anew. Those who understand, rather than memorize, have something that will sustain them throughout life. You'll find that every problem taught in this book has a purpose which is explained to you. Memorizing isn't necessary to your successful dialogue. Understanding will definitely improve your achievement level and your satisfaction.

Contents

Preface   III

Directions to the Student   V

Chapter 1 Objectives		XII
Chapter 1	**Building a Vocabulary for Intermediate Algebra**	**1**
	Unit 1: Basic Terminology	1
	Unit 2: The Set of Real Numbers	9
	Unit 3: Properties of the Real Numbers	14
	Unit 4: Simplifying Open Expressions	20
	Unit 5: Solving Linear Equations	28
	Unit 6: Applications	40
	Summary	43
	Mastery Test	45
Chapter 2 Objectives		46
Chapter 2	**Simplifying Radical Expressions**	**47**
	Unit 1: Counting Number Exponents	47
	Unit 2: Nth Roots	56
	Unit 3: Multiplying and Adding Nth Roots	68
	Unit 4: Simplifying Nth Roots	77
	Unit 5: The Imaginary Number, i	85
	Unit 6: Applications	94
	Summary	98
	Mastery Test	99

Chapter 3 Objectives — 100
Chapter 3   Simplifying Radical Fractions — 101
- Unit 1: Simplifying a Division of Square Roots — 101
- Unit 2: Reducing Radical Fractions with Indices Greater Than 2 — 114
- Unit 3: Simplifying Radical Fractions — 121
- Unit 4: Radical Fractions with Binomial Denominators — 128
- Unit 5: Applications — 133
- Summary — 137
- Mastery Test — 138

Chapter 4 Objectives — 140
Chapter 4   Rational Number Exponents — 141
- Unit 1: The Meaning of Rational Number Exponents — 141
- Unit 2: Multiplying Like Bases with Rational Number Exponents — 147
- Unit 3: Division of Like Bases with Rational Number Exponents — 152
- Unit 4: Raising a Power to a Power with Rational Number Exponents — 156
- Unit 5: Applications — 160
- Summary — 164
- Mastery Test — 165

Chapter 5 Objectives — 166
Chapter 5   Polynomials — 167
- Unit 1: Addition and Subtraction of Polynomials — 167
- Unit 2: Multiplication of Polynomials — 176
- Unit 3: Division of Polynomials — 186
- Unit 4: Applications — 193
- Summary — 200
- Mastery Test — 201

Chapter 6 Objectives		202
Chapter 6	**Factoring Polynomials**	**203**
	Unit 1: Factoring by the Common Factor Method	203
	Unit 2: Factoring Trinomials of the Form $x^2 + bx + c$	210
	Unit 3: Factoring Quadratics of the Form $ax^2 + bx + c$	218
	Unit 4: Factoring Sums/Differences of Two Cubes	228
	Unit 5: Reviewing and Extending Factoring Skills	233
	Unit 6: Factoring Polynomials with More Than Two Factors	242
	Unit 7: Applications	245
	Summary	251
	Mastery Test	252
Chapter 7 Objectives		254
Chapter 7	**Polynomial Fractions**	**255**
	Unit 1: Simplifying Fractions	255
	Unit 2: Multiplying/Dividing Polynomial Fractions	260
	Unit 3: Writing Polynomial Fractions with a Common Denominator	268
	Unit 4: Adding/Subtracting Polynomial Fractions	274
	Unit 5: Applications	281
	Summary	286
	Mastery Test	287

Chapter 8 Objectives — 288
Chapter 8 Equation Solving — 289
- Unit 1: Equivalent Linear Equations — 289
- Unit 2: Generating Linear Equations — 296
- Unit 3: Polynomial Equations with Rational Solutions — 304
- Unit 4: Completing the Square — 316
- Unit 5: Polynomial Equations with Complex Number Solutions — 324
- Unit 6: Solving Equations with Two Variables — 332
- Unit 7: Applications — 341
- Summary — 348
- Mastery Test — 349

Chapter 9 Objectives — 350
Chapter 9 Inequalities and Absolute Values — 351
- Unit 1: Graphing Number Line Inequalities — 351
- Unit 2: Equivalent Inequalities — 361
- Unit 3: Inequalities with Two Variables — 368
- Unit 4: Solving Absolute Value Equations — 378
- Unit 5: Interpreting Absolute Value as Distance — 386
- Unit 6: Absolute Value Inequalities — 392
- Unit 7: Applications — 398
- Summary — 408
- Mastery Test — 409

Chapter 10 Objectives — 410
Chapter 10 Linear Functions — 411
- Unit 1: Relations and Functions — 411
- Unit 2: Graphing Linear Functions — 421
- Unit 3: The Slope of a Linear Function — 431
- Unit 4: y-Intercepts of Linear Functions — 443
- Unit 5: Writing the Equation of a Line — 451
- Unit 6: Applications — 461
- Summary — 468
- Mastery Test — 469

CHAPTER 11 OBJECTIVES	470
CHAPTER 11 QUADRATIC FUNCTIONS AND RELATIONS	**473**
UNIT 1: INTRODUCTION TO QUADRATIC FUNCTIONS	473
UNIT 2: GRAPHS OF QUADRATIC FUNCTIONS OF THE FORM $f(x) = ax^2 + bx + c$	478
UNIT 3: GRAPHING QUADRATIC FUNCTIONS BY TABLES OF SOLUTIONS	482
UNIT 4: FUNCTIONS OF THE FORM $f(x) = (x - h)^2 + k$	493
UNIT 5: GRAPHING CIRCLES	502
UNIT 6: APPLICATIONS	511
SUMMARY	519
MASTERY TEST	520
CHAPTER 12 OBJECTIVES	522
CHAPTER 12 SOLVING SYSTEMS OF EQUATIONS	**523**
UNIT 1: SOLVING PAIRS OF LINEAR EQUATIONS	523
UNIT 2: SYSTEMS OF 2 LINEAR EQUATIONS	527
UNIT 3: THE COMMON SOLUTION OF A LINEAR AND A QUADRATIC EQUATION	538
UNIT 4: SYSTEMS OF 3 LINEAR EQUATIONS	551
UNIT 5: APPLICATIONS	560
SUMMARY	575
MASTERY TEST	576
SQUARE ROOT TABLE	**579**
ANSWERS FOR ALL TESTS AND FEEDBACK EXERCISES	**580**
INDEX	**621**

Chapter 1 Objectives

The following problems illustrate the objectives of this chapter. At this time you are not expected to know how to do these problems. However, if all these problems are thoroughly understood, proceed directly to the Chapter Mastery Test. The number in parentheses which follows each problem indicates the unit in which it can be learned.

1. The next consecutive counting number to 412 is _____. (1)

2. Show the set of odd numbers between 47 and 52. (1)

3. Write the set of composite numbers between 19 and 24. (1)

4. What is the HCF of 42 and 28? (1)

5. What is the LCM of 42 and 28? (1)

6. Are $9(5x)$ and $(9 \cdot 5)x$ equivalent expressions? (1)

7. Evaluate. $9 + 4 \cdot 2$ (1)

8. Which of the following is a conditional equation?
   a. $7x + 5 = 31$
   b. $9 + 13 = 22$ (1)

9. Which of the following is an identity?
   a. $x + y = y + x$
   b. $2 \cdot 11 = 22$ (1)

10. $\{\ldots, -3, -2, -1, 0, 1, 2, 3, \ldots\}$ is the set of _____. (2)

11. Is every point of the real number line represented by a rational number? (2)

12. What is the opposite of $\frac{-1}{2}$? (2)

13. What is the reciprocal of $\frac{5}{12}$? (2)

14. Do $5 - 13$ and $-13 + 5$ have the same evaluation? (3)

15. Because addition is an associative operation, $7x + (5y + 4)$ is equivalent to _____. (3)

16. Because multiplication is a commutative operation, $(5x) \cdot 4$ is equivalent to _____. (3)

17. Simplify. $3(4x - 3) - 5x$ (4)

18. Simplify. $\frac{1}{4}x - 2 - \frac{1}{6}x + \frac{3}{4}$ (4)

19. Solve. $5x - 6 = 3x + 1$ (5)

20. Solve. $\frac{5}{6}x + \frac{1}{3} = \frac{2}{3}x - \frac{5}{12}$ (5)

Chapter 1
Building a Vocabulary for Intermediate Algebra

Unit 1: Basic Terminology

Many students overlook the value of words in a mathematics course, but vocabulary is an essential part of mathematics achievement. Clearly, it is difficult to respond to questions involving words that are not understood. Also, any explanation that involves words outside the learner's vocabulary will not be well understood.

The value of a good vocabulary, however, goes beyond the obvious. A good vocabulary enables the learner to efficiently store new information, retrieve it more effectively, and, most importantly, assimilate it into his/her own knowledge base. Students who learn the vocabulary remember more easily, understand more quickly, and apply their learning more effectively. Learn the vocabulary of Algebra and take a major step toward successfully learning Intermediate Algebra.

2 CHAPTER 1

Each unit of this text begins with a listing of vocabulary words like the one shown below. Each listed term will be carefully described in the unit and will be crucial to an understanding of the unit.

Counting numbers	Set of counting numbers
Consecutive	Member
Element	Even numbers
Odd numbers	Subset
Between	Empty set
Factor	Prime number
Composite number	Common factor
Highest common factor (HCF)	Multiples
Least common multiple (LCM)	Numerical expression
Evaluate	Parentheses
Square brackets	Order of operations
Open expression	Variable
Coefficient	Equivalent open expression
Statement	Equation
Conditional equation	Identity

1
The first numbers that a child learns are: 1, 2, 3, etc. These numbers are **counting numbers**. The smallest counting number is 1. Is there a largest counting number?　　　　　No

2
　　{1, 2, 3, . . . } is the **set of counting numbers**. The three dots (. . .) indicate that the numbers go on forever. 4 is the next **consecutive** counting number to 3. Is 4 a counting number?　　　　　Yes

3
Any number used to count the grains of sand on a beach is a **member** or **element** of the set {1, 2, 3, . . . }. Is 913 an element of {1, 2, 3, . . . }?　　　　　Yes

4
　　{2, 4, 6, . . .} is the set of **even** numbers. Is 462 an even number?　　　　　Yes, it can be evenly divided by 2.

5

{1, 3, 5, . . .} is the set of **odd** numbers. Is 917 an odd number?

Yes

6

Because every even number is also a counting number, {2, 4, 6, . . . } is a **subset** of {1, 2, 3, . . . }. This is indicated by: {2, 4, 6, . . .} ⊂ {1, 2, 3, . . . }. Use the subset symbol to show that every odd number is a counting number.

{1, 3, 5, . . . } ⊂ {1, 2, 3, . . . }

7

42, 43, and 44 are the counting numbers **between** 41 and 45. What counting numbers are between 17 and 22?

18, 19, 20, 21

8

The **empty set**, { }, is the set which has no objects or numbers in it. Is 0 an element of the empty set, { }?

No, it has no elements

9

The set of numbers between 10 and 20 that are greater than 25 is the empty set because there are no numbers that meet the specifications. What is the set of odd numbers between 20 and 30 that are less than 16?

{ }

10

10 is a counting number **factor** of 30 because 10 • 3 = 30. Is 5 also a counting number factor of 30? [Note: The symbol (•) indicates multiplication.]

Yes, 5 • 6 = 30

11

The set of counting number factors of 35 is {1, 5, 7, 35}. What is the set of counting number factors of 15?

{1, 3, 5, 15}

12

17 is a **prime number** because it has exactly two counting number factors. Which number shown below is a prime number?
    (a) 39   (b) 1   (c) 23

23

13

42 is a **composite number** because it has more than two counting number factors. Which number shown below is a composite number?

    (a) 79   (b) 1   (c) 51   51

14

The number 1 is neither prime nor composite because it has exactly one factor — itself. Is there any other counting number that is neither prime nor composite?   No

15

2 and 6 are **common factors** of 12 and 18 because each is a factor of both 12 and 18. Find another common factor of 12 and 18.   1 or 3

16

6 is the **highest common factor (HCF)** of 12 and 18. What is the HCF of 15 and 25?   5

17

The highest common factor (HCF) of 24 and 36 is 12. Find the HCF of 8 and 14.   2

18

What is the HCF of 13 and 9?   1

19

When the HCF of two counting numbers is not obvious,

1. Find the factors of the smaller number, and
2. Check each as a factor of the larger number.

Find the HCF of 42 and 66.   6

20

Find the HCF of 51 and 39.   3

21

Find the HCF of 85 and 34.   17

A Vocabulary for Algebra

22
    10, 30, and 55 are **multiples** of 5. Is 80 also a multiple of 5?   Yes

23
    24, 36, 48, etc., are multiples of both 3 and 4. Is every multiple of 12 a multiple of both 3 and 4?   Yes

24
    12 is the **least common multiple (LCM)** of 3 and 4. Find the LCM of 8 and 6 by finding the smallest counting number that is a multiple of both 8 and 6.   24

25
The LCM of 10 and 14 is 70. Find the LCM of 6 and 14.   42

26
What is the LCM of 23 and 2?   46

27
When the LCM of two counting numbers is not obvious,

1. Find their HCF,
2. Multiply the numbers and divide by the HCF.

Find the LCM of 12 and 30.   HCF is 6. LCM is $\frac{12 \cdot 30}{6} = 60$

28
Find the LCM of 24 and 40. $\frac{24 \cdot 40}{8} = 120$

29
Find the LCM of 35 and 60.   420

30
The problem $5 + 7 \cdot 3$ is a **numerical expression**. Is $6 + 4 \cdot 7$ a numerical expression?   Yes

31
To **evaluate** a numerical expression, find its answer. Evaluate. $8 + 10 + 3$ 21

32

The **parentheses** in 8 • (4 + 7) indicate that 4 + 7 is to be evaluated first. Evaluate 8 • (4 + 7) which is usually written as 8(4 + 7).

88

33

In evaluating 5 + (2 + [6 • 2]), grouping symbols such as parentheses and **square brackets** indicate the order in which to perform the operations.

$$5 + (2 + [6 \cdot 2])$$
$$5 + (2 + 12)$$
$$5 + 14$$
$$19$$

Evaluate. 10 + (6 + [3 • 5])

31

34

In a numerical expression with no grouping symbols, all multiplication is performed before any addition.
Evaluate. 5 • 7 + 2

37

35

The **Order of Operations** is: In evaluating a numerical expression with no grouping symbols, all multiplication is performed before addition.
In evaluating 8 + 2 • 3, the first step is to _____.

multiply 2 and 3

36

A numerical expression like 4 + 5 • 9 can be evaluated, but an **open expression** like x + 3 cannot be evaluated until a number replaces x. What kind of an expression is 4(x + 8)?

open

37

The open expression 7 • x is normally written without the dot as 7x. Similarly, 4 • y can be written without the dot as _____.

4y

38
The open expression 9 + 4x becomes 9 + 4 • 3 when x is replaced by 3. The letter x is a **variable** and may be replaced by any element in its replacement set. What is the variable in 7k + 4?

k

39
The 3 in 3x is a **coefficient**. 3x means to multiply 3 and x. In 7x the 7 is a _____ and 7x means to _____ 7 and x.

coefficient, multiply

40
The coefficient of x in 5 + x is 1 and the expression can be written as 5 + 1x. What is the coefficient of y in y + 4 • 7?

1

41
Two open expressions are equivalent if they have the same evaluation for any replacement of the variable. x + 5 and 5 + x are **equivalent open expressions**. Are 7x and x • 7 equivalent?

Yes

42
Are 5 + x and 5 + 1x equivalent?

Yes

43
 7 + 2 = 9 is a **statement** with "equals" as its verb. A statement might be true or false, but it must be one or the other. Is 6 + 4 = 2 a true statement?

No

44
 x + 4 = 7 is an **equation**. Until x is replaced, x + 4 = 7 is neither true nor false. Is 4x = 12 an equation or a statement?

equation

45
An equation becomes a statement when its variable is replaced. If x is replaced by 5, the equation x + 4 = 9 becomes a _____ statement.

true

46

If x is replaced by 8, the equation x + 4 = 9 becomes a _____ statement.

false

47

Some equations are **conditional equations** because they become true statements for some variable replacements and false statements for other replacements. Is x + 4 = 9 a conditional equation?

Yes

48

Some equations are **identities** because they become true statements for all acceptable replacements of the variable. Is x + 3 = 3 + x an identity for any counting number replacement of x?

Yes

FEEDBACK UNIT 1

This quiz reviews the preceding unit. Answers are at the back of the book.

1. The next consecutive counting number to 813 is _____.

2. Find the set of even numbers between 33 and 38.

3. Is {5, 10, 15, . . .} a subset of the set of counting numbers?

4. What is the set of counting number factors of 30.

5. What is the set of prime numbers between 22 and 30.

6. Is 1 a common factor for 54 and 97?

7. Is 1 a common multiple for 31 and 45?

8. What is the HCF of 17 and 10?

9. What is the HCF of 18 and 32?

10. What is the LCM of 15 and 20?

11. What is the LCM of 6 and 18?

12. Evaluate. $8 + 4 \cdot 7$

13. Evaluate. $4(9 + 2) + 6$

14. Are 9 + x and x + 9 equivalent open expressions?

15. 8 + 9 = 72 is a(n) _____ (statement, equation).

16. x + 7 = 12 is a(n) _____ (conditional equation, identity).

Unit 2: The Set of Real Numbers

The following mathematical terms are crucial to an understanding of this unit.

Set of real numbers
Set of integers
Next consecutive integer
Terminating decimal
Dense
Irrational number
Reciprocal

Real number line
Negative integers
Set of rational numbers
Repeating decimal
Infinite
Opposite

1
The **set of real numbers**, *R*, is of great importance to the study of calculus and is the major replacement set of this text. It contains four major subsets. One of those subsets is the set of counting numbers. Is every counting number also a real number?

Yes

2
The figure below shows the **real number line**. To visualize the real numbers correctly, think of the number line as a set of separate, distinct points. What real number is associated with the point marked D?

4

3
Each real number is associated with a unique point on the real number line. Does each point on the real number line represent a real number?

Yes

4

On the real number line shown below, points for the 12 smallest counting numbers have been designated. Between two consecutive counting numbers, are there other points associated with counting numbers?

No

5

Between two consecutive counting numbers, are there other points associated with real numbers?

Yes

6

The **set of integers**, { . . . , -3, -2, -1, 0, 1, 2, 3, . . . }, is another important subset of the set of real numbers, **R**. The set contains all the counting numbers, 1, 2, 3, . . . , and these are positive integers. 8 is both a counting number and a _____ integer.

positive

7

The set of integers contains the counting numbers, 0, and the **negative integers** -1, -2, -3, ... The integer -9 is not a counting number. It is a _____ integer.

negative

8

The integers -6 and -5 are consecutive integers. The **next consecutive integer** to -19 is -18 (-19 + 1). The next consecutive integer to -4 is ___.

-3 (-4 + 1)

9

On the real number line shown below, points for the integers -8 to 12 have been labeled. Between two consecutive integers, are there other points associated with integers?

No

10

Between two consecutive integers, are there other points associated with real numbers?

Yes

11

The **set of rational numbers** contains all the ratios $\frac{x}{y}$ where x and y are integers and y is not 0. Each of the ratios $\frac{9}{8}, \frac{-2}{11}, \frac{3}{4}$ is a rational number. Is $\frac{-9}{5}$ a rational number?

Yes

12

The set of rational numbers can be shown as

$$\left\{ \frac{x}{y} \mid x \text{ and } y \text{ integers}, y \neq 0 \right\}$$

Is $\frac{7}{0}$ a rational number?

No, denominator cannot be 0.

13

Any rational number can be written as a decimal numeral. The ratio $\frac{1}{2}$ can be written as .5 which is a **terminating decimal**. Write the terminating decimal for $\frac{1}{4}$.

.25

14

The ratio $\frac{1}{3}$ can be written as $.33\overline{3}$ (the bar over 3 indicates the digit 3 repeats forever) which is a **repeating decimal**. Write the repeating decimal for $\frac{2}{3}$.

$.6\overline{66}$

15

Each rational number can be written as a terminating or repeating decimal. Does the repeating decimal $.454\overline{545}$ represent a rational number?

Yes, it is $\frac{5}{11}$

16

The number line shown below has some points labeled with rational numbers. Between 0 and 1 are there other rational numbers that have not been labeled?

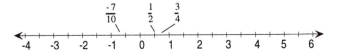

Yes

17

The set of rational numbers is **dense** which means that between any two rational numbers there are other rational numbers. Is the set of counting numbers dense?

No, between consecutive counting numbers there are no counting numbers.

18

The set of rational numbers is dense. Between any two rational numbers there are other rational numbers. Is the set of integers dense?

No

19

The density of the rational numbers means there are an **infinite** (no limit) number of rational numbers between 0 and 1. Are there an infinite number of rational numbers between .1 and .11?

Yes

20

Even though the set of rational numbers is dense, they do not "fill up" the real number line. In other words, there are _____ (no, some) points on the real number line that are not rational numbers.

some (actually an infinite number)

21

On the real number line, there are an infinite number of points that are not associated with rational numbers. The most common of these other numbers that is encountered in arithmetic is Pi, π. The number π is used in formulas involving circles. Is π a rational number?

No

22

π is an **irrational number**. Although it is often approximated as $\frac{22}{7}$ or 3.14, the irrational number π actually has a decimal numeral that neither terminates nor repeats. Irrational numbers _____ (are, are not) real numbers.

are

23

Irrational numbers are associated with decimal numerals which neither terminate nor repeat. These real numbers that are not rational (irrational) have non-terminating, non-repeating decimals. Since $\sqrt{2}$ is an irrational number, its decimal numeral is _____.

non-terminating, non-repeating

24

Every real number is either a rational number or an irrational number. If z is a real number that is not rational, then it is a(n) _____ number.

irrational

25

The set of real numbers contains every rational number and every irrational number. If w is an irrational number, must w also be a real number?

Yes

26

The real number line can be thought of as a set of points with no empty spaces between points. Does the set of real numbers "fill up" the real number line?

Yes

27

In the set of real numbers, every element has an **opposite**. The opposite of 7 is -7. The opposite of $\frac{5}{7}$ is _____.

$\frac{-5}{7}$

28

In the set of real numbers, every element except 0 has a **reciprocal**. The reciprocal of 7 is $\frac{1}{7}$. The reciprocal of $\frac{5}{7}$ is _____.

$\frac{7}{5}$

Feedback Unit 2

This quiz reviews the preceding unit. Answers are at the back of the book.

1. Is -6 a counting number?

2. What is the next consecutive integer to -8?

3. Is the set of integers dense?

4. If a real number has a repeating decimal, then it is a(n) _____ number.

5. If a real number has a non-terminating, non-repeating decimal, then it is a(n) _____ number.

6. If a real number has a terminating decimal, then it is a(n) _____ number.

7. Is every point on the real number line associated with a rational number?

8. Is every point on the real number line associated with a real number?

9. What is the opposite of $\frac{2}{3}$?

10. What is the reciprocal of $\frac{7}{9}$?

Unit 3: Properties of the Real Numbers

The following mathematical terms are crucial to an understanding of this unit.

Binary operation
Division sign
Associative Law of Addition
Inverse Law of Addition
Associative Law of Multiplication
Inverse Law of Multiplication
Distributive Law of Addition
    Over Multiplication

Minus sign
Commutative Law of Addition
Identity Element for Addition
Commutative Law of Multiplication
Identity Element for Multiplication
Completeness Property

1

Addition is a **binary operation** for the set of real numbers because:
    a.   Any two real numbers can be added, and
    b.   There is exactly one result (answer) for each pair.
When π and -37 are added, is the result a real number?   Yes

2

Multiplication is another binary operation for the set of counting numbers because:
    a.   Any two real numbers can be multiplied, and
    b.   There is exactly one result (answer) for each pair.
When -47 and $\frac{5}{7}$ are multiplied is the result a real number?   Yes

3

Subtraction is another binary operation for the real numbers. However, subtraction can be completely defined as addition, and that process is valuable because of some special properties of addition.
For example, 7 – 3 is the same as 7 + (-3).
Is 8 – 4 the same as 8 + (-4)?   Yes

4

Whenever a **minus sign** is placed between two real numbers, it means to add the opposite of the second number to the first. 9 – 11 means 9 plus -11 or 9 + (-11).
4 – 7 means _____.   4 + (-7)

5

Using variables x and y for real numbers, subtraction can be defined by the following identity: $x - y = x + (-y)$
    -4 – 6 = -4 + (-6)   8 – 9 = _____   8 + (-9)

6

    -11 – (-5) = -11 + 5   13 – (-1) = _____   13 + 1

7

Division would be a binary operation for the real numbers except for the fact that 0 cannot be a divisor. Which of the following problems has no real number answer?

$7 \div 0$ $\qquad$ $0 \div 7$

$7 \div 0$ is undefined.
$0 \div 7 = 0$

8

Division for all real numbers, except 0 as a divisor, can be completely defined in terms of multiplication. That process is valuable because of some special properties of multiplication. For example, $7 \div 3$ is the same as $7 \cdot \frac{1}{3}$. Is $8 \div 4$ the same as $8 \cdot \frac{1}{4}$?

Yes

9

Whenever a **division sign** is placed between two real numbers, it means to multiply the reciprocal of the second number times the first. $9 \div 11$ means $9 \cdot \frac{1}{11}$.

$4 \div 7$ means _____.

$4 \cdot \frac{1}{7}$

10

Using variables x and y for real numbers, division can be defined by the following identity:

$$\text{when } y \neq 0, \; x \div y = x \cdot \frac{1}{y}$$

$12 \div -6 = 12 \cdot \frac{-1}{6}$ $\qquad$ $8 \div -9 =$ _____

$8 \cdot \frac{-1}{9}$

11

$\frac{7}{2} \div \frac{3}{8} = \frac{7}{2} \cdot \frac{8}{3}$ $\qquad$ $\frac{8}{9} \div \frac{3}{5} =$ _____

$\frac{8}{9} \cdot \frac{5}{3}$

12

There are 4 important properties of addition in the real number system. One property assures that the order of two addends has no effect on the evaluation of an addition. For example, 6 + 15 has the same evaluation as 15 + 6. Does -74 + π have the same evaluation as π + (-74)?

Yes

13

The **Commutative Law of Addition** states that x + y = y + x is an identity for all real number replacements of x and y. Is k + r equivalent to r + k for all real numbers?

Yes

14

Addition is an associative operation in the real number system. The grouping of three addends has no effect on the evaluation of an addition. For example, $(2 + 7) + \frac{1}{2}$ has the same evaluation as $2 + \left(7 + \frac{1}{2}\right)$. Does -13 + (π + 5) have the same evaluation as (-13 + π) + 5?

Yes

15

The **Associative Law of Addition** states that x + (y + z) = (x + y) + z is an identity for all real number replacements of x, y, and z. Is (w + k) + m equivalent to w + (k + m) for all real numbers?

Yes

16

Zero is the **Identity Element for Addition** in the real number system. This means that x + 0 = x is an identity for all real number replacements of x. Is p + 0 equivalent to p for all real numbers?

Yes

17

Every real number has an opposite in the set of real numbers. This property is the **Inverse Law of Addition**. It states: For each real number x there exists a real number -x such that x + (-x) = 0. According to this law, what is the sum of two opposites?

0

18

There are 4 important properties of multiplication in the real number system. One property assures that the order of two factors has no effect on the evaluation of a multiplication. For example, -9 • 4 has the same evaluation as 4 • -9. Does $\sqrt{2} \cdot \sqrt{3}$ have the same evaluation as $\sqrt{3} \cdot \sqrt{2}$?   Yes

19

The **Commutative Law of Multiplication** states that xy = yx is an identity for all real number replacements of x and y. Is kr equivalent to rk for all real numbers?   Yes

20

Multiplication is an associative operation in the real number system. The grouping of three factors has no effect on the evaluation of a multiplication. For example, $(5 \cdot -3) \cdot \frac{2}{3}$ has the same evaluation as $5 \cdot \left(-3 \cdot \frac{2}{3}\right)$. Does -6 (4π) have the same evaluation as (-6 • 4)π?   Yes

21

The **Associative Law of Multiplication** states that x(yz) = (xy)z is an identity for all real number replacements of x, y, and z. Is (wk)m equivalent to w(km) for all real numbers?   Yes

22

One is the **Identity Element for Multiplication** in the real number system. This means that 1x = x is an identity for all real number replacements of x. Is 1p equivalent to p for all real numbers?   Yes

23

Every real number except 0 has a reciprocal in the set of real numbers. This property is the **Inverse Law of Multiplication**. It states: For each real number x, x ≠ 0, there exists a real number $\frac{1}{x}$ such that x • $\frac{1}{x}$ = 1. According to this law, what is the product of two reciprocals?   1

24

There is one important property involving both addition and multiplication. This property assures that adding first with the numerical expression 5(8 + 9) gives the same result as multiplying first with the numerical expression 5 • 8 + 5 • 9. Do 5(8 + 9) and 5 • 8 + 5 • 9 have the same evaluation?   Yes

25

The **Distributive Law of Addition Over Multiplication** states that x(y + z) = xy + xz is an identity for all real number replacements of x, y, and z. Is k(r + m) equivalent to kr + km for all real numbers?   Yes

26

The content of the **Completeness Property** was presented in Unit 2. If a line is envisioned as a continuous set of points without any gaps or breaks between points, the Completeness Property assures that each of those points represents a unique real number and each real number is represented by exactly one point. Do the real numbers "fill up" the number line?   Yes

Feedback Unit 3

This quiz reviews the preceding unit. Answers are at the back of the book.

1. The evaluation of 463 • 57 will be the same as the evaluation of 57 • 463 because multiplication is a _____ operation.

2. The order of numbers in a subtraction problem _____ (can, cannot) be reversed without changing the evaluation.

3. Is division a commutative operation?

4. Does the grouping of three numbers in an addition change the evaluation?

5. The evaluation of (43 • 51) • 904 will be the same as the evaluation of 43 • (51 • 904) because multiplication is a(n) _____ operation.

6. Do the numerical expressions 5(7 + 4) and 5 • 7 + 5 • 4 have the same evaluation?

7. What is the identity element for addition?

8. Is there a real number that has no reciprocal?

9. If a line is thought of as a continuous set of points, does every point represent a rational number?

10. What is the identity element for multiplication?

Unit 4: Simplifying Open Expressions

The following mathematical terms are crucial to an understanding of this unit.

- Simplify an open expression
- Like terms
- Terms
- Remove the parentheses

1

To **simplify an open expression** means to write an equivalent open expression that involves less additions and/or multiplications. 9 + z + 7 is simplified as:

$$9 + z + 7$$
$$z + (9 + 7)$$
$$z + 16$$

Is 9 + z + 7 equivalent to z + 16?　　　　　　　　　　　　　　　Yes

2

The open expression, 6 + y + 2, has three addends or **terms**: 6, y, and 2. How many terms are in its simplification y + 8?　　　　　　　　　　　　　　　2, y and 8

3

To simplify x – 8 + 6, add -8 and 6.

$$x - 8 + 6$$
$$x + (-8 + 6)$$
$$x + (-2)$$
$$x - 2$$

Simplify. x – 8 – 7　　　　　　　　　　　　　　　x – 15

4

Simplify. -7 + x + 2　　　　　　　　　　　　　　　x – 5

5

To simplify $\left[\frac{3}{4}x + \frac{5}{6}\right] + \frac{-2}{3}$, the following steps are used:

$$\left[\frac{3}{4}x + \frac{5}{6}\right] + \frac{-2}{3}$$

$$\frac{3}{4}x + \left[\frac{5}{6} + \frac{-2}{3}\right]$$

$$\frac{3}{4}x + \frac{1}{6}$$

Simplify. $\left(\frac{7}{8}x + \frac{2}{5}\right) - \frac{1}{4}$　　　　　　　　　　　$\frac{7}{8}x + \frac{3}{20}$

6

Complete the simplification:

$$\left(\tfrac{2}{3}x - \tfrac{1}{3}\right) - \tfrac{2}{7} = \tfrac{2}{3}x + \left(\tfrac{-1}{3} - \tfrac{2}{7}\right) = \underline{\qquad}$$

$\tfrac{2}{3}x - \tfrac{13}{21}$

7

Simplify. $\left(\tfrac{3}{5}x - \tfrac{5}{9}\right) + \tfrac{1}{3}$

$\tfrac{3}{5}x - \tfrac{2}{9}$

8

Simplify. $\left(\tfrac{3}{8}x + \tfrac{2}{3}\right) + \tfrac{3}{4}$

$\tfrac{3}{8}x + \tfrac{17}{12}$

9

The open expression 5x + 3x can be simplified to 8x because 5x and 3x are **like terms**. To add like terms, just add the coefficients. Simplify 9y + 7y.

16y

10

To simplify 6x + 4 + 2x + 5, first change the ordering and grouping.

$$6x + 4 + 2x + 5$$
$$(6x + 2x) + (4 + 5)$$
$$8x + 9$$

Simplify. 9 + 3x + 4 + 8x

11x + 13 or 13 + 11x

11

Simplify. x + 12 + 6x + 9
[Note: The coefficient of x is 1.]

7x + 21

12

To simplify 4x – 9 – 6x + 12, like terms are added.

$$4x - 9 - 6x + 12$$
$$(4x - 6x) + (-9 + 12)$$
$$-2x + 3$$

Simplify. 12x – 3 + 8 + 2x

14x + 5

13

To simplify $\frac{3}{4}x - \frac{2}{3}x$, the following steps are used:

$$\frac{3}{4}x - \frac{2}{3}x$$

$$\frac{3}{4}x + \frac{-2}{3}x$$

$$\left(\frac{3}{4} + \frac{-2}{3}\right)x$$

$$\frac{1}{12}x$$

Simplify. $\frac{7}{8}x - \frac{1}{5}x$ $\qquad$ $\frac{27}{40}x$

14

Simplify. $2x - \frac{5}{6}x$ $\qquad$ $\left(\frac{2}{1} + \frac{-5}{6}\right)x = \frac{7}{6}x$

15

Simplify. $\frac{2}{3}x - \frac{1}{4}x$ $\qquad$ $\frac{5}{12}x$

16

To simplify $8 \cdot (4x)$, the following steps are used:

$$8 \cdot (4x)$$
$$(8 \cdot 4)x$$
$$32x$$

Simplify. $7 \cdot (3y)$ $\qquad$ $(7 \cdot 3)y$
$\qquad\qquad\qquad\qquad\qquad\qquad\qquad\qquad$ $21y$

17

To simplify $(-4x)3$, the order and the grouping of the factors is altered.

$$(-4x)3$$
$$3(-4x)$$
$$(3 \cdot -4)x$$
$$-12x$$

Simplify. $(5x) \cdot -4$ $\qquad$ $-20x$

18
Simplify. (-8x) • -2 16x

19
To simplify $\frac{-3}{8} \cdot \left(\frac{5}{9}x\right)$, the following steps are used:

$$\frac{-3}{8} \cdot \left(\frac{5}{9}x\right) = \left(\frac{-3}{8} \cdot \frac{5}{9}\right)x = \frac{-5}{24}x$$

Notice that cancellation was used in multiplying $\frac{-3}{8} \cdot \frac{5}{9}$.

Simplify. $-3 \cdot \left(\frac{5}{6}x\right)$ $\frac{-5}{2}x$

20
Complete the simplification:

$$\frac{-5}{12} \cdot \left(\frac{2}{3}x\right) = \left(\frac{-5}{12} \cdot \frac{2}{3}\right)x = \underline{}$$

$\frac{-5}{18}x$

21
Simplify. $\frac{5}{6} \cdot (6x)$ 5x

22
Simplify. $\frac{7}{2}\left(\frac{2}{7}z\right)$ 1z or z

23
Simplify. $8\left(\frac{3}{8}x\right)$ 3x

24
To **remove the parentheses** from 5(2x + 3), 5 is multiplied by 2x and 5 is also multiplied by 3.

$$5(2x + 3)$$
$$5 \cdot 2x + 5 \cdot 3$$
$$10x + 15$$

Remove the parentheses from 4(3x + 7). 12x + 28

25

To simplify 5 + 3(2x + 1), the parentheses are removed first.

$$5 + 3(2x + 1)$$
$$5 + 3 \cdot 2x + 3 \cdot 1$$
$$5 + 6x + 3$$
$$6x + 8$$

Simplify. 7 + 2(3x + 5)

7 + 6x + 10
6x + 17

26

The parentheses of 7x + (4x + 9) are preceded by a plus sign and a multiplier of 1 is used.

$$7x + (4x + 9)$$
$$7x + 1(4x + 9)$$
$$7x + 4x + 9$$
$$11x + 9$$

Simplify. 6x + (8x + 5)

6x + 8x + 5
14x + 5

27

To simplify 3(2x + 5) + 2(5x + 1), first remove both pairs of parentheses.

$$3(2x + 5) + 2(5x + 1)$$
$$6x + 15 + 10x + 2$$
$$16x + 17$$

Simplify. 5(4x + 1) + 4(6x + 8)

44x + 37

28

The parentheses of -6(2x + 5) are removed by multiplying both 2x and 5 by -6.

$$-6(2x + 5)$$
$$-6 \cdot 2x + -6 \cdot 5$$
$$-12x + (-30)$$
$$-12x - 30$$

Remove the parentheses from -7(2x + 5).

-14x - 35

29
Remove the parentheses from -8(2x – 3). $-16x + 24$

30
To simplify 5 – (3x – 6), first remove the parentheses using -1 as the multiplier.

$$5 - (3x - 6)$$
$$5 - 1(3x - 6)$$
$$5 - 3x + 6$$
$$-3x + 11$$

Simplify. 6x – (2x – 7) $6x - 2x + 7 = 4x + 7$

31
Simplify. 8 – (3x + 2) $8 - 3x - 2 = -3x + 6$

32
The first step in simplifying 5 – 2(3x + 4) is to remove the parentheses.

$$5 - 2(3x + 4)$$
$$5 - 6x - 8$$
$$-6x - 3$$

Simplify. 7 – 3(2x – 4) $7 - 6x + 12 = -6x + 19$

33
Complete the simplification below where the parentheses are removed as the first step.

$$2x + 5\left(\tfrac{2}{5}x + \tfrac{1}{3}\right) = 2x + 2x + \tfrac{5}{3} = \underline{\qquad}$$

$4x + \tfrac{5}{3}$

34
Simplify. $3\left(2x - \tfrac{1}{4}\right) - 2(x + 3)$ $4x - \tfrac{27}{4}$

35
Simplify. $3\left(\tfrac{1}{4}x + \tfrac{1}{5}\right) - 2\left(\tfrac{1}{3}x + \tfrac{1}{5}\right)$ $\tfrac{1}{12}x + \tfrac{1}{5}$

36
Simplify. $\tfrac{1}{3}\left(\tfrac{1}{4} - \tfrac{2}{5}x\right) + \tfrac{5}{8}x$ $\tfrac{59}{120}x + \tfrac{1}{12}$

37
Simplify. $\frac{1}{2}\left(\frac{5}{6}x - \frac{1}{3}\right) + \frac{3}{4}x$ $\frac{7}{6}x - \frac{1}{6}$

FEEDBACK UNIT 4

This quiz reviews the preceding unit. Answers are at the back of the book.

Simplify the following expressions.

1. $4 + x + 11$

2. $5x \cdot 7$

3. $6y + 2(5y + 3)$

4. $7(3 + 2x)$

5. $8x + 1 + 6x + 4$

6. $4 + 3x + 5(2x + 6)$

7. $4x + 7 - x + 6$

8. $-8 + 3x - 9 - 4x$

9. $5 - 2(3x - 4)$

10. $-3x + 4(2x - 5)$

11. $-(2x + 5) + (4 - 5x)$

12. $-(3x - 2) - (5x + 9)$

13. $\left(x - \frac{2}{5}\right) - \frac{1}{4}$

14. $\frac{5}{6} + \left(\frac{1}{4}x - \frac{2}{3}\right)$

15. $-8\left(\frac{5}{8}x\right)$

16. $\frac{-3}{4}\left(\frac{8}{3}x\right)$

17. $\frac{3}{4}x - 3 - \frac{5}{8}x + \frac{1}{5}$

18. $\frac{-2}{5} + 3x - \frac{7}{10}x + \frac{7}{5}$

19. $\frac{2}{3}(6x - 9) - 4x$

20. $\frac{-1}{2}(4x - 10) - \frac{9}{2}$

Unit 5: Solving Linear Equations

The following mathematical terms are crucial to an understanding of this unit.

    Linear equation   Solution
    Truth set   Checking solutions

1
In the **linear equation** $x + 4 = 7$, if x is replaced by 5 the resulting statement, $5 + 4 = 7$, is false. If x is replaced by 9 in $x + 4 = 7$, the statement $9 + 4 = 7$ is _____. (true or false)
    false

2
In the linear equation $x + 2 = 9$, if x is replaced by 7 the statement $7 + 2 = 9$ is _____. (true or false)
    true

3
The linear equation $x + 3 = 7$ can be translated into a question:
    What number can be added to 3 and equal 7?
The answer to the question is _____.
    4

4
The linear equation $x - 8 = 5$ can be translated into a question:
    What number can be added to -8 and equal 5?
The answer to the question is _____.
    13

5
The equation $5x = 10$ translates into a multiplication question.
    What number can be multiplied by 5 and equal 10?
The answer to the question is _____.
    2

6
The equation 3x = 18 translates into a multiplication question.
What number can be multiplied by 3 and equal 18?
The answer to the question is _____.

6

7
The equation -4x = 20 translates into a multiplication question.
What number can be multiplied by -4 and equal 20?
The answer to the question is _____.

-5

8
Some equations, like 4x + 7 = -13 do not translate easily into simple questions. The fact that every rational number has an opposite is a powerful tool for solving non-simple equations. What is the opposite of -9?

9

9
Two equations are equivalent if they become true statements for the same number replacements of the variable. Are x + 3 = 5 and 4x = 8 equivalent equations?

Yes, 2 makes both true.

10
Any number may be added to both sides of an equation to generate a new, equivalent equation.
Is x + 4 = 9 equivalent to x + 4 + 12 = 9 + 12?

Yes, 12 was added to both sides of x + 4 = 9.

11
To solve 2x + 5 = 13, generate an equivalent equation by adding -5 to both sides of the equation.

$$\begin{aligned} 2x + 5 &= 13 \\ -5 & -5 \\ \hline 2x + 0 &= 8 \\ 2x &= 8 \end{aligned}$$

2x + 5 = 13 and 2x = 8 are equivalent equations and both become true statements when x = _____.

4

12

To solve $3x - 7 = 8$, first generate an equivalent equation by adding 7 to both sides.

$$\begin{array}{rl} 3x - 7 =& 8 \\ +7 & +7 \\ \hline 3x + 0 =& 15 \\ 3x =& 15 \end{array}$$

$3x - 7 = 8$ and $3x = 15$ are equivalent equations and both become true statements when x = _____.

5

13

To solve $3 + 2x = -9$, first generate an equivalent equation.

$$\begin{array}{rl} 3 + 2x =& -9 \\ -3 & -3 \\ \hline 0 + 2x =& -12 \\ 2x =& -12 \end{array}$$

$3 + 2x = -9$ and $2x = -12$ are equivalent equations and both become true statements when x = _____.

-6

14

To solve $5x - 2 = 13$, the following steps are used.

$$\begin{array}{rl} 5x - 2 =& 13 \\ +2 & +2 \\ \hline 5x + 0 =& 15 \\ 5x =& 15 \end{array}$$

The number that will make $5x - 2 = 13$ true is _____.

3

15

To solve $15 - 4x = 7$, the following steps are used.

$$\begin{array}{rl} 15 - 4x =& 7 \\ -15 & -15 \\ \hline 0 - 4x =& -8 \\ -4x =& -8 \end{array}$$

The number that will make $15 - 4x = 7$ true is _____.

2

16

To find the **solution** for $x + \frac{5}{7} = \frac{3}{4}$, the opposite of $\frac{5}{7}$ is added to both sides of the equation.

$$x + \frac{5}{7} = \frac{3}{4}$$

$$\left(x + \frac{5}{7}\right) - \frac{5}{7} = \left(\frac{3}{4}\right) - \frac{5}{7}$$

$$x + \left(\frac{5}{7} - \frac{5}{7}\right) = \left(\frac{3}{4} - \frac{5}{7}\right)$$

$$x + 0 = \left(\frac{21}{28} - \frac{20}{28}\right)$$

$$x = \frac{1}{28}$$

Therefore, the solution of $x + \frac{5}{7} = \frac{3}{4}$ is _____.

$\frac{1}{28}$

17

To find the solution for $x - \frac{1}{3} = \frac{2}{9}$, first add the opposite of $\frac{-1}{3}$ to both sides of the equation.

$$x - \frac{1}{3} = \frac{2}{9}$$

$$\left(x - \frac{1}{3}\right) + \frac{1}{3} = \left(\frac{2}{9}\right) + \frac{1}{3}$$

$$x + \left(\frac{-1}{3} + \frac{1}{3}\right) = \left(\frac{2}{9} + \frac{1}{3}\right)$$

$$x + 0 = \left(\frac{2}{9} + \frac{3}{9}\right)$$

$$x = \frac{5}{9}$$

Find the solution for $x - \frac{1}{3} = \frac{2}{9}$.

$\frac{5}{9}$

18

Find the solution for $x + \frac{3}{4} = \frac{2}{3}$ by adding the opposite of $\frac{3}{4}$ to both sides of the equation.

$$x + 0 = \frac{2}{3} - \frac{3}{4}$$

$$x = \frac{-1}{12}$$

19

Find the solution for $x - \frac{5}{9} = \frac{3}{2}$ by adding the opposite of $\frac{-5}{9}$ to both sides of the equation.

$$x + 0 = \frac{3}{2} + \frac{5}{9}$$
$$x = \frac{37}{18}$$

20

Any nonzero number may be multiplied by both sides of an equation to generate a new, equivalent equation. To solve $7x = 17$, both sides of the equation are multiplied by the reciprocal of 7.

$$7x = 17$$
$$\tfrac{1}{7} \cdot 7x = \tfrac{1}{7} \cdot 17$$
$$1x = \tfrac{17}{7}$$
$$x = \tfrac{17}{7}$$

What is the solution of $7x = 17$?

$$\frac{17}{7}$$

21

The fact that every rational number except 0 has a reciprocal is a powerful tool in equation solving. To solve $-5x = 11$, an equivalent equation is generated by multiplying both sides by $\frac{-1}{5}$.

$$-5x = 11$$
$$\tfrac{-1}{5} \cdot -5x = \tfrac{-1}{5} \cdot 11$$
$$1x = \tfrac{-11}{5}$$
$$x = \tfrac{-11}{5}$$

What is the solution of $-5x = 11$?

$$\frac{-11}{5}$$

22
Find the solution of 3x = 23 by multiplying both sides by the reciprocal of 3.

Solution is $\frac{23}{3}$

23
Find the solution of -4x = 5 by multiplying both sides by the reciprocal of -4.

Solution is $\frac{-5}{4}$

24
Find the solution of 8x = -7 by generating an equivalent equation.

Solution is $\frac{-7}{8}$

25
Find the solution of -9x = -4 by generating an equivalent equation.

Solution is $\frac{4}{9}$

26
Multiply both sides of $\frac{3}{4}$x = 2 by $\frac{4}{3}$. Simplify both sides and determine the solution for the equation.

x = $\frac{8}{3}$

27
To solve the equation $\frac{3}{4}$x = 2, the following steps are used.

$$\frac{3}{4}x = 2$$

$$\frac{4}{3} \cdot \frac{3}{4}x = \frac{4}{3} \cdot 2$$

$$1x = \frac{8}{3}$$

$$x = \frac{8}{3}$$

What is the solution of $\frac{5}{3}$x = 7?

$\frac{21}{5}$

28

To find a solution of 4x + 3 = 21, the following steps are used:

$$4x + 3 = 21$$
$$4x + 3 - 3 = 21 - 3$$
$$4x = 18$$
$$\frac{1}{4} \cdot 4x = \frac{1}{4} \cdot 18$$
$$x = \frac{9}{2}$$

The solution of 4x + 3 = 21 is _____.

$\frac{9}{2}$

29

Find a solution for 7x – 5 = 19 by first adding the opposite of -5 to both sides of the equation.

7x = 24

$\frac{24}{7}$

30

Find a solution for -9x – 7 = 4.

$\frac{-11}{9}$

31

Find a solution for 6x + 9 = 5.

$\frac{-2}{3}$

32

The variable x appears on both sides of the equation 5x + 3 = 2x – 8. To solve, first add the opposite of 2x to both sides of the equation.

$$5x + 3 = 2x - 8$$
$$5x + 3 - 2x = 2x - 8 - 2x$$
$$3x + 3 = -8$$
$$3x + 3 - 3 = -8 - 3$$
$$3x = -11$$
$$x = \frac{-11}{3}$$

Solve 7x – 3 = 5x + 9 by first adding -5x to both sides.

2x – 3 = 9
solution is 6

33
Solve 4x – 8 = 6 – 3x by first adding 3x to both sides.

7x – 8 = 6
solution is 2

34
Solve 2x + 9 = 3 + 5x by first adding -5x to both sides.

-3x + 9 = 3
solution is 2

35
Solve. 5x + 2 = 3x – 34

-18

36
To solve $\frac{3}{2}x - \frac{5}{6} = \frac{1}{4}$, first multiply each term by the LCM of the denominators.

$$\frac{3}{2}x - \frac{5}{6} = \frac{1}{4}$$

$$12 \cdot \frac{3}{2}x - 12 \cdot \frac{5}{6} = 12 \cdot \frac{1}{4}$$

$$6 \cdot \frac{3}{1}x - 2 \cdot \frac{5}{1} = 3 \cdot \frac{1}{1}$$

$$18x - 10 = 3$$

Are $\frac{3}{2}x - \frac{5}{6} = \frac{1}{4}$ and 18x – 10 = 3 equivalent equations?

Yes

37
Since $\frac{3}{2}x - \frac{5}{6} = \frac{1}{4}$ and 18x – 10 = 3 are equivalent equations, solve one of them to find a solution for both.

18x = 13

$x = \frac{13}{18}$

38

To solve $\frac{2}{3}x - \frac{4}{9} = \frac{5}{6}$, first multiply each term by the common denominator, 18.

$$\frac{2}{3}x - \frac{4}{9} = \frac{5}{6}$$

$$18 \cdot \frac{2}{3}x - 18 \cdot \frac{4}{9} = 18 \cdot \frac{5}{6}$$

$$6 \cdot \frac{2}{1}x - 2 \cdot \frac{4}{1} = 3 \cdot \frac{5}{1}$$

$$12x - 8 = 15$$

Find a solution for $\frac{2}{3}x - \frac{4}{9} = \frac{5}{6}$.

$12x = 23$

$x = \frac{23}{12}$

39

To solve $\frac{1}{3}x + \frac{3}{4} = \frac{1}{6}$, each term is multiplied by the common denominator, 12.

$$\frac{1}{3}x + \frac{3}{4} = \frac{1}{6}$$

$$12 \cdot \frac{1}{3}x + 12 \cdot \frac{3}{4} = 12 \cdot \frac{1}{6}$$

$$4 \cdot \frac{1}{1}x + 3 \cdot \frac{3}{1} = 2 \cdot \frac{1}{1}$$

$$4x + 9 = 2$$

Is $4x + 9 = 2$ equivalent to $\frac{1}{3}x + \frac{3}{4} = \frac{1}{6}$?

Yes

40

Since $\frac{1}{3}x + \frac{3}{4} = \frac{1}{6}$ and $4x + 9 = 2$ are equivalent equations, solve one of them to find a solution for both.

$4x = -7$

$x = \frac{-7}{4}$

41

Solve. $6x - \frac{3}{10} = \frac{4}{5}$

$x = \frac{11}{60}$

42

Solve. $\frac{5}{12}x - \frac{3}{4} = \frac{7}{8}$

$x = \frac{39}{10}$

A VOCABULARY FOR ALGEBRA   37

43
Solve. $\frac{5}{6}x - 9 = \frac{2}{3}$

$x = \frac{58}{5}$

44

The **truth set** of $\frac{1}{3}x - \frac{3}{4} = \frac{1}{2}$ is $\{\frac{15}{4}\}$ because when x is replaced by $\frac{15}{4}$, the equation becomes the true statement $\frac{1}{2} = \frac{1}{2}$. Replace x by 3 in $\frac{2}{5}x - \frac{1}{2} = \frac{3}{10}$ and determine whether it is a solution.

$\frac{6}{5} - \frac{1}{2} = \frac{3}{10}$

$\frac{7}{10} = \frac{3}{10}$ is false.

3 is not a solution.

45

Find the truth set of $\frac{2}{5}x - \frac{1}{2} = \frac{3}{10}$.

Then **check** the solution by showing that both sides of the equation have the same evaluation.

$\{2\}$

$\frac{4}{5} - \frac{1}{2} = \frac{3}{10}$

$\frac{8}{10} - \frac{5}{10} = \frac{3}{10}$

$\frac{3}{10} = \frac{3}{10}$ is true.

46

Determine if $\frac{3}{5}$ is a solution of $\frac{5}{6}x - \frac{2}{9} = \frac{2}{3}$ by evaluating the numerical expression on the left side of the equality below.

$$\frac{5}{6}x - \frac{2}{9} = \frac{2}{3}$$

If $x = \frac{3}{5}$ $\frac{5}{6} \cdot \frac{3}{5} - \frac{2}{9} = \frac{2}{3}$

$\underline{\qquad} = \frac{2}{3}$

$\frac{1}{2} - \frac{2}{9} = \frac{2}{3}$

$\frac{9}{18} - \frac{4}{18} = \frac{2}{3}$

$\frac{5}{18} = \frac{2}{3}$ is false.

$\frac{3}{5}$ is not a solution.

47

Solve and check. $\frac{5}{6}x - \frac{2}{9} = \frac{2}{3}$

$\left\{\frac{16}{15}\right\}$

$\frac{5}{6} \cdot \frac{16}{15} - \frac{2}{9} = \frac{2}{3}$

$\frac{8}{9} - \frac{2}{9} = \frac{2}{3}$

$\frac{6}{9} = \frac{2}{3}$ is true.

48

To determine if $\frac{2}{5}$ is a solution of $\frac{-3}{10}x + \frac{7}{15} = \frac{3}{5} - \frac{19}{30}x$, each side of the equality is separately evaluated. Evaluate the numerical expression on the right side of the equality below and state whether $\frac{2}{5}$ is a solution.

$$\frac{-3}{10}x + \frac{7}{15} = \frac{3}{5} - \frac{19}{30}x,$$

If $x = \frac{2}{5}$

$$\frac{-3}{10} \cdot \frac{2}{5} + \frac{7}{15} = \frac{3}{5} - \frac{19}{30} \cdot \frac{2}{5}$$

$$\frac{-3}{25} + \frac{7}{15} = \underline{\qquad}$$

$$\frac{-9}{75} + \frac{35}{75} = \underline{\qquad}$$

$$\frac{26}{75} = \underline{\qquad}$$

$\frac{3}{5} - \frac{19}{75}$

$\frac{45}{75} - \frac{19}{75}$

$\frac{26}{75}$

$\frac{26}{75} = \frac{26}{75}$ is true.

$\frac{2}{5}$ is a solution.

49

Find the truth set and check. $\frac{2}{3}x - \frac{7}{10} = x - \frac{1}{6}$

$\left\{\frac{-8}{5}\right\}$

Check:

$\frac{-53}{30} = \frac{-53}{30}$ is true.

Feedback Unit 5

This quiz reviews the preceding unit. Answers are at the back of the book.

For problems 1-16, solve the equation.

1. $x + \frac{2}{3} = 3$

2. $x - \frac{3}{11} = \frac{-2}{3}$

3. $x + \frac{4}{5} = \frac{7}{8}$

4. $\frac{-7}{8}x = -3$

5. $\frac{3}{7}x = \frac{6}{5}$

6. $3x = -11$

7. $-5x = -16$

8. $4x - 3 = 19$

9. $7x + 1 = 29$

10. $-2x + 7 = -13$

11. $-9 + x = -2$

12. $7x - 3 = 2x + 12$

13. $-3x + 4 = x - 16$

14. $\frac{2}{5}x - \frac{7}{10} = \frac{1}{2}$

15. $\frac{-7}{8}x + \frac{3}{4} = \frac{-1}{2}x + 3$

16. $\frac{1}{6}x + \frac{3}{4} = \frac{5}{12}x - \frac{1}{3}$

For problems 17-19, solve and check.

17. $\frac{1}{2}x + \frac{2}{3} = \frac{5}{6}$

18. $\frac{3}{4}x - \frac{7}{8} = \frac{1}{2}x - 2$

19. $\frac{2}{3}x - \frac{1}{6} = \frac{1}{2}x + \frac{5}{12}$

UNIT 6: APPLICATIONS

In this Applications Section, the format of the text has been altered. Answers for the word problems will appear beneath the problems rather than in the right-hand column. The emphasis with these word problems is to select the correct procedure to follow with the numbers. For that reason, once the procedure is selected a calculator may be used to calculate the answer.

1
Discounts, mark-ups, and commissions are often stated as percents. For example, a discount for a sales item may be stated as: 15% off of regular price. This is equivalent to: A sales price of ____% of the regular price.

Answer: 85%. The regular price is 100%. With a discount of 15%, the sales price is the difference, 85%.

2
Find the sales price for an item which is advertised as 20% off of its regular price of $59.95.

Answer: The sales price is 80% of $59.95 (.80 • 59.95) or $47.96.

3
A retail merchant buys an item for its wholesale price (usually stated as its cost) and sells it for a higher amount (usually stated as its retail price). The difference between the cost and retail price is the mark-up, and this is often stated as a percent of the cost. If a mark-up is 40%, what percent is the retail price of the cost?

Answer: 140%. The cost is 100%. The mark-up is 40%. Therefore, the retail price is 140% of the cost.

4
A retail merchant buys an item for $37.55 and applies a 35% mark-up. What is the retail price?

> Answer: The retail price is $50.6925 which must be rounded off to $50.69. The retail price is found as 135% of $37.55 (1.35 • 37.55).

5
Many salespeople work on a commission basis. That is, they are paid a percent of their sales rather than an hourly wage. For example, a realtor may be paid 2% of each property sale. If such a realtor sells a house for $125,000, the commission would be _____.

> Answer: $2,500. The commission is 2% of $125,000 (.02 • 125,000).

6
Car salespersons commonly work on a commission basis. If such a salesman is paid 6% on each sale, what is his commission on the sale of a Lexus for $52,500?

> Answer: $3,150. The commission is 6% of $52,500 (.06 • 52,500).

7
If a store advertises a 15% discount off regular price and now offers coats for $42.46, what was the regular price before the sale?

> Answer: $49.95. To find the regular price, use the equation .85R = 42.46 because 85% of the regular price is the sales price.

8

A manufacturer sells tennis shoes to the retailer for $34.85 a pair. If the retailer prices them at $59.95, what is the percent of mark-up?

> Answer: 72%. To find the percent of mark-up, use the equation 34.85M = 59.95 which gives the result 1.72 or 172%. Since the retail price is 172% of the cost, the mark-up is 72%.

FEEDBACK UNIT 6 FOR APPLICATIONS

1. Find the sales price for an item which is advertised as 30% off of its regular price of $39.95.

2. A retail merchant buys an item for $13.66 and applies a 40% mark-up. What is the retail price?

3. A car saleswoman earns a commission of 5% on selling a Buick for $32,500. What was her commission?

4. A store advertises a 20% discount off regular price and has suits for $87.75. What was the regular price before the sale?

5. An electronics manufacturer sells TVs to the retailer for $385.62 each. If the retailer prices them at $549.95, what is the percent of mark-up?

Summary for Chapter 1

The following mathematical terms are crucial to an understanding of this chapter.

- Counting numbers
- Consecutive
- Element
- Even numbers
- Between
- Factor
- Composite number
- Highest common factor (HCF)
- Least common multiple (LCM)
- Evaluate
- Square brackets
- Open expression
- Coefficient
- Statement
- Conditional equation
- Set of real numbers
- Set of integers
- Next consecutive integer
- Terminating decimal
- Dense
- Irrational number
- Reciprocal
- Binary operation
- Division sign
- Associative Law of Addition
- Inverse Law of Addition
- Associative Law of Multiplication
- Inverse Law of Multiplication
- Distributive Law of Addition Over Multiplication
- Set of counting numbers
- Member
- Subset
- Odd numbers
- Empty set
- Prime number
- Common factor
- Multiples
- Numerical expression
- Parentheses
- Order of operations
- Variable
- Equivalent open expression
- Equation
- Identity
- Real number line
- Negative integers
- Set of rational numbers
- Repeating decimal
- Infinite
- Opposite
- Minus sign
- Commutative Law of Addition
- Identity Element for Addition
- Commutative Law of Multiplication
- Identity Element for Multiplication
- Completeness Property
- Terms
- Remove the parentheses

Continued Next Page

Simplify an open expression
Like terms
Linear equation

Solution
Checking solutions
Truth set

This chapter reviews and extends the vocabulary needed for success in Intermediate Algebra.

Also included here were the evaluation of numerical expressions involving integers and rational numbers, the skills of simplifying open expressions, and techniques for solving linear equations.

With a mastery of this chapter, the student is ready to learn Intermediate Algebra.

Chapter 1 Mastery Test

The following questions test the objectives of Chapter 1. Answers are at the back of the book. The number in parentheses which follows each problem indicates the unit in which it can be learned.

1. The next consecutive counting number to 351 is _____. (1)

2. Show the set of even numbers between 61 and 66. (1)

3. Write the set of prime numbers between 1 and 14. (1)

4. What is the HCF of 39 and 26? (1)

5. What is the LCM of 21 and 28? (1)

6. Are 6(4x + 3) and 6 • 4x + 6 • 3 equivalent expressions? (1)

7. Evaluate. 7 + 3 • 5 (1)

8. Which of the following is a conditional equation?
    a. x + 3 = 3 + x
    b. 2x − 7 = 5 (1)

9. Which of the following is an identity?
    a. x(yz) = (xy)z
    b. 8 − 5 = 1 • 3 (1)

10. {1, 2, 3, . . .} is the set of _____ _____. (2)

11. Is every point of the real number line represented by a real number? (2)

12. What is the opposite of $\frac{3}{4}$? (2)

13. What is the reciprocal of $\frac{-1}{2}$? (2)

14. Do 3 − 19 and -19 + 3 have the same evaluation? (3)

15. Because multiplication is an associative operation, (3x)y is equivalent to _____. (3)

16. Because addition is a commutative operation, 4 + 5x is equivalent to _____. (3)

17. Simplify. 2x − 5(x − 4) (4)

18. Simplify. $\frac{-3}{8}x - 3 + \frac{5}{12}x + \frac{7}{3}$ (4)

19. Solve. 7x − 4 = 3x + 5 (5)

20. Solve. $\frac{1}{4}x + \frac{5}{6} = \frac{2}{3}x + \frac{7}{3}$ (5)

21. Find the sales price for an item which is advertised as 30% off of its regular price of $79.90. (6)

22. A retail merchant buys an item for $42.55 and applies a 40% mark-up. What is the retail price? (6)

Chapter 2 Objectives

The following problems illustrate the objectives of this chapter. At this time you are not expected to know how to do these problems. However, if all these problems are thoroughly understood, proceed directly to the Chapter Mastery Test. The number in parentheses which follows each problem indicates the unit in which it can be learned.

1. $7^5 \cdot 7^8 =$ _____ (1)

2. $(5^4)^7 =$ _____ (1)

3. $\dfrac{3^8}{3^2} =$ _____ (1)

4. Is 216 a perfect cube? (1)

5. $\sqrt[4]{7^4} =$ _____ (2)

6. $\sqrt[3]{13^3} =$ _____ (2)

7. $\sqrt[3]{-64} =$ _____ (2)

8. $\sqrt[4]{9^{28}} =$ _____ (2)

9. $\sqrt[3]{z^{12}} =$ _____ (2)

10. $\sqrt{7} \cdot \sqrt{5} =$ _____ (3)

11. $3\sqrt{5} + \sqrt{5} =$ _____ (3)

12. $\sqrt[4]{7} \cdot 2\sqrt[4]{6} =$ _____ (3)

13. $13\sqrt[5]{4x^3} - 9\sqrt[5]{4x^3} =$ _____ (3)

14. $\sqrt{5}\,(6\sqrt{2} - \sqrt{5}) =$ _____ (3)

15. $(3\sqrt{2} - \sqrt{5})(2\sqrt{2} - 4\sqrt{3}) =$ _____ (3)

16. $\sqrt{84} =$ _____ (4)

17. $\sqrt{90} =$ _____ (4)

18. $\sqrt{96} =$ _____ (4)

19. $\sqrt[5]{243} =$ _____ (4)

20. $\sqrt[3]{128} =$ _____ (4)

21. $\sqrt[8]{5^8 \cdot 7} =$ _____ (4)

22. $\sqrt[5]{x^9} =$ _____ (4)

23. $-\sqrt{5} + 4\sqrt{24} - \sqrt{125} =$ _____ (4)

24. $2i \cdot -i =$ _____ (5)

25. $5i \cdot i\sqrt{25} =$ _____ (5)

26. $-13 + 2i - 3 - 5i =$ _____ (5)

Chapter 2
Simplifying Radical Expressions

Unit 1: Counting Number Exponents

The following mathematical terms are crucial to an understanding of this unit.

- Factors
- Squared
- To the fourth power
- Perfect squares
- Perfect fourth powers
- Power expression
- Simplifying power multiplications
- Dividing power expressions
- Exponent
- Cubed
- Base
- Perfect cubes
- Perfect fifth powers
- Multiplication of power expressions
- Raising a power to a power

1
In the multiplication expression 2 • -5 • 3 the numbers 2, -5, and 3 are **factors**. The factors of 5 • -3 are 5 and _____.

-3

2

When a multiplication expression uses the same factor repeatedly, it is often written with an **exponent.** 7 • 7 • 7 may be written 7^3 and the small "3" is an _____.   exponent

3

6^2 means 6 • 6 and is read "6 to the second power or 6 **squared**."
Use an exponent to write: 5 squared. 5^2

4

7^3 means 7 • 7 • 7 and is read "7 to the third power or 7 **cubed**."
Use an exponent to write: 2 cubed. 2^3

5

8^4 means 8 • 8 • 8 • 8 and is read "8 **to the fourth power**."
Use an exponent to write: 6 to the fourth power. 6^4

6

Use an exponent to write: 3 to the eighth power. 3^8

7

Use an exponent to write: 7 to the twelfth power. 7^{12}

8

The expression 3 • 3 • 3 • 3 is equivalent to 3^4.
The expression 5 • 5 • 5 • 5 is equivalent to _____. 5^4

9

The expression 10 • 10 • 10 is equivalent to 10^3.
The expression 4 • 4 • 4 • 4 • 4 is equivalent to _____. 4^5

10

-5 • -5 • -5 must be written with parentheses: $(-5)^3$.
-3 • -3 • -3 • -3 • -3 is equivalent to _____. $(-3)^5$

11

-6 • -6 • -6 • -6 is equivalent to _____. $(-6)^4$, but not -6^4

SIMPLIFYING RADICAL EXPRESSIONS   49

12
In the expression 2^3 the 2 is the **base** and the 3 is the exponent. The number that is a factor is the _____ and the number of times it is used is the exponent.

base

13
In the expression $(-23)^4$ the base is _____ and it is used ____ times as a factor.

-23, 4

14
The base of 5^6 is ____ and it is used as a factor _____ times.

5, 6

15
When a negative number is used as the base, it must always be indicated with parentheses or some grouping symbol.

$(-2)^3$ means $-2 \cdot -2 \cdot -2$
-2^3 means $-1 \cdot 2 \cdot 2 \cdot 2$

Use an exponent to show that -3 is used 5 times as a factor.

$(-3)^5$

16
To evaluate 5^3, the following steps are used:

$5^3 = 5 \cdot 5 \cdot 5 = 125$

Evaluate 4^3.

64

17
Complete the evaluation of 7^3.

$7^3 = 7 \cdot 7 \cdot 7 = $ _____

343

18
Complete the evaluation of 1^5.

$1^5 = 1 \cdot 1 \cdot 1 \cdot 1 \cdot 1 = $ _____

1

19
Complete the evaluation of $(-2)^3$.

$(-2)^3 = (-2)(-2)(-2) = $ _____

-8

20

$(-9)^2 = $ _____ 81

21

$2^4 = $ _____ 16

22

$10^3 = $ _____ 1,000

23

To evaluate x^2 when $x = 5$, use parentheses to replace x by 5.

$$x^2 = (5)^2 = 5 \cdot 5 = 25$$

Evaluate x^2 when $x = 9$. 81

24

When x^2 is evaluated the results are **perfect squares**. Find the perfect squares associated with the elements of $\{1, 2, 3, \ldots, 10\}$.

1, 4, 9, 16, 25, 36, 49, 64, 81, 100

25

When x^3 is evaluated the results are **perfect cubes**. Find the perfect cubes associated with the elements of $\{1, 2, 3, \ldots, 10\}$.

1, 8, 27, 64, 125, 216, 343, 512, 729, 1000

26

When x^4 is evaluated the results are **perfect fourth powers**. Find the perfect fourth powers associated with the elements of $\{1, 2, 3, 4, 5\}$.

1, 16, 81, 256, 625

27

When x^5 is evaluated the results are **perfect fifth powers**. Find the perfect fifth powers associated with the elements of $\{1, 2, 3\}$.

1, 32, 243

SIMPLIFYING RADICAL EXPRESSIONS   51

28
The numbers found as answers in the four preceding frames will be most useful in later units of this chapter. An ability to recognize perfect squares, cubes, fourth, and fifth powers is _____ (valuable, useless) later in this chapter.

valuable

29
Is 625 a perfect fourth power?

Yes, $625 = 5^4$

30
x^3 is a **power expression** with base x and exponent 3.
Write the power expression with base y and exponent 5.

y^5

31
$x^4 \cdot x^3$ is the **multiplication of power expressions**.

$$x^4 \cdot x^3 \text{ means } xxxx \cdot xxx$$

How many factors of x are in $x^4 \cdot x^3$?

7

32
$x^2 \cdot x^3$ is the multiplication of power expressions.

$$x^2 \cdot x^3 \text{ means } xx \cdot xxx$$

How many factors of x are in $x^2 \cdot x^3$?

5

33
To **simplify** the multiplication of power expressions with the same base, add the exponents.

$$x^3 \cdot x^4 = x^{3+4} = x^7$$

Simplify. $x^2 \cdot x^3$

$x^{2+3} = x^5$

34
$x^8 \cdot x^3 = x^{8+3} = x^{11}$
$x^4 \cdot x^5 =$ _____

x^9

35
$x \cdot x^4 =$ _____

x^5

36

$x^3 \cdot x^6 =$ _____

x^9

37

$x^7 \cdot x^4 =$ _____

x^{11}

38

$x^8 \cdot x^2 =$ _____

x^{10}

39

To simplify $(x^2)^3$ first notice that two exponents are involved. Begin with the "inside" exponent which is 2.

$(x^2)^3$ means $\quad (xx)^3$
$(xx)^3$ means $\quad (xx)(xx)(xx)$
$\quad\quad\quad\quad\quad\quad (xxxxxx)$
$\quad\quad\quad\quad\quad\quad x^6$

Simplify. $(x^4)^2$

xxxxxxxx = x^8

40

When like bases are multiplied the exponents are added.

$$x^a \cdot x^b = x^{a+b}$$

When the base of an exponent is itself a power expression the exponents are multiplied.

$$(x^a)^b = x^{ab}$$

$(x^3)^5 = x^{15}$ and $x^3 \cdot x^5 =$ _____

$x^{3+5} = x^8$

41

Both of the simplification facts stated in the previous frame are a direct result of the meaning of an exponent. If there is a doubt how to simplify a power expression, use the meaning of the exponent.

$x^3 \cdot x^5$ means $xxx \cdot xxxxx = x^8$
$(x^3)^5$ means $(x^3)(x^3)(x^3)(x^3)(x^3) = x^{15}$

$x^4 \cdot x^3 = x^7$ and $(x^4)^3 =$ _____

$(x^4)(x^4)(x^4) = x^{12}$

42

$(x^7)^4 = x^{28}$ and $x^7 \cdot x^4 =$ _____

$x^{7+4} = x^{11}$

SIMPLIFYING RADICAL EXPRESSIONS   53

43

$x^5 \cdot x^4 = x^9$ and $(x^5)^4 = $ _____

$x^{5 \cdot 4} = x^{20}$

44

$(x^3)^9 = x^{27}$ and $x^3 \cdot x^9 = $ _____

$x^{3+9} = x^{12}$

45

In **raising a power to a power**, multiply the exponents.

$(x^2)^8 = $ _____

x^{16}

46

$(x^9)^2 = $ _____

x^{18}

47

$(x^7)^6 = $ _____

x^{42}

48

$(x^4)^3 = $ _____

x^{12}

49

To simplify $\dfrac{x^2}{x^3}$, two factors of x are divided out of both the numerator and denominator.

$$\dfrac{x^2}{x^3} = \dfrac{x \cdot x}{x \cdot x \cdot x} = \dfrac{\cancel{x} \cdot \cancel{x}}{\cancel{x} \cdot \cancel{x} \cdot x} = \dfrac{1}{x}$$

Simplify $\dfrac{x^3}{x^4}$ by dividing three factors of x out of both the numerator and denominator.

$\dfrac{x \cdot x \cdot x}{x \cdot x \cdot x \cdot x} = \dfrac{1}{x}$

50

To simplify $\dfrac{x^5}{x^2}$, two factors of x are divided out of both the numerator and denominator.

$$\dfrac{x^5}{x^2} = \dfrac{xxxxx}{xx} = \dfrac{\cancel{xx} \cdot xxx}{\cancel{xx}} = \dfrac{x^3}{1} = x^3$$

Simplify $\dfrac{x^5}{x^3}$ by dividing out three factors of x.

$\dfrac{xxx \cdot xx}{xxx} = \dfrac{x^2}{1} = x^2$

51

In **dividing power expressions** with the same base the exponents can be subtracted.

$$\frac{x^5}{x^2} = \frac{x^{5-2}}{1} = x^3 \quad \text{and} \quad \frac{x^3}{x^9} = \frac{1}{x^{9-3}} = \frac{1}{x^6}$$

When k is greater than m, $\dfrac{x^k}{x^m} = \dfrac{x^{k-m}}{1} = x^{k-m}$

When k is less than m, $\dfrac{x^k}{x^m} =$ _____ $\dfrac{1}{x^{m-k}}$

52

The simplification of $\dfrac{x^9}{x^2}$ is shown below.

$$\frac{x^9}{x^2} = \frac{x^{9-2}}{1} = \frac{x^7}{1} = x^7$$

$\dfrac{x^{10}}{x^7} =$ _____ $\dfrac{x^{10-7}}{1} = \dfrac{x^3}{1} = x^3$

53

$\dfrac{x^8}{x^4} =$ _____ $\dfrac{x^{8-4}}{1} = \dfrac{x^4}{1} = x^4$

54

$\dfrac{x^7}{x^6} =$ _____   x

55

The simplification of $\dfrac{x^2}{x^8}$ is shown below.

$$\frac{x^2}{x^8} = \frac{1}{x^{8-2}} = \frac{1}{x^6}$$

$\dfrac{x^{12}}{x^{14}} =$ _____ $\dfrac{1}{x^{14-12}} = \dfrac{1}{x^2}$

56

$\dfrac{x^5}{x^8} =$ _____ $\dfrac{1}{x^{8-5}} = \dfrac{1}{x^3}$

57

$\dfrac{x^8}{x^{10}} =$ _____ $\dfrac{1}{x^2}$

58. $\dfrac{x^4}{x^7} =$ _____ $\dfrac{1}{x^3}$

Feedback Unit 1

This quiz reviews the preceding unit. Answers are at the back of the book.

1. $6^3 =$ _____
2. $(-3)^4 =$ _____
3. Is 27 a perfect cube?
4. Is 32 a perfect fourth power?
5. Is 64 a perfect cube?
6. $x^6 \cdot x^9 =$ _____
7. $x^5 \cdot x^3 =$ _____
8. $x^7 \cdot x^4 =$ _____
9. $(x^2)^7 =$ _____
10. $(x^5)^4 =$ _____
11. $(y^4)^3 =$ _____
12. $\dfrac{x^8}{x^2} =$ _____
13. $\dfrac{x^5}{x^9} =$ _____
14. $\dfrac{x^7}{x^6} =$ _____

UNIT 2: Nth Roots

In Chapter 1, the irrational numbers were shown as an important subset of the real numbers. In this unit, a special class of these irrational numbers is the major topic.

The following mathematical terms are crucial to an understanding of this unit.

Square root	Radical sign, $\sqrt{}$
Radicand	Table of square roots
Perfect square integer	Simplest form (of a radical)
Cube	Cube root
Cube-root symbol	Perfect cube
Fifth root	Index of the radical
Nth root	

1
The **square root** of 17 is indicated by $\sqrt{17}$.
Show the square root of 29.

$\sqrt{29}$

2
The square root of 43 is indicated by $\sqrt{43}$. The symbol $\sqrt{}$ is a **radical sign**. Use a radical sign to indicate the square root of 37.

$\sqrt{37}$

3
The square root of 8 is shown by $\sqrt{8}$. 8 is the **radicand** of $\sqrt{8}$. What is the radicand of $\sqrt{37}$?

37

4
The square root of 19 is shown by $\sqrt{19}$. What is the number under the radical sign called?

Radicand

5

$\sqrt{7}$ is the **positive** number that multiplied by itself will give 7.

$$\sqrt{7} \cdot \sqrt{7} = 7$$

What will be the answer when $\sqrt{11}$ is multiplied by itself?

$\sqrt{11} \cdot \sqrt{11} = 11$

6

$$\sqrt{17} \cdot \sqrt{17} = (\sqrt{17})^2 = 17$$

What will be the result when $\sqrt{41}$ is squared (raised to the second power)?

$(\sqrt{41})^2 = $ _____.

41

7

$\sqrt{53}$ is the positive real number that multiplied by itself will give 53.

$(\sqrt{53})^2 = $ _____.

53

8

$\sqrt{100}$ is the positive real number that multiplied by itself will give 100.

$(\sqrt{100})^2 = 10^2 = $ _____.

100

9

A positive real number, x, and its square root (another positive real number), y, are related by the identity:

If $\sqrt{x} = y$, then $x = y^2$

The square root of 16 is 4. The identity showing the square-root relationship between 16 and 4 is:

If $\sqrt{16} = 4$, then _____.

$16 = 4^2$

10

The square root of 16 can be shown as $\sqrt{16}$ or as 4. The square root of 25 can also be shown in two ways. The square root of 25 is $\sqrt{25}$ or _____.

5

11
For some real numbers there is more than one way to indicate the square root. The square root of 49 can be shown as $\sqrt{49}$ or _____.

7

12
For some real numbers there is only one way to show the square root accurately. The square root of 41 can be shown accurately in only one way, $\sqrt{41}$. Is there more than one way to show the square root of 81?

Yes, $\sqrt{81}$ or 9

13
The square root of 100 can be shown in two ways:

    10 or $\sqrt{100}$

How many ways can the square root of 64 be shown?

Two, $\sqrt{64}$ or 8

14
The square root of 93 can be approximated by a terminating or repeating decimal, but accurately shown only as $\sqrt{93}$. There is no rational number that is equal to the square root of 93. Is there more than one way to accurately show the square root of 55?

No

15
Tables of square roots are frequently found in mathematics texts. A **table of square roots** is shown on page 579. What is the decimal numeral given in that table for the square root of 55?

7.416

16
The square-root table gives the decimal numeral 4.796 as the square root of 23. If 4.796 is squared, will it give 23?

No (it will be close to 23)

17
Square-root tables, such as the one on page 579, give rational number approximations for their numbers. The square root of 23 is not a rational number, so the decimal numeral 4.796 cannot be exactly equal to it. Is 4.796 a rational number?

Yes (terminating decimals name rational numbers)

SIMPLIFYING RADICAL EXPRESSIONS   59

18
Use the square-root table on page 579. Find the
rational number approximation for $\sqrt{85}$.

9.220

19
Is $\sqrt{85}$ exactly equal to 9.220?

No (they are approximately equal)

20
Use the square-root table on page 579.
Find the entry for $\sqrt{81}$.

9.000

21
Is $\sqrt{81}$ exactly equal to 9?

Yes, 81 is a **perfect square integer**

22
Most square roots shown in the table on page 579
are rational number approximations, but some
real numbers have square roots that are equal
to rational numbers. Is the square root of 4, $\sqrt{4}$,
a rational number?

Yes ($\sqrt{4} = 2$)

23
Some entries in the square-root table on page 579
are rational numbers that are exactly equal to a
square root. For example, the square root of 49, $\sqrt{49}$,
has an entry of 7.000. Is $\sqrt{49}$ exactly equal to 7.000?

Yes

24
Most entries in the square-root table on page 579 are
rational number approximations for the square roots.
For example, the table gives 6.856 as the square root
of 47. Is $\sqrt{47}$ exactly equal to 6.856?

No, 6.856 approximates $\sqrt{47}$

25

For the objectives of this book, there is little value to decimal approximations. Consequently, we will consider $\sqrt{29}$ to be the **simplest form** of the square root of 29. What is the simplest form for the square root of 31?

$\sqrt{31}$

26

The simplest form of the square root of 36 is 6. Both 6 and $\sqrt{36}$ are forms of the square root of 36, but _____ is the simplest form.

6

27

The simplest form of the square root of 71 is $\sqrt{71}$. What is the simplest form for the square root of 64?

8

28

Neither the radicand nor the square root of a real number can be negative. $\sqrt{x^2} = x$ is an identity only if x is replaced by 0 or a positive. Which of the following is true?

    a. $\sqrt{9} = 3$ b. $\sqrt{9} = -3$

$\sqrt{9} = 3$ is true
$\sqrt{9} = -3$ is false

29

Which of the following is true?

    a. $\sqrt{(-3)^2} = -3$ b. $\sqrt{(3)^2} = 3$

$\sqrt{9} = -3$ is false
$\sqrt{9} = 3$ is true

30

$\sqrt{x^2} = x$ is an identity only when x is non-negative. Complete the following, assuming that y is positive.

$\sqrt{y^2} = $ _____ .

y

31

Neither the radicand nor the square root can be negative. If z is a positive then $\sqrt{z^2} = $ _____ .

z

Simplifying Radical Expressions   61

32
$\sqrt{13^2} = $ _____   13

33
$\sqrt{(-103)^2} = $ _____   103

34
$\sqrt{(-942)^2} = $ _____   942

35
$\sqrt{58^2} = $ _____   58

36
The **cube** of 5 is $5^3 = 125$. The cube of 2 is $2^3 = $ _____.   8

37
The **cube root** of 125 is 5 because $5^3 = 125$.
The cube root of 8 is _____ because (___)^3 = 8.   2, 2

38
The cube root of 17 is indicated by $\sqrt[3]{17}$ in which
$\sqrt[3]{}$ is the **cube-root symbol** and 17 is the radicand.
What is the radicand of $\sqrt[3]{29}$?   29

39
The cube root of 43 is indicated by $\sqrt[3]{43}$.
Show the cube root of 37. $\sqrt[3]{37}$

40
$\sqrt[3]{7}$ is the real number, positive or negative, that
used three times as a factor will give 7.
$\sqrt[3]{7} \cdot \sqrt[3]{7} \cdot \sqrt[3]{7} = 7$
What will be the answer when $\sqrt[3]{11}$ is used three times
as a factor? $(\sqrt[3]{11})^3 = 11$

41
What will be the result when $\sqrt[3]{41}$ is cubed (raised
to the third power)?   41

42

$\sqrt[3]{53}$ is the positive real number that raised to the third power (cubed) will give 53.

$(\sqrt[3]{53})^3 = $ _____   53

43

$\sqrt[3]{100}$ is the positive real number that raised to the third power will give 100.

$(\sqrt[3]{100})^3 = $ _____   100

44

Notice that with a cube root the radicand and the root may be either positive or negative. If y and x are real numbers, y is the cube root of x if

$y^3 = x$ or, equivalently, $\sqrt[3]{x} = y$

The cube root of 8 ($\sqrt[3]{8}$) is 2 because $2^3 = 8$.
What is the cube root of -8?

$\sqrt[3]{-8} = -2$ because $(-2)^3 = -8$

45

A real number, x, and its cube root, y, are related by the identity: If $\sqrt[3]{x} = y$ then $x = y^3$.
The cube root of 125 is 5 because $5^3 = $ _____   125

46

The cube root of 8 can be shown as $\sqrt[3]{8}$ or as 2.
The cube root of 27 can also be shown in two ways.
The cube root of 27 is $\sqrt[3]{27}$ or _____.

3 because $3^3 = 27$

47

The cube root of 64 may be shown as either $\sqrt[3]{64}$ or 4 because 64 is a **perfect cube**. $64 = 4^3$
For most real numbers there is only one way to show their cube root. 41 is not a perfect cube and the only way to accurately show its cube root is _____.

$\sqrt[3]{41}$

48
The cube root of 29 is $\sqrt[3]{29}$. The relationship is:
$$(\sqrt[3]{29})^3 = 29 \quad \text{or} \quad \sqrt[3]{29^3} = 29$$
Complete the identity. $\sqrt[3]{41^3} = $ _____

41

49
If x is a real number, then $\sqrt[3]{x^3} = x$ is an identity.
Complete the identity. $\sqrt[3]{y^3} = $ _____.

y

50
A square root cannot be negative, but a cube root always has the same sign as the radicand. The cube root of a positive is positive and the cube root of a negative is _____.

negative

51
If z is a real number, then $\sqrt[3]{z^3} = $ _____.

z

52
A cube root always has the same sign as the radicand.
$\sqrt[3]{(-13)^3} = $ _____

-13

53
A cube root always has the same sign as the radicand.
$\sqrt[3]{103^3} = $ _____

103

54
$\sqrt[3]{942^3} = $ _____

942

55
$\sqrt[3]{(-58)^3} = $ _____

-58

56
The **fifth root** of 17 is indicated by $\sqrt[5]{17}$. The defining characteristic of $\sqrt[5]{17}$ is:
$$(\sqrt[5]{17})^5 = 17$$
The symbol $\sqrt[7]{43}$ indicates the _____ root of 43.

seventh

57

The fourth root of 47 is indicated by $\sqrt[4]{47}$. Write a symbol for the eighth root of 37.

$\sqrt[8]{37}$

58

The fifth root of -8 is indicated by $\sqrt[5]{-8}$. Write a symbol for the ninth root of -13.

$\sqrt[9]{-13}$

59

The sixth root of 19 is shown by $\sqrt[6]{19}$. Write a symbol for the fourth root of 59.

$\sqrt[4]{59}$

60

The symbolism $\sqrt[r]{x}$ represents the rth root of x. The counting number r is the **index of the radical**. What is the index of $\sqrt[3]{8}$?

3

61

$\sqrt{5}$ is the same as $\sqrt[2]{5}$ because an index of 2 is understood. What is the index of $\sqrt{7}$?

2

62

The index of $\sqrt[5]{13}$ is 5. What is the index of $\sqrt[4]{7^3}$?

4

63

The index of $\sqrt{7^5}$ is 2 and the exponent of 7 is 5. What is the index of $\sqrt[3]{5^2}$?

3

64

$\sqrt[5]{71}$ is the real number that used five times as a factor will give 71.

$(\sqrt[5]{71})^5 = 71$

What will be the answer when $\sqrt[6]{87}$ is used six times as a factor?

$(\sqrt[6]{87})^6 = 87$

65

When the index is an even number, the radicand must be non-negative for the radical expression to name a real number. For example, $\sqrt[4]{-16}$ cannot name a real number because the radicand is negative. Is $\sqrt[6]{-32}$ an expression that names a real number?

No, index is even and radicand negative

66

When the index is an odd number, any real number radicand names a real number. For example, $\sqrt[7]{-21}$ names a negative real number. Is $\sqrt[9]{-47}$ an expression that names a real number?

Yes, index is odd

It is important here to understand that negative radicands with even indices (plural of index) do not name real numbers. In the remainder of this unit, if variables are used for radicands with even indices, then the replacement set for those variables will be restricted to positive real numbers.

67

If x is a real number, $\sqrt[5]{x^5} = x$ is an identity.
Complete the identity. $\sqrt[9]{y^9} = $ _____

y

68

In general, the **nth root** of x is shown as $\sqrt[n]{x}$ and its defining identity is $\sqrt[n]{x^n} = x$. Complete the identity.
$\sqrt[13]{x^{13}} = $ _____

x

69

$\sqrt[5]{2^5} = 2$ $\sqrt[4]{19^4} = $ _____

19

70

The simplification of radical expressions depends on the identity $\sqrt[n]{x^n} = x$. Apply this identity and simplify $\sqrt[7]{z^7}$.

z

66 CHAPTER 2

71
When the index of a radical is the same as the exponent on its radicand, the simplification gives the base of the exponent.

$\sqrt[8]{y^8}$ = _____

y

72
To simplify $\sqrt[5]{x^{20}}$ the following steps are used.

$\sqrt[5]{x^{20}} = \sqrt[5]{(x^4)^5} = x^4$

x^4 is one of the _____ equal factors of x^{20}.

five

73
Complete the simplification. $\sqrt[3]{y^{21}} = \sqrt[3]{(y^7)^3}$ = _____

y^7

74
Complete the simplification. $\sqrt[6]{z^{24}} = \sqrt[6]{(z^4)^6}$ = _____

z^4

75
The index of $\sqrt[4]{x^{28}}$ is ___ and the first step of its simplification is to write the radicand as _____.

4
$(x^7)^4$

76
Complete the simplification. $\sqrt[4]{x^{28}} = \sqrt[4]{(x^7)^4}$ = _____

x^7

77
The index of $\sqrt{x^{12}}$ is ___ and the first step of its simplification is to write the radicand as _____.

2
$(x^6)^2$

78
Complete the simplification. $\sqrt{x^{12}} = \sqrt{(x^6)^2}$ = _____

x^6

79
$\sqrt[6]{y^{30}}$ = _____

$\sqrt[6]{(y^5)^6} = y^5$

80
$\sqrt[7]{x^{35}}$ = _____

$\sqrt[7]{(x^5)^7} = x^5$

81
$\sqrt[4]{x^{32}}$ = _____

$\sqrt[4]{(x^8)^4} = x^8$

82. $\sqrt{z^{42}} =$ _____ z^{21}

83. $\sqrt[3]{6^{15}} =$ _____ 6^5

84. $\sqrt[5]{9^{25}} =$ _____ 9^5

85. $\sqrt[7]{4^{42}} =$ _____ 4^6

Feedback Unit 2

This quiz reviews the preceding unit. Answers are at the back of the book.

1. $\sqrt[3]{(-15)^3} =$ _____

2. $\sqrt{3^2} =$ _____

3. $\sqrt[4]{312^4} =$ _____

4. $\sqrt{(-21)^2} =$ _____

5. $\sqrt[5]{927^5} =$ _____

6. $\sqrt[5]{-32} =$ _____

7. $\sqrt[4]{16} =$ _____

8. $\sqrt[3]{-125} =$ _____

9. $\sqrt{(x^3)^2} =$ _____

10. $\sqrt[5]{a^{25}} =$ _____

11. $\sqrt[4]{z^{12}} =$ _____

12. $\sqrt[7]{57^{14}} =$ _____

13. $\sqrt[3]{9^{15}} =$ _____

14. $\sqrt{13^6} =$ _____

15. $\sqrt[8]{a^{24}} =$ _____

16. $\sqrt[3]{21^{36}} =$ _____

Unit 3: Multiplying and Adding Nth Roots

The following mathematical terms are crucial to an understanding of this unit.

> Multiply nth roots Coefficient
> Nth root addition Like terms

1
To multiply $\sqrt{5} \cdot \sqrt{3}$, multiply the radicands 5 and 3.

$\sqrt{5} \cdot \sqrt{3} = \sqrt{15}$ and $\sqrt{7} \cdot \sqrt{3} =$ _____ $\sqrt{21}$

2
$\sqrt{3} \cdot \sqrt{13} = \sqrt{39}$ and $\sqrt{2} \cdot \sqrt{13} =$ _____ $\sqrt{26}$

3
The radicands of $\sqrt[3]{7}$ and $\sqrt[5]{6}$ cannot be multiplied because the indices are different. The best that can be done is to write the multiplication as $\sqrt[3]{7} \cdot \sqrt[5]{6}$. Write the multiplication of $\sqrt[4]{11}$ and $\sqrt[7]{5}$. $\sqrt[4]{11} \cdot \sqrt[7]{5}$

4
To **multiply nth roots**,
    a.   they must have the same index,
    b.   then multiply the radicands.

$\sqrt[4]{3} \cdot \sqrt[4]{13} = \sqrt[4]{39}$ and $\sqrt[8]{2} \cdot \sqrt[8]{17} =$ _____ $\sqrt[8]{34}$

5
The **coefficient** of $3\sqrt{5}$ is 3. The coefficient of $-\sqrt{11}$ is ____. -1

Simplifying Radical Expressions

6
To multiply $3\sqrt{5}$ and $-\sqrt{11}$, the coefficients and nth roots are separately multiplied.

$3\sqrt{5} \cdot -\sqrt{11} = (3 \cdot -1)(\sqrt{5} \cdot \sqrt{11}) = -3\sqrt{55}$
$8\sqrt{11} \cdot \sqrt{5} = $ _____ $8\sqrt{55}$

7
$2\sqrt[3]{5} \cdot 3\sqrt[3]{13} = (2 \cdot 3) \cdot (\sqrt[3]{5} \cdot \sqrt[3]{13}) = 6\sqrt[3]{65}$
$5\sqrt[6]{2} \cdot 7\sqrt[6]{3} = $ _____ $35\sqrt[6]{6}$

8
$5\sqrt{10} \cdot 7\sqrt{3} = $ _____ $35\sqrt{30}$

9
$8\sqrt[5]{7} \cdot \sqrt[5]{6} = $ _____ $8\sqrt[5]{42}$

10
$-\sqrt{5} \cdot 2\sqrt{3} = $ _____ $-2\sqrt{15}$

11
$-\sqrt{19} \cdot -\sqrt{3} = $ _____ $\sqrt{57}$

12
Nth roots with the same index are multiplied by placing the product of the radicands under a radical with the same index.

$\sqrt[4]{3x^2} \cdot \sqrt[4]{5x} = \sqrt[4]{15x^3}$ and $\sqrt[6]{7r} \cdot \sqrt[6]{5s} = $ _____ $\sqrt[6]{35rs}$

13
$\sqrt[3]{5} \cdot \sqrt[3]{7} = \sqrt[3]{35}$ and $\sqrt[4]{9} \cdot \sqrt[4]{7} = $ _____ $\sqrt[4]{63}$

14
$\sqrt[5]{x} \cdot \sqrt[5]{x^3} = \sqrt[5]{x^4}$ and $\sqrt[6]{4x^2} \cdot \sqrt[6]{13x^3} = $ _____ $\sqrt[6]{52x^5}$

15

When nth roots have coefficients, the example below shows how to multiply them.

$$9\sqrt[5]{2x^2} \cdot 5\sqrt[5]{7x^2} = 45\sqrt[5]{14x^4}$$

$$6\sqrt[7]{8y^3} \cdot 9\sqrt[7]{5y^2} = \underline{}$$

$54\sqrt[7]{40y^5}$

16

$$\sqrt[9]{5^4} \cdot \sqrt[9]{5^3} = \underline{}$$

$\sqrt[9]{5^7}$

17

$$\sqrt[7]{3x} \cdot \sqrt[4]{5x} = \underline{}$$

Indices are different.

18

$$\sqrt[5]{3y^2} \cdot \sqrt[5]{y^2} = \underline{}$$

$\sqrt[5]{3y^4}$

19

$$2\sqrt[6]{5x^2y^4} \cdot 6\sqrt[6]{4x^2y} = \underline{}$$

$12\sqrt[6]{20x^4y^5}$

20

$$-7\sqrt[9]{3x^3y^5} \cdot 5\sqrt[9]{7x^4y} = \underline{}$$

$-35\sqrt[9]{21x^7y^6}$

21

Nth root addition is possible only when both the indices and radicands are the same. Can $2\sqrt{6}$ and $5\sqrt{7}$ be added?

No, radicands are different.

22

Nth roots are like terms when they have the same index and radicand. Are $4\sqrt{13}$ and $5\sqrt{13}$ like terms?

Yes

23

The sum of nth roots that are **like terms** are found by adding the coefficients.

$$7\sqrt{6} + 5\sqrt{6} = 12\sqrt{6} \quad \text{and} \quad 8\sqrt{5} - 6\sqrt{5} = \underline{}$$

$2\sqrt{5}$

24

Are $-13\sqrt[4]{5x^3}$ and $6\sqrt[4]{5x^3}$ like terms?

Yes, index and radicand are the same.

SIMPLIFYING RADICAL EXPRESSIONS

25
The sum of $-13\sqrt[4]{5x^3}$ and $6\sqrt[4]{5x^3}$ are found by adding their coefficients.

$-13\sqrt[4]{5x^3} + 6\sqrt[4]{5x^3} = -7\sqrt[4]{5x^3}$

$8\sqrt[5]{6x^2} + 9\sqrt[5]{6x^2} = $ _____ $17\sqrt[5]{6x^2}$

26
$3\sqrt{3} + 4\sqrt{3} = 7\sqrt{3}$ and $5\sqrt{3} + 3\sqrt{3} = $ _____ $8\sqrt{3}$

27
$4\sqrt[7]{3x} + 8\sqrt[7]{3x} = $ _____ $12\sqrt[7]{3x}$

28
$-3\sqrt{3} + 5\sqrt{3} - 6\sqrt{3} = $ _____ $-4\sqrt{3}$

29
Simplify $4\sqrt{7} + 5\sqrt{3} + 2\sqrt{7}$ by adding like terms. $6\sqrt{7} + 5\sqrt{3}$

30
Simplify $11\sqrt[8]{5} + 7\sqrt[7]{5} - 8\sqrt[8]{5}$ by adding like terms. $3\sqrt[8]{5} + 7\sqrt[7]{5}$

31
Simplify $2\sqrt[5]{8} - 9\sqrt[5]{16} + 5\sqrt[5]{16}$ by adding like terms. $2\sqrt[5]{8} - 4\sqrt[5]{16}$

32
Simplify $\sqrt{3} - 5 + 3\sqrt{3} + 7$ by adding like terms. $4\sqrt{3} + 2$

33
$6\sqrt[4]{11} + 9 + 5\sqrt[4]{11} - 7 = $ _____ $11\sqrt[4]{11} + 2$

34
$6 - 4\sqrt{3} + 9 - \sqrt{3} = $ _____ $15 - 5\sqrt{3}$

35
$-3\sqrt{5} + 4\sqrt{3} + \sqrt{5} - 3\sqrt{3} = $ _____ $-2\sqrt{5} + \sqrt{3}$

36
$-3\sqrt[6]{5} - 5\sqrt{10} + \sqrt[6]{5} - 2\sqrt{10} = $ _____ $-2\sqrt[6]{5} - 7\sqrt{10}$

37

To multiply $2\sqrt{3}$ and $(\sqrt{5}+5\sqrt{7})$, each term of $(\sqrt{5}+5\sqrt{7})$ is multiplied by $2\sqrt{3}$.

$$2\sqrt{3}(\sqrt{5}+5\sqrt{7})$$
$$2\sqrt{3}\cdot\sqrt{5}+2\sqrt{3}\cdot 5\sqrt{7}$$
$$2\sqrt{15}+10\sqrt{21}$$

$6(2\sqrt{5}-3\sqrt{11}) = $ _____ $12\sqrt{5}-18\sqrt{11}$

38

The example below shows the multiplication of $\sqrt{3}$ by $(\sqrt{3}+\sqrt{7})$.

$$\sqrt{3}(\sqrt{3}+\sqrt{7})$$
$$\sqrt{3}\cdot\sqrt{3}+\sqrt{3}\cdot\sqrt{7}$$
$$3+\sqrt{21}$$

$\sqrt{5}(\sqrt{3}+\sqrt{2}) = $ _____ $\sqrt{15}+\sqrt{10}$

39

The example below contains some cube roots.

$$2\sqrt[3]{3}(7+3\sqrt[3]{7})$$
$$14\sqrt[3]{3}+6\sqrt[3]{21}$$

$\sqrt[4]{7}(9-5\sqrt[4]{2}) = $ _____ $9\sqrt[4]{7}-5\sqrt[4]{14}$

40

Remember that nth roots can be multiplied when they have the same index.

$$2\sqrt[7]{3}(5\sqrt[7]{11}-7)$$
$$(10\sqrt[7]{33}-14\sqrt[7]{3}$$

$5\sqrt[8]{6}(3\sqrt[8]{4}-2) = $ _____ $15\sqrt[8]{24}-10\sqrt[8]{6}$

41

$2\sqrt{5}(\sqrt{6}-8\sqrt{2}) = $ _____ $2\sqrt{30}-16\sqrt{10}$

42

$\sqrt{3}(2\sqrt{5}+5) = $ _____ $2\sqrt{15}+5\sqrt{3}$

43
$3(7\sqrt[6]{5} - \sqrt{2}) = $ _____

$21\sqrt[6]{5} - 3\sqrt{2}$

44
$5\sqrt[9]{11}(3\sqrt[9]{2} - 4\sqrt[9]{5}) = $ _____

$15\sqrt[9]{22} - 20\sqrt[9]{55}$

45
$-2\sqrt{6}(3\sqrt{5} + 7\sqrt{7}) = $ _____

$-6\sqrt{30} - 14\sqrt{42}$

46
$7\sqrt[3]{2}(10 - 3\sqrt[3]{2}) = $ _____

$70\sqrt[3]{2} - 21\sqrt[3]{4}$

47
$\sqrt[5]{3}(6 - 4\sqrt[5]{7}) = $ _____

$6\sqrt[5]{3} - 4\sqrt[5]{21}$

48
$4\sqrt[7]{3}(\sqrt[7]{9} - 2\sqrt[7]{15}) = $ _____

$4\sqrt[7]{27} - 8\sqrt[7]{45}$

49
To multiply $(\sqrt{7} + \sqrt{5})(\sqrt{3} + \sqrt{6})$ each term of $(\sqrt{3} + \sqrt{6})$ must be multiplied by each term of $(\sqrt{7} + \sqrt{5})$. This means there are four separate multiplications.

$$(\sqrt{7} + \sqrt{5})(\sqrt{3} + \sqrt{6})$$
$$\sqrt{7} \cdot \sqrt{3} + \sqrt{7} \cdot \sqrt{6} + \sqrt{5} \cdot \sqrt{3} + \sqrt{5} \cdot \sqrt{6}$$
$$\sqrt{21} + \sqrt{42} + \sqrt{15} + \sqrt{30}$$

Does this result contain any like terms?

No, radicands are different.

50
To multiply $(6 + \sqrt{2})(2 + \sqrt{5})$ there are four separate multiplications.

$$(6 + \sqrt{2})(2 + \sqrt{5})$$
$$6 \cdot 2 + 6 \cdot \sqrt{5} + \sqrt{2} \cdot 2 + \sqrt{2} \cdot \sqrt{5}$$
$$12 + 6\sqrt{5} + 2\sqrt{2} + \sqrt{10}$$

Does this result contain any like terms?

No

51

$(\sqrt{2} - \sqrt{5})(\sqrt{7} + \sqrt{3}) =$ _____

$\sqrt{14} + \sqrt{6} - \sqrt{35} - \sqrt{15}$

52

$(\sqrt{7} + \sqrt{2})(\sqrt{5} - \sqrt{3}) =$ _____

$\sqrt{35} - \sqrt{21} + \sqrt{10} - \sqrt{6}$

53
The following problem involves cube roots.

$$(4 - \sqrt[3]{3})(7 + \sqrt[3]{3})$$
$$4 \cdot 7 + 4 \cdot \sqrt[3]{3} - \sqrt[3]{3} \cdot 7 - \sqrt[3]{3} \cdot \sqrt[3]{3}$$
$$28 + 4\sqrt[3]{3} - 7\sqrt[3]{3} - \sqrt[3]{9}$$

Does this result contain any like terms?

Yes

54
Simplify $28 + 4\sqrt[3]{3} - 7\sqrt[3]{3} - \sqrt[3]{9}$ by adding its like terms.

$28 - 3\sqrt[3]{3} - \sqrt[3]{9}$

55
One of the terms obtained in the multiplication below is the square root of a perfect square. Notice how that effects the problem.

$$(6 + \sqrt{7})(5 + \sqrt{7})$$
$$6 \cdot 5 + 6 \cdot \sqrt{7} + \sqrt{7} \cdot 5 + \sqrt{7} \cdot \sqrt{7}$$
$$30 + 6\sqrt{7} + 5\sqrt{7} + \sqrt{49}$$
$$30 + 6\sqrt{7} + 5\sqrt{7} + 7$$

Does this result contain any like terms?

Yes

56
Simplify $30 + 6\sqrt{7} + 5\sqrt{7} + 7$ by adding its like terms.

$37 + 11\sqrt{7}$

57
Complete the multiplication and simplification shown below.

$$(9 + \sqrt{5})(2 - \sqrt{5})$$
$$9 \cdot 2 - 9 \cdot \sqrt{5} + \sqrt{5} \cdot 2 - \sqrt{5} \cdot \sqrt{5}$$
$$18 - 9\sqrt{5} + 2\sqrt{5} - \sqrt{25}$$

_____

$18 - 9\sqrt{5} + 2\sqrt{5} - 5$
$13 - 7\sqrt{5}$

SIMPLIFYING RADICAL EXPRESSIONS   75

58
Complete the multiplication and simplification shown below.

$$(2\sqrt{3} + \sqrt{5})(4\sqrt{3} + \sqrt{5})$$
$$2\sqrt{3} \cdot 4\sqrt{3} + 2\sqrt{3} \cdot \sqrt{5} + \sqrt{5} \cdot 4\sqrt{3} + \sqrt{5} \cdot \sqrt{5}$$
$$8\sqrt{9} + 2\sqrt{15} + 4\sqrt{15} + \sqrt{25}$$
$$8 \cdot 3 + 2\sqrt{15} + 4\sqrt{15} + 5$$
_____

$24 + 6\sqrt{15} + 5$
$29 + 6\sqrt{15}$

59
Complete the multiplication and simplification shown below.

$$(2\sqrt{5} - \sqrt{3})(7\sqrt{5} + \sqrt{2})$$
$$14\sqrt{25} + 2\sqrt{10} - 7\sqrt{15} - \sqrt{6}$$
_____

$70 + 2\sqrt{10} - 7\sqrt{15} - \sqrt{6}$

60
Complete the multiplication and simplification shown below.

$$(\sqrt{6} - \sqrt{5})(\sqrt{6} + \sqrt{5})$$
$$\sqrt{36} + \sqrt{30} - \sqrt{30} - \sqrt{25}$$
_____

$6 - 5 = 1$

61 $(3\sqrt[4]{5} - 4)(2\sqrt[4]{5} - 3) = $ _____

$6\sqrt[4]{25} - 17\sqrt[4]{5} + 12$

62 $(\sqrt[3]{7} + 4)(2\sqrt[3]{7} - 4) = $ _____

$2\sqrt[3]{49} + 4\sqrt[3]{7} - 16$

63 $(7\sqrt{3} - 2)(8\sqrt{3} + 5) = $ _____

$56\sqrt{9} + 35\sqrt{3} - 16\sqrt{3} - 10$
$158 + 19\sqrt{3}$

64 $(6\sqrt[4]{5} - 5)(2\sqrt[4]{5} + 7) = $ _____

$12\sqrt[4]{25} + 32\sqrt[4]{5} - 35$

65 $(5\sqrt{3} - 2)(5\sqrt{3} + 2) = $ _____

71

66 $(\sqrt{3} - \sqrt{11})(\sqrt{3} + \sqrt{11}) = $ _____

-8

Feedback Unit 3

This quiz reviews the preceding unit. Answers are at the back of the book.

1. $6\sqrt{5} \cdot 7\sqrt{3} =$ _____

2. $\sqrt{7} \cdot 2\sqrt{10} =$ _____

3. $5\sqrt{10} \cdot -3\sqrt{3} =$ _____

4. $\sqrt[3]{7} \cdot 3\sqrt[3]{6} =$ _____

5. $-\sqrt[5]{9} \cdot -2\sqrt[5]{10} =$ _____

6. $3\sqrt{3} + 8\sqrt{3} =$ _____

7. $3\sqrt{17} - 5\sqrt{3} - 7\sqrt{17} =$ _____

8. $-\sqrt{5} - 2\sqrt{6} + \sqrt{6} - \sqrt{5} =$ _____

9. $5\sqrt[3]{7} - 2\sqrt[3]{3} - 4\sqrt[3]{7} =$ _____

10. $3(\sqrt{5} - \sqrt{3}) =$ _____

11. $2\sqrt{7}(\sqrt{3} + \sqrt{6}) =$ _____

12. $\sqrt{5}(6\sqrt{3} - \sqrt{5}) =$ _____

13. $2\sqrt[6]{3}(4 - 3\sqrt[6]{5}) =$ _____

14. $(\sqrt{5} - 4)(2\sqrt{5} + 3) =$ _____

15. $(\sqrt{6} - 2)(\sqrt{5} + 7) =$ _____

16. $(2\sqrt{3} - 5)(\sqrt{7} + \sqrt{10}) =$ _____

17. $(\sqrt[7]{5} + 6)(3\sqrt[7]{8} + 1) =$ _____

18. $(5\sqrt[8]{7} - 4)(2\sqrt[8]{7} + 9) =$ _____

Unit 4: Simplifying Nth Roots

The following mathematical terms are crucial to an understanding of this unit.

    Simplify a square root   Simplify a cube root

1
The square root $\sqrt{20}$ is simplified by first writing it as a multiplication in which one of the radicands is a perfect square.

$$\sqrt{20} = \sqrt{4} \cdot \sqrt{5} = 2 \cdot \sqrt{5} = 2\sqrt{5}$$

The simplification is possible because $\sqrt{4} =$ ____.

    2

2
The simplification of a square root is made possible by the identity $\sqrt{x^2} = x$ for all positive real numbers x.

$$\sqrt{9} = 3 \quad \sqrt{16} = 4 \quad \sqrt{25} = 5 \quad \sqrt{36} = 6$$

The simplifications shown above are possible because 9, 16, 25, and 36 are _____ _____.

    perfect squares

3
To **simplify a square root**, find a factor of the radicand that is a perfect square. Each of the following is a perfect square.

    4, 9, 16, 25, 36, ____, ____, ____, and ____

    49, 64, 81, 100

4
Find a perfect square factor of 50.

    25

5
The simplification of $\sqrt{50}$ is shown below.

$$\sqrt{50} = \sqrt{25} \cdot \sqrt{2} = 5\sqrt{2}$$

Find a perfect square factor of 45.

    9

6

The simplification of $\sqrt{45}$ is shown below.

$$\sqrt{45} = \sqrt{9} \cdot \sqrt{5} = 3\sqrt{5}$$

Find a perfect square factor of 72.   4, 9, and 36

7

To simplify $\sqrt{72}$, use the largest perfect square factor of 72.

$$\sqrt{72} = \sqrt{36} \cdot \sqrt{2} = 6\sqrt{2}$$

Find a perfect square factor of 48.   4 and 16

8

To simplify $\sqrt{48}$, use the largest perfect square factor of 48.

$$\sqrt{48} = \sqrt{16} \cdot \sqrt{3} = 4\sqrt{3}$$

Find a perfect square factor of 60.   4

9

Simplify. $\sqrt{60}$ $\sqrt{4} \cdot \sqrt{15} = 2\sqrt{15}$

10

Simplify $\sqrt{200}$ using the largest perfect-square factor of 200. $\sqrt{100} \cdot \sqrt{2} = 10\sqrt{2}$

11

Simplify. $\sqrt{54}$ $\sqrt{9} \cdot \sqrt{6} = 3\sqrt{6}$

12

Simplify. $\sqrt{24}$ $\sqrt{4} \cdot \sqrt{6} = 2\sqrt{6}$

13

Simplify. $\sqrt{80}$ $\sqrt{16} \cdot \sqrt{5} = 4\sqrt{5}$

14

Simplify. $\sqrt{63}$ $3\sqrt{7}$

15
Simplify. $\sqrt{75}$

$5\sqrt{3}$

16
Simplify. $\sqrt{98}$

$7\sqrt{2}$

17
The example below shows how to handle a coefficient when simplifying a square root.

$$5\sqrt{18} = 5\sqrt{9} \cdot \sqrt{2} = 5 \cdot 3\sqrt{2} = \underline{\qquad}$$

$15\sqrt{2}$

18
Simplify. $6\sqrt{25}$

30

19
Simplify. $4\sqrt{125}$

$20\sqrt{5}$

20
Simplify. $5\sqrt{12}$

$10\sqrt{3}$

21
Simplify. $2\sqrt{300}$

$20\sqrt{3}$

22
Simplify. $-5\sqrt{24}$

$-10\sqrt{6}$

23
Simplify. $-3\sqrt{20}$

$-6\sqrt{5}$

24
The cube root $\sqrt[3]{40}$ is simplified by first writing it as a multiplication in which one of the radicands is a perfect cube.

$$\sqrt[3]{40} = \sqrt[3]{8} \cdot \sqrt[3]{5} = 2 \cdot \sqrt[3]{5} = 2\sqrt[3]{5}$$

The simplification is possible because $\sqrt[3]{8} = \underline{\qquad}$.

2

25

The simplification of a cube root is made possible by the identity $\sqrt[3]{x^3} = x$ for all real numbers x.

$$\sqrt[3]{27} = 3 \quad \sqrt[3]{64} = 4 \quad \sqrt[3]{125} = 5 \quad \sqrt[3]{216} = 6$$

The simplifications shown above are possible because 27, 64, 125, and 216 are _____ _____.

perfect cubes

26

To **simplify a cube root**, find a factor of the radicand that is a perfect cube. Each of the following is a perfect cube.

8, 27, 64, 125, 216, ____, ____, ____, and ____

343, 512, 729, 1000

27

Find a perfect cube factor of 54.

27

28

The simplification of $\sqrt[3]{54}$ is shown below.

$$\sqrt[3]{54} = \sqrt[3]{27} \cdot \sqrt[3]{2} = 3\sqrt[3]{2}$$

Find a perfect cube factor of 7000.

1000

29

The simplification of $\sqrt[3]{7000}$ is shown below.

$$\sqrt[3]{7000} = \sqrt[3]{1000} \cdot \sqrt[3]{7} = 10\sqrt[3]{7}$$

Find a perfect cube factor of 48.

8

30

Simplify $\sqrt[3]{48}$ using 8 as the largest perfect cube factor of 48.

$\sqrt[3]{8} \cdot \sqrt[3]{6} = 2\sqrt[3]{6}$

31

Simplify. $\sqrt[3]{1080}$

$\sqrt[3]{216} \cdot \sqrt[3]{5} = 6\sqrt[3]{5}$

SIMPLIFYING RADICAL EXPRESSIONS   81

32
In general, the simplification of nth roots is dependent upon the identity $\sqrt[n]{x^n} = x$ for any positive real number x. To simplify $\sqrt[4]{80}$ it is necessary to find a perfect 4th power factor of 80. To simplify $\sqrt[5]{192}$ it is necessary to find a _____ _____ power factor of 192.

perfect 5th

Because most of the perfect cubes, 4th powers, etc., are very large numbers, the remainder of the problems in this section show the radicands in factored form. The process for simplifying remains the same. In those cases where the index is even, assume that variables represent positive real numbers.

33
The simplification of $\sqrt[4]{32}$ depends upon the fact that $32 = 2^5$.
$$\sqrt[4]{32} = \sqrt[4]{2^5} = \sqrt[4]{2^4} \cdot \sqrt[4]{2} = 2\sqrt[4]{2}$$
Simplify $\sqrt[5]{192}$ using the fact that $192 = 32 \cdot 6 = 2^5 \cdot 6$.

$2\sqrt[5]{6}$

34
Simplify $\sqrt[6]{5^8}$ using the fact that $\sqrt[6]{5^6} = 5$.

$\sqrt[6]{5^6} \cdot \sqrt[6]{5^2} = 5\sqrt[6]{5^2}$

35
To simplify $\sqrt[3]{162}$, a perfect third power of 162 is needed. Since $162 = 2 \cdot 3^4$, what is a perfect third-power factor of 162?

3^3

36
Complete the simplification:
$$\sqrt[3]{162} = \sqrt[3]{2 \cdot 3^4} = \sqrt[3]{3^3} \cdot \sqrt[3]{2 \cdot 3} = \underline{\qquad}$$

$3\sqrt[3]{6}$

37
To simplify $\sqrt[7]{5 \cdot 3^8}$, a perfect 7th power of $5 \cdot 3^8$ is needed. What is a perfect 7th power factor of $5 \cdot 3^8$?

3^7

38
Simplify. $\sqrt[7]{5 \cdot 3^8}$

$\sqrt[7]{3^7} \cdot \sqrt[7]{5 \cdot 3^1} = 3\sqrt[7]{15}$

39
To simplify $\sqrt[4]{5^7 \cdot 6}$, a perfect 4th power of $5^7 \cdot 6$ is needed. What is a perfect 4th power factor of $5^7 \cdot 6$?

5^4

40
Simplify. $\sqrt[4]{5^7 \cdot 6}$

$\sqrt[4]{5^4} \cdot \sqrt[4]{5^3 \cdot 6}$
$5\sqrt[4]{5^3 \cdot 6}$

41
To simplify $\sqrt[7]{x^{23}}$, it is necessary to find a factor of x^{23} that is a perfect 7th power. x^7, x^{14}, and x^{21} are perfect 7th power factors of x^{23} because $(x)^7 = x^7$, $(x^2)^7 = x^{14}$, and $(x^3)^7 =$ _____

x^{21}

42
Complete the simplification.
$\sqrt[7]{x^{23}} = \sqrt[7]{x^{21}} \cdot \sqrt[7]{x^2} = \sqrt[7]{(x^3)^7} \cdot \sqrt[7]{x^2} =$ _____

$x^3 \sqrt[7]{x^2}$

43
What is the largest 4th power factor of x^{13}?

$x^{12} = (x^3)^4$

44
Complete the simplification.
$\sqrt[4]{x^{13}} = \sqrt[4]{x^{12}} \cdot \sqrt[4]{x^1} = \sqrt[4]{(x^3)^4} \cdot \sqrt[4]{x} =$ _____

$x^3 \sqrt[4]{x}$

45
What is the largest 6th power factor of x^{47}?

$x^{42} = (x^6)^7$

46
Complete the simplification.
$\sqrt[6]{x^{47}} = \sqrt[6]{x^{42}} \cdot \sqrt[6]{x^5} = \sqrt[6]{(x^7)^6} \cdot \sqrt[6]{x^5} =$ _____

$x^7 \sqrt[6]{x^5}$

47
Complete the simplification:
$\sqrt[8]{x^{39}} = \sqrt[8]{x^{32}} \cdot \sqrt[8]{x^7} =$ _____.

$\sqrt[8]{(x^4)^8} \cdot \sqrt[8]{x^7} = x^4 \sqrt[8]{x^7}$

48
Simplify $\sqrt[3]{x^{25}}$ by first finding the largest 3rd power factor of x^{25}.

$\sqrt[3]{x^{24}} \cdot \sqrt[3]{x} = x^8 \sqrt[3]{x}$

Simplifying Radical Expressions

49
Simplify $\sqrt[5]{x^{32}}$ by first finding the largest 5th power factor of x^{32}.

$\sqrt[5]{x^{30}} \cdot \sqrt[5]{x^2} = x^6 \sqrt[5]{x^2}$

50
Simplify. $\sqrt[4]{x^{19}}$

$x^4 \sqrt[4]{x^3}$

51
Simplify. $\sqrt[6]{x^{35}}$

$x^5 \sqrt[6]{x^5}$

52
In Unit 3, nth root expressions were added when the index and radicand were identical. Now that process can be expanded to some addition expressions which first require simplification. The sum of $7\sqrt{162} - 12\sqrt{2}$ can be found by first simplifying $\sqrt{162}$.

$$7\sqrt{162} - 12\sqrt{2}$$
$$7\sqrt{81} \cdot \sqrt{2} - 12\sqrt{2}$$
$$7 \cdot 9 \cdot \sqrt{2} - 12\sqrt{2}$$
$$63\sqrt{2} - 12\sqrt{2}$$
$$51\sqrt{2}$$

Add $5\sqrt{12} - 6\sqrt{3}$ by first simplifying $5\sqrt{12}$.

$10\sqrt{3} - 6\sqrt{3} = 4\sqrt{3}$

53
Add $3\sqrt{2} + \sqrt{8}$ by first simplifying $\sqrt{8}$.

$3\sqrt{2} + 2\sqrt{2} = 5\sqrt{2}$

54
Add $3\sqrt{5} + \sqrt{45}$ by first simplifying $\sqrt{45}$.

$6\sqrt{5}$

55
$\sqrt{18} + \sqrt{50} = $ _____

$8\sqrt{2}$

56
$\sqrt{27} - \sqrt{12} = $ _____

$\sqrt{3}$

57
$\sqrt{18} - \sqrt{2} + \sqrt{16} = $ _____

$2\sqrt{2} + 4$

58 $\sqrt{50} - \sqrt{45} - \sqrt{32} + \sqrt{80} =$ _____ $\sqrt{2} + \sqrt{5}$

59 $5\sqrt[3]{32} + 3\sqrt[3]{4} =$ _____ $13\sqrt[3]{4}$

Feedback Unit 4

This quiz reviews the preceding unit. Answers are at the back of the book.

Simplify the following expressions.

1. $\sqrt{225} =$ _____
2. $\sqrt{98} =$ _____
3. $-\sqrt{40} =$ _____
4. $\sqrt{128} =$ _____
5. $\sqrt{56} =$ _____
6. $\sqrt{8} \cdot \sqrt{8} =$ _____
7. $\sqrt{6^2} =$ _____
8. $\sqrt{8} =$ _____
9. $\sqrt{28} =$ _____
10. $\sqrt{63} =$ _____
11. $5\sqrt{18} =$ _____
12. $-\sqrt{20} =$ _____
13. $\sqrt{40} =$ _____
14. $\sqrt{80} =$ _____
15. $\sqrt[3]{375} =$ _____
16. $\sqrt[4]{162} =$ _____
17. $\sqrt[5]{96} =$ _____
18. $\sqrt[5]{2^5 \cdot 3} =$ _____
19. $\sqrt[4]{3^7 \cdot 5} =$ _____
20. $\sqrt[9]{7^9 \cdot 3} =$ _____
21. $\sqrt[5]{x^9} =$ _____
22. $\sqrt[6]{x^{19}} =$ _____
23. $\sqrt[7]{x^{38}} =$ _____
24. $3\sqrt{8} + \sqrt{32} =$ _____
25. $\sqrt{18} - \sqrt{48} - \sqrt{32} + \sqrt{27} =$ _____

UNIT 5: THE IMAGINARY NUMBER, i

The following mathematical terms are crucial to an understanding of this unit.

    Imaginary number i   Multiply imaginary numbers
    Add imaginary numbers

1
Radical signs with even-number indices must have non-negative radicands if they are to name real numbers. Is $\sqrt{51}$ a real number?

    Yes

2
The square of any positive or negative real number is a positive real number. Is $\sqrt{-85}$ a real number?

    No, a real number squared cannot be negative.

3
Radical signs with odd-number indices and real-number radicands always name real numbers. Is $\sqrt[7]{-13}$ a real number?

    Yes, a negative number raised to the 7th power can be negative.

4
$\sqrt{59}$ is the positive real number, k, such that k • k or $k^2 = 59$. k is between the integers 7 and 8, because $7^2 = 49$ and $8^2 = 64$. $\sqrt{97}$ is a _____ (positive, negative) real number.

    positive

5
$-\sqrt{47}$ is the negative real number which is between -6 and -7, because $-\sqrt{36} = -6$ and $-\sqrt{49} = -7$. $-\sqrt{15}$ is a _____ (positive, negative) real number.

    negative

86 CHAPTER 2

6

$\sqrt{46}$ is a positive real number between 6 and 7.
$\sqrt{78}$ is between _____ and _____.

8, 9

7

$-\sqrt{85}$ is a negative real number between -9 and -10.
$-\sqrt{35}$ is between _____ and _____.

-5, -6

8

$\sqrt{-38}$ is not a real number. There is no real number that can be squared to give -38. Is $\sqrt{-91}$ a real number?

No

9
Which of the following is a positive real number?

$\sqrt{59}$ $-\sqrt{13}$ $\sqrt{-43}$

$\sqrt{59}$

10
Which of the following is a negative real number?

$-\sqrt{37}$ $\sqrt{-41}$ $\sqrt{51}$

$-\sqrt{37}$

11
Which of the following is not a real number?

$-\sqrt{93}$ $\sqrt{-46}$ $\sqrt{95}$

$\sqrt{-46}$

12
Is $\sqrt{-1}$ a real number?

No

13
The **imaginary number i** is defined as: $i = \sqrt{-1}$.
Is i a real number?

No

14
By using the imaginary number i, square roots such as $\sqrt{-13}$ can be simplified.

$$\sqrt{-13} = \sqrt{-1 \cdot 13} = \sqrt{-1} \cdot \sqrt{13} = i\sqrt{13}$$

Simplify $\sqrt{-47}$ using the same process.

$\sqrt{-1} \cdot \sqrt{47} = i\sqrt{47}$

15
Is $\sqrt{-5}$ a real number?   No

16
Simplify $\sqrt{-5}$ using the fact that $\sqrt{-5} = \sqrt{-1} \cdot \sqrt{5}$. $i\sqrt{5}$

17
Simplify. $\sqrt{-91}$ $i\sqrt{91}$

18
Simplify. $\sqrt{-87}$ $i\sqrt{87}$

19
Simplify. $\sqrt{-19}$ $i\sqrt{19}$

20
Simplify. $\sqrt{-2}$ $i\sqrt{2}$

21
$\sqrt{-4}$ is simplified as follows:
$\sqrt{-4} = \sqrt{-1} \cdot \sqrt{4} = i \cdot 2 = 2i$
Simplify. $\sqrt{-25}$ $5i$

22
Simplify. $\sqrt{-49}$ $7i$

23
Simplify. $\sqrt{-81}$ $9i$

24
To simplify $\sqrt{-24}$, the following steps are used:
$\sqrt{-24} = \sqrt{-1} \cdot \sqrt{24} = i \cdot \sqrt{4} \cdot \sqrt{6} = i \cdot 2\sqrt{6} = 2i\sqrt{6}$
Simplify. $\sqrt{-28}$ $2i\sqrt{7}$

25
Simplify. $\sqrt{-45}$ $3i\sqrt{5}$

26
Simplify. $\sqrt{-8}$ $2i\sqrt{2}$

27
Simplify. $\sqrt{-27}$

$3i\sqrt{3}$

28
Simplify. $\sqrt{-20}$

$2i\sqrt{5}$

29
To simplify $-\sqrt{-50}$, use -1 as the coefficient of the radical.

$-\sqrt{-50} = -\sqrt{-1} \cdot \sqrt{50} = -1 \cdot i \cdot \sqrt{25} \cdot \sqrt{2} = -5i \cdot \sqrt{2}$

Simplify. $-\sqrt{-27}$

$-3i\sqrt{3}$

30
To simplify $-3\sqrt{-98}$, use -3 as the coefficient of the radical.

$-3\sqrt{-98} = -3\sqrt{-1} \cdot \sqrt{49} \cdot \sqrt{2} = -3i \cdot 7 \cdot \sqrt{2} = \underline{}$

$-21i\sqrt{2}$

31
Simplify. $4\sqrt{-75}$

$20i\sqrt{3}$

32
Simplify. $5\sqrt{-32}$

$20i\sqrt{2}$

33
Simplify. $-7\sqrt{-48}$

$-28i\sqrt{3}$

34
Simplify. $12\sqrt{-18}$

$36i\sqrt{2}$

35
To simplify $6\sqrt{-16}$, the following steps are used:

$6\sqrt{-16} = 6 \cdot \sqrt{-1} \cdot \sqrt{16} = 6 \cdot i \cdot 4 = 24i$

Simplify. $-2\sqrt{-9}$

$-6i$

36
Simplify. $15\sqrt{-25}$

$75i$

37
Simplify. $-4\sqrt{-44}$

$-8i\sqrt{11}$

38
Simplify. $10\sqrt{-20}$

$20i\sqrt{5}$

39
Simplify. $8\sqrt{-81}$

$72i$

40
Simplify. $-3\sqrt{-24}$

$-6i\sqrt{6}$

41
The square of any real number cannot be negative.

$(5)^2 = 25$
$(-4)^2 = 16$
$0^2 = 0$

Can the square of a real number be a negative number?

No

42
The square of any real number cannot be
a _____ number.

negative

43
The imaginary number i can be squared to give -1.
$i^2 = -1$
The square of a real number can never be negative.
Since $i^2 = -1$, is i a real number?

No

44
The imaginary number i is not a real number. i has no position on the real number line because i is not
a _____ number.

real

45
i is an imaginary number. $i^2 =$ _____.

-1

46
i is not a real number. i is a(n) _____
(real, imaginary) number.

imaginary

90 CHAPTER 2

47

$\sqrt{-1}$ = _____   i

48

i^2 = _____   -1

49

i is an imaginary number. What number can be squared to give -1?   i

50

$i^2 = -1$. To **multiply imaginary numbers**, the following steps are used:

$3i \cdot 2i = 6i^2 = 6 \cdot -1 = -6$

Complete the multiplication problem:

$5i \cdot 4i = 20i^2 = 20 \cdot -1 =$ _____   -20

51

Complete the multiplication problem:

$7i \cdot 5i = 35i^2 = 35 \cdot -1 =$ _____   -35

52

$6i \cdot 3i =$ _____   -18

53

To multiply -5i and 3i, the following steps are used:

$-5i \cdot 3i = -15i^2 = -15 \cdot -1 = 15$

Complete the multiplication problem:

$-7i \cdot 2i = -14i^2 = -14 \cdot -1 =$ _____   14

54

Complete the multiplication problem:

$-4i \cdot 6i = -24i^2 = -24 \cdot -1 =$ _____   24

55

Complete the multiplication problem:

$4i \cdot -9i = -36i^2 = -36 \cdot -1 =$ _____   36

56

$-3i \cdot 8i =$ _____   24

57

$-6i \cdot 8i =$ _____   48

58
To multiply -3i and -6i, the following steps are used:

$-3i \cdot -6i = 18i^2 = 18 \cdot -1 =$ _____   -18

59
Complete the multiplication problem:

$-5i \cdot -8i = 40i^2 = 40 \cdot -1 =$ _____   -40

60

$-2i \cdot -5i =$ _____   -10

61

i = 1i and -i = -1i
$4i \cdot -i =$ _____   4

62

i = 1i and -i = -1i
$-i \cdot i =$ _____   1

63

$-5i \cdot -i =$ _____   -5

64

$i \cdot i =$ _____   -1

65

$4i \cdot 3i =$ _____   -12

66

$-3i \cdot 9i =$ _____   27

67

$4i \cdot -8i =$ _____   32

68

$i\sqrt{3} \cdot i\sqrt{3}$ is simplified as
$i\sqrt{3} \cdot i\sqrt{3} = i^2\sqrt{9} = -1 \cdot 3 = -3$
$i\sqrt{5} \cdot i\sqrt{5} = $ _____   -5

69

$i\sqrt{7} \cdot i\sqrt{7} = i^2\sqrt{49} = -1 \cdot 7 = -7$
$i\sqrt{6} \cdot i\sqrt{6} = $ _____   -6

70

$i\sqrt{10} \cdot i\sqrt{10} = $ _____   -10

71

4i • -3i = _____   12

72

$i\sqrt{5} \cdot i\sqrt{5} = $ _____   -5

73

To **add imaginary numbers**, add the coefficients of i.

5i + -8i = -3i   6i + 13i = 19i

Add. 7i + 3i   10i

74

To simplify 8i – 13i, add the coefficients.

8i – 13i = -5i   and   12i – 5i = _____   7i

75

3i – 8i + 5i – 10i = _____   -10i

76

To simplify 5 + 3i – 2 – 7i, add like terms.

5 + 3i – 2 – 7i = (5 – 2) + (3i – 7i) = 3 – 4i

Simplify. 4 – 9i – 7 + 13i   -3 + 4i

77

13 – 5i + 5 – i = _____   18 – 6i

78. $-4i + 14i - 7i + 3 =$ _____   3i + 3 or 3 + 3i

79. $-4 + 8i - 4i + 9 =$ _____   5 + 4i

80. $i + 3 - i + 7i - 3 =$ _____   7i

Feedback Unit 5

This quiz reviews the preceding unit. Answers are at the back of the book.

1. The square of any real number is never _____ (positive, negative).

2. Which of the following are imaginary numbers?
   $\sqrt{4}$, 3, $\sqrt{-2}$, $-\sqrt{4}$, i, 3i

3. $2i \cdot 4i =$ _____

4. $-i \cdot 3i =$ _____

5. $i \cdot i =$ _____

6. $i\sqrt{3} \cdot i\sqrt{3} =$ _____

7. $4i \cdot -i =$ _____

8. $-13 + 5i - 4 - 8i =$ _____

9. $8i - 3i - 17i =$ _____

10. $7 - 4i - 9 - 13i =$ _____

11. $7 - 3i + 8 + 3i =$ _____

12. $-14 + 6i + 14 - 6i =$ _____

13. $4 - 7i + 14 - 12i =$ _____

14. $-13i + 5 - i - 5 =$ _____

15. $7 - 3i - 5 + 2i =$ _____

UNIT 6: APPLICATIONS

In this Applications Section, the format of the text has been altered. Answers for the problems appear beneath them rather than in the right-hand column. Your studying emphasis should be on learning the best procedures to follow with word problems. For that reason, once the procedure is learned a calculator may be used to complete the answer.

1

In solving word problems, it is often helpful to make drawings or tables that describe and/or organize the known and unknown facts. For example, if 2 trains leave the same town and travel in opposite directions, the drawing below may be helpful.

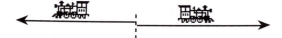

Suppose that the situation is changed. The 2 trains leave the same town and travel in the same direction. What would be an appropriate drawing?

Answer: The arrows indicating the paths of the 2 trains need to have the same starting point and go in the same direction. The arrows should be of different lengths because one train is likely to go further than the other.

2

Suppose Train #1 leaves Town A and heads toward Town B and Train #2 leaves Town B and heads toward Town A. Make a drawing that would describe the situation?

Answer: The arrows indicating the paths of the 2 trains need to go in opposite directions. One arrow starts at A and the other at B.

3

Distance problems usually involve the formula D = RT where D is the distance, R is the uniform rate of speed, and T is the time. For example, the distance traveled by a car going 50 miles per hour (mph) for 3 hours is D = 50 • 3 or 150 miles. Find the distance traveled by a car going 45 mph for 4 hours.

Answer: 180 miles. Use the formula D = RT with R = 45 and T = 4. D = 45 • 4 = 180.

4

A table may be set up for showing information in a distance problem. For example, the table below includes all the information given in this situation: A hiker walks 5 hours at a rate of 4 mph.

	Rate (R)	Time (T)	Distance (D = RT)
The hiker	4 mph	5 hours	20 miles

Construct a table like the one above for the following situation: An airplane travels at a rate of 650 mph for 3 hours.

Answer:

	Rate (R)	Time (T)	Distance (D = RT)
The airplane	*650 mph*	*3 hours*	*1950 miles*

5

A table like the one used in frame 4 can be constructed even when some of the information is unknown. For example, the table shown below organizes the information for the following situation: A train travels at a uniform rate of 48 mph between two cities.

	Rate (R)	Time (T)	Distance (D = RT)
The train	48 mph	T hours	48T miles

Notice that the number of hours traveled was not stated. Since T is unknown it is represented by its variable. The distance, D, is also unknown, but another variable is not necessary because the distance can be represented as 48T.

Construct a table like the one shown above for the following situation. A car travels 6 hours at a uniform speed.

Answer:

	Rate (R)	Time (T)	Distance (D = RT)
The car	R mph	6 hours	6R miles

6

The table below illustrates how the distance information for 2 trains may be organized. The situation is: Two trains leave the same town at the same time traveling in opposite directions. Train #1 goes 37 mph for 4 hours and stops for an hour. Train #2 goes 23 mph for 5 hours.

	Rate (R)	Time (T)	Distance (D = RT)
Train #1	37 mph	4 hours	148 miles
Train #2	23 mph	5 hours	115 miles

Make a drawing of the situation above and find the distance between the 2 trains after Train #2 has traveled 5 hours.

Answer: The drawing must show the distance arrows for the 2 trains going in opposite directions.

The distance arrow of Train #1 is longer than that of Train #2. Each is labeled by the distance traveled. The distance between the trains at that time will be the sum of 148 and 115, so it is 263 miles.

Simplifying Radical Expressions

Feedback Unit 6 for Applications

This quiz reviews the preceding unit. Answers are at the back of the book.

1. Two cars leave the same town and travel in opposite directions. What would be an appropriate drawing for this situation?

2. Two river boats leave the same dock and both travel downstream. What would be an appropriate drawing for this situation?

3. A hiker leaves camp Monday morning. Late that afternoon she stops and camps. The next day she continues her hike in the same direction. What would be an appropriate drawing for this situation?

4. A table like the one shown below may be set up for showing information in a distance problem.

	Rate (R)	Time (T)	Distance (D = RT)
The hiker	4 mph	5 hours	20 miles

   Construct a table like the one above for the following situation: An airboat travels at a rate of 18 mph for 2 hours.

5. Construct a distance table for showing the following information. A car travels 8 hours at a uniform speed.

6. Two trains leave the same town and travel on parallel tracks in the same direction. Train#1 travels at a uniform speed of 45 mph. Train #2 travels at a uniform speed of 62 mph. Construct a table for showing this information and find the distance between the 2 trains after 3 hours.

Summary for Chapter 2

The following mathematical terms are crucial to an understanding of this chapter.

Factors	Exponent
Squared	Cubed
To the fourth power	Base
Perfect squares	Perfect cubes
Perfect fourth powers	Perfect fifth powers
Power expression	Multiplication of power expressions
Simplifying power multiplications	Raising a power to a power
Dividing power expressions	Square root
Radical sign, $\sqrt{}$	Radicand
Table of square roots	Perfect square integer
Simplest form (of a radical)	Cube
Cube root	Cube-root symbol
Perfect cube	Fifth root
Index of the radical	Nth root
Coefficient	Multiply nth roots
Like terms	Nth root addition
Simplify a cube root	Simplify a square root
Multiply imaginary numbers	Imaginary number i
Add imaginary numbers	

This chapter begins with a review of exponents and their related terminology, but the emphasis is upon radical expressions.

The radical expression $\sqrt[7]{13}$ has an index of 7 with a radicand of 13.
The radical expression $\sqrt[6]{5 \cdot 7^6}$ can be simplified because a factor of the radicand is the same power as the index.
$$\sqrt[6]{5 \cdot 7^6} = 7\sqrt[6]{5}$$
Two radical expressions can be multiplied when they have the same index.
$$\sqrt[4]{5} \cdot \sqrt[4]{11} = \sqrt[4]{55}$$
Two radical expressions can be added when the radicals are identical.
$$8\sqrt[5]{7} + 9\sqrt[5]{7} = 17\sqrt[5]{7}$$

The square root of a negative number cannot be a real number. This gives rise to the concept of the imaginary number i which has $i^2 = -1$ as its defining attribute. Using i, square roots with negative radicands can be written as the product of i with a real number. $\sqrt{-7} = i\sqrt{7}$

Simplifying Radical Expressions 99

Chapter 2 Mastery Test

The following questions test the objectives of Chapter 2. Answers are at the back of the book. The number in parentheses which follows each problem indicates the unit in which it can be learned.

1. $6^4 \cdot 6^9 =$ _____ (1)

2. $(7^5)^3 =$ _____ (1)

3. $\dfrac{5^6}{5^2} =$ _____ (1)

4. Is 729 a perfect cube? (1)

5. $\sqrt[6]{5^6} =$ _____ (2)

6. $\sqrt[9]{(-19)^9} =$ _____ (2)

7. $\sqrt[3]{-216} =$ _____ (2)

8. $\sqrt[5]{7^{40}} =$ _____ (2)

9. $\sqrt[3]{y^{18}} =$ _____ (2)

10. $\sqrt{3} \cdot \sqrt{5} =$ _____ (3)

11. $\sqrt{3} - 2\sqrt{3} =$ _____ (3)

12. $2\sqrt[3]{6} \cdot -5\sqrt[3]{2} =$ _____ (3)

13. $4\sqrt[6]{5x^2} - 7\sqrt[6]{5x^2} =$ _____ (3)

14. $-2(4\sqrt{3} - 7\sqrt{2}) =$ _____ (3)

15. $(2\sqrt{3} - \sqrt{5})(2\sqrt{3} + 3\sqrt{5}) =$ _____ (3)

16. $\sqrt{45} =$ _____ (4)

17. $3\sqrt{99} =$ _____ (4)

18. $-2\sqrt{48} =$ _____ (4)

19. $\sqrt[4]{81} =$ _____ (4)

20. $\sqrt[3]{192} =$ _____ (4)

21. $\sqrt[5]{7^5 \cdot 9} =$ _____ (4)

22. $\sqrt[7]{x^{24}} =$ _____ (4)

23. $4\sqrt{7} + 9\sqrt{75} - \sqrt{147} =$ _____ (4)

24. $2i \cdot 3i =$ _____ (5)

25. $-i \cdot 4i =$ _____ (5)

26. $-13 - i + 14 + 2i =$ _____ (5)

27. Two cars leave two different towns and travel toward each other. What would be an appropriate drawing for this situation? (6)

28. A bicycler leaves home at 8 a.m. traveling due east. At noon she turns around and heads for home. What would be an appropriate drawing for this situation at 2 p.m.? (6)

Chapter 3 Objectives

The following problems illustrate the objectives of this chapter. At this time you are not expected to know how to do these problems. However, if all these problems are thoroughly understood, proceed directly to the Chapter Mastery Test. The number in parentheses which follows each problem indicates the unit in which it can be learned.

Simplify each of the following.

1. $\dfrac{\sqrt{18}}{\sqrt{50}}$ (1)

2. $\dfrac{\sqrt{6}}{\sqrt{30}}$ (1)

3. $\dfrac{6\sqrt{7}}{\sqrt{3}}$ (1)

4. $\dfrac{5\sqrt{11}}{10\sqrt{5}}$ (1)

5. $\dfrac{14\sqrt{48}}{8\sqrt{6}}$ (1)

6. $\sqrt{\dfrac{11}{3}}$ (1)

7. $\dfrac{2}{9}\sqrt{\dfrac{3}{2}}$ (1)

8. $\dfrac{\sqrt[7]{32}}{\sqrt[7]{4}}$ (3)

9. $\dfrac{10\sqrt[3]{6}}{\sqrt[3]{125}}$ (3)

10. $\dfrac{\sqrt[9]{5^5 \cdot 7^4}}{\sqrt[9]{5^3 \cdot 7^{13}}}$ (3)

11. $\dfrac{28\sqrt[6]{3^3}}{21\sqrt[6]{3^5}}$ (3)

12. $\dfrac{9\sqrt[4]{7}}{15\sqrt[4]{81}}$ (3)

13. $\dfrac{-6\sqrt[3]{4}}{\sqrt[3]{36}}$ (3)

14. $\dfrac{6\sqrt[3]{4000}}{5\sqrt[3]{6}}$ (3)

15. $\dfrac{4}{5-\sqrt{6}}$ (4)

16. $\dfrac{\sqrt{6}}{1-\sqrt{2}}$ (4)

17. $\dfrac{6-5\sqrt{3}}{2+\sqrt{3}}$ (4)

18. $\dfrac{7+3\sqrt{7}}{7-3\sqrt{7}}$ (4)

19. $\dfrac{\sqrt{2}+4\sqrt{5}}{\sqrt{2}-2\sqrt{5}}$ (4)

20. $\dfrac{2\sqrt{3}+\sqrt{5}}{3\sqrt{3}+2\sqrt{5}}$ (4)

Chapter 3
Simplifying Radical Fractions

Unit 1: Simplifying a Division of Square Roots

The following mathematical terms are crucial to an understanding of this unit.

    Reduce a fraction   Rationalizing the denominator

1
Any nonzero number divided by itself has a quotient of 1.

$$\frac{7}{7} = 1 \qquad \frac{15}{15} = 1 \qquad \frac{109}{109} = \underline{\qquad} \qquad\qquad 1$$

2
When a radical expression is divided by itself, the quotient is 1.

$$\frac{\sqrt{22}}{\sqrt{22}} = 1 \qquad \frac{\sqrt{7}}{\sqrt{7}} = 1 \qquad \frac{\sqrt{47}}{\sqrt{47}} = \underline{\qquad} \qquad\qquad 1$$

3

$$\frac{\sqrt{15}}{\sqrt{15}} = 1 \quad \text{and} \quad \frac{\sqrt{91}}{\sqrt{91}} = \underline{\qquad}$$

1

4

To **reduce a fraction**, divide out a common factor.

$$\frac{\sqrt{15}}{\sqrt{21}} = \frac{\sqrt{3} \cdot \sqrt{5}}{\sqrt{3} \cdot \sqrt{7}} = \frac{\cancel{\sqrt{3}} \cdot \sqrt{5}}{\cancel{\sqrt{3}} \cdot \sqrt{7}} = \frac{\sqrt{5}}{\sqrt{7}}$$

What was the common factor in this example?

$\sqrt{3}$

5

To reduce $\frac{\sqrt{10}}{\sqrt{15}}$, divide out the common factor $\sqrt{5}$.

$$\frac{\sqrt{10}}{\sqrt{15}} = \frac{\sqrt{5} \cdot \sqrt{2}}{\sqrt{5} \cdot \sqrt{3}} = \frac{\cancel{\sqrt{5}} \cdot \sqrt{2}}{\cancel{\sqrt{5}} \cdot \sqrt{3}} = \frac{\sqrt{2}}{\sqrt{3}}$$

Reduce $\frac{\sqrt{22}}{\sqrt{77}}$ by dividing out the common factor $\sqrt{11}$.

$\frac{\sqrt{2}}{\sqrt{7}}$

6

Reduce $\frac{\sqrt{15}}{\sqrt{6}}$ by dividing out the common factor $\sqrt{3}$.

$\frac{\sqrt{5}}{\sqrt{2}}$

7

In some cases, reducing a fraction will eliminate the radical from the numerator.

$$\frac{\sqrt{19}}{\sqrt{38}} = \frac{\cancel{\sqrt{19}} \cdot \sqrt{1}}{\cancel{\sqrt{19}} \cdot \sqrt{2}} = \frac{\sqrt{1}}{\sqrt{2}} = \frac{1}{\sqrt{2}}$$

Reduce. $\frac{\sqrt{5}}{\sqrt{30}}$

$\frac{1}{\sqrt{6}}$

8

Reduce. $\frac{\sqrt{10}}{\sqrt{30}}$

$\frac{1}{\sqrt{3}}$

9

Reduce. $\frac{\sqrt{42}}{\sqrt{35}}$

$\frac{\sqrt{6}}{\sqrt{5}}$

10

Reduce. $\frac{\sqrt{7}}{\sqrt{14}}$

$\frac{1}{\sqrt{2}}$

SIMPLIFYING RADICAL FRACTIONS 103

11
In some cases, reducing a fraction will eliminate the radical from the denominator.

$$\frac{\sqrt{22}}{\sqrt{11}} = \frac{\sqrt{11} \cdot \sqrt{2}}{\sqrt{11} \cdot \sqrt{1}} = \frac{\sqrt{2}}{\sqrt{1}} = \frac{\sqrt{2}}{1} = \sqrt{2}$$

Reduce. $\dfrac{\sqrt{14}}{\sqrt{2}}$
$\sqrt{7}$

12
Reduce. $\dfrac{\sqrt{30}}{\sqrt{6}}$
$\sqrt{5}$

13
Reduce. $\dfrac{\sqrt{26}}{\sqrt{22}}$
$\dfrac{\sqrt{13}}{\sqrt{11}}$

14
Reduce. $\dfrac{\sqrt{11}}{\sqrt{77}}$
$\dfrac{1}{\sqrt{7}}$

15
To reduce $\dfrac{5\sqrt{22}}{10\sqrt{35}}$ separate the coefficients from the radical expressions as shown in the following steps.

$$\frac{5\sqrt{22}}{10\sqrt{35}} = \frac{5 \cdot \sqrt{22}}{10 \cdot \sqrt{35}} = \frac{1}{2} \cdot \frac{\sqrt{22}}{\sqrt{35}} = \frac{\sqrt{22}}{2\sqrt{35}}$$

Use this process to complete the reduction below.

$$\frac{7\sqrt{15}}{5\sqrt{21}} = \frac{7 \cdot \sqrt{15}}{5 \cdot \sqrt{21}} = \frac{7}{5} \cdot \frac{\sqrt{5}}{\sqrt{7}} = \underline{\qquad}$$
$\dfrac{7\sqrt{5}}{5\sqrt{7}}$

16
Notice that $\dfrac{7\sqrt{5}}{5\sqrt{7}}$ should not be further reduced. 7 is a factor of the numerator, but $\sqrt{7}$ is a factor of the denominator.
Should $\dfrac{\sqrt{5}}{5}$ be reduced?
No, $\sqrt{5} \neq 5$

17

In reducing $\dfrac{8\sqrt{6}}{3\sqrt{14}}$ keep the coefficients and radical expressions separated.

$$\dfrac{8\sqrt{6}}{3\sqrt{14}} = \dfrac{8}{3} \cdot \dfrac{\sqrt{6}}{\sqrt{14}} = \dfrac{8}{3} \cdot \dfrac{\sqrt{3}}{\sqrt{7}} = \dfrac{8\sqrt{3}}{3\sqrt{7}}$$

Use the same process to reduce the fraction below.

$$\dfrac{4\sqrt{10}}{6\sqrt{22}} = \dfrac{4}{6} \cdot \dfrac{\sqrt{10}}{\sqrt{22}} = \underline{\hspace{2cm}} \qquad \dfrac{2}{3} \cdot \dfrac{\sqrt{5}}{\sqrt{11}} = \dfrac{2\sqrt{5}}{3\sqrt{11}}$$

18

In reducing a radical fraction do not divide a coefficient into a radicand or a radicand into a coefficient.
Complete the following.

$$\dfrac{9\sqrt{15}}{12\sqrt{6}} = \dfrac{9}{12} \cdot \dfrac{\sqrt{15}}{\sqrt{6}} = \underline{\hspace{2cm}} \qquad \dfrac{3}{4} \cdot \dfrac{\sqrt{5}}{\sqrt{2}} = \dfrac{3\sqrt{5}}{4\sqrt{2}}$$

19

Reduce. $\dfrac{6\sqrt{10}}{5\sqrt{30}}$ $\qquad \dfrac{6\sqrt{1}}{5\sqrt{3}} = \dfrac{6}{5\sqrt{3}}$

20

Reduce. $\dfrac{8\sqrt{14}}{10\sqrt{7}}$ $\qquad \dfrac{4\sqrt{2}}{5\sqrt{1}} = \dfrac{4\sqrt{2}}{5}$

21

Reduce. $\dfrac{4\sqrt{15}}{10\sqrt{65}}$ $\qquad \dfrac{2\sqrt{3}}{5\sqrt{13}}$

22

Reduce. $\dfrac{3\sqrt{33}}{10\sqrt{22}}$ $\qquad \dfrac{3\sqrt{3}}{10\sqrt{2}}$

23

The fraction $\dfrac{\sqrt{13}}{\sqrt{6}}$ has been reduced, but it is not completely simplified because its denominator contains the radical expression, $\sqrt{6}$. Does reducing a fraction always produce its most simplified form?

No, radical may still be in the denominator.

24
To simplify $\frac{\sqrt{13}}{\sqrt{6}}$, the fact that $\sqrt{6} \cdot \sqrt{6} = \sqrt{36} = 6$ is used.

$$\frac{\sqrt{13}}{\sqrt{6}} = \frac{\sqrt{13}}{\sqrt{6}} \cdot \frac{\sqrt{6}}{\sqrt{6}} = \frac{\sqrt{78}}{\sqrt{36}} = \frac{\sqrt{78}}{6}$$

Is $\frac{\sqrt{78}}{6}$ in its most simplified form?   Yes, there is no radical in the denominator.

25
Is the fraction $\frac{5}{\sqrt{7}}$ in its most simplified form?   No

26
To simplify $\frac{5}{\sqrt{7}}$ multiply numerator and denominator by $\sqrt{7}$.

$$\frac{5}{\sqrt{7}} = \frac{5}{\sqrt{7}} \cdot \frac{\sqrt{7}}{\sqrt{7}} = \frac{5\sqrt{7}}{\sqrt{49}} = \frac{5\sqrt{7}}{7}$$

Is $\frac{5\sqrt{7}}{7}$ in its most simplified form?   Yes, there is no radical in the denominator.

27
Is the fraction $\frac{\sqrt{3}}{\sqrt{10}}$ in its most simplified form?   No

28
To completely simplify $\frac{\sqrt{3}}{\sqrt{10}}$ the radical must be eliminated from its denominator.

$$\frac{\sqrt{3}}{\sqrt{10}} = \frac{\sqrt{3}}{\sqrt{10}} \cdot \frac{\sqrt{10}}{\sqrt{10}} = \frac{\sqrt{30}}{\sqrt{100}} = \frac{\sqrt{30}}{10}$$

Is $\frac{\sqrt{30}}{10}$ in its most simplified form?   Yes

29

To completely simplify $\dfrac{\sqrt{31}}{\sqrt{2}}$ the radical must be eliminated from its denominator.

$$\dfrac{\sqrt{31}}{\sqrt{2}} = \dfrac{\sqrt{31}}{\sqrt{2}} \cdot \dfrac{\sqrt{2}}{\sqrt{2}} = \dfrac{\sqrt{62}}{\sqrt{4}} = \dfrac{\sqrt{62}}{2}$$

What number can be multiplied by both numerator and denominator of $\dfrac{8}{\sqrt{11}}$ to produce a fraction with a rational number denominator?

$\sqrt{11}$

30

To completely simplify $\dfrac{8}{\sqrt{11}}$ the following steps are used.

$$\dfrac{8}{\sqrt{11}} = \dfrac{8}{\sqrt{11}} \cdot \dfrac{\sqrt{11}}{\sqrt{11}} = \dfrac{8\sqrt{11}}{\sqrt{121}} = \dfrac{8\sqrt{11}}{11}$$

The process illustrated above is called "**rationalizing the denominator**." The denominator of a fraction is changed from an irrational number to a _____ number.

rational

31

The process of rationalizing a denominator depends on two facts.
   a. Any nonzero number divided by itself is 1.

$\dfrac{\sqrt{3}}{\sqrt{3}} = 1$ and $\dfrac{\sqrt{5}}{\sqrt{5}} = 1$

   b. Any square root expression multiplied by itself gives its radicand.

$\sqrt{3} \cdot \sqrt{3} = 3$ and $\sqrt{5} \cdot \sqrt{5} =$ _____

5

32

In rationalizing the denominator of a fraction the process shown below is followed.

$$\dfrac{\sqrt{7}}{\sqrt{3}} = \dfrac{\sqrt{7}}{\sqrt{3}} \cdot 1 = \dfrac{\sqrt{7}}{\sqrt{3}} \cdot \dfrac{\sqrt{3}}{\sqrt{3}} = \dfrac{\sqrt{21}}{3}$$

Use the rationalizing the denominator process to simplify $\dfrac{\sqrt{10}}{\sqrt{3}}$.

$\dfrac{\sqrt{30}}{3}$

33
The example below shows the rationalizing the denominator process.

$$\frac{\sqrt{3}}{\sqrt{5}} = \frac{\sqrt{3}}{\sqrt{5}} \cdot \frac{\sqrt{5}}{\sqrt{5}} = \frac{\sqrt{15}}{5}$$ and $$\frac{\sqrt{7}}{\sqrt{2}} = \underline{\hspace{2cm}}$$

$$\frac{\sqrt{14}}{2}$$

34
A fraction is not completely simplified until its denominator contains no radical expression.

$$\frac{\sqrt{15}}{\sqrt{2}} = \frac{\sqrt{15}}{\sqrt{2}} \cdot \frac{\sqrt{2}}{\sqrt{2}} = \frac{\sqrt{30}}{2}$$ and $$\frac{\sqrt{11}}{\sqrt{2}} = \underline{\hspace{2cm}}$$

$$\frac{\sqrt{22}}{2}$$

35
The example below involves a coefficient for the numerator's radical. Notice how the coefficient is handled.

$$\frac{17\sqrt{6}}{\sqrt{5}} = \frac{17\sqrt{6}}{\sqrt{5}} \cdot \frac{\sqrt{5}}{\sqrt{5}} = \frac{17\sqrt{6} \cdot \sqrt{5}}{5} = \frac{17\sqrt{30}}{5}$$

Rationalize the denominator of $\frac{11\sqrt{7}}{\sqrt{10}}$.

$$\frac{11\sqrt{70}}{10}$$

36
Rationalize the denominator of $\frac{4\sqrt{7}}{\sqrt{11}}$.

$$\frac{4\sqrt{77}}{11}$$

37
The example below involves a coefficient for the denominator's radical. Notice how the coefficient is handled.

$$\frac{\sqrt{15}}{2\sqrt{7}} = \frac{\sqrt{15}}{2\sqrt{7}} \cdot \frac{\sqrt{7}}{\sqrt{7}} = \frac{\sqrt{105}}{2 \cdot 7} = \frac{\sqrt{105}}{14}$$

Rationalize the denominator of $\frac{\sqrt{13}}{5\sqrt{2}}$.

$$\frac{\sqrt{26}}{10}$$

38
Rationalize the denominator of $\frac{\sqrt{22}}{4\sqrt{3}}$.

$$\frac{\sqrt{66}}{12}$$

39

Sometimes, the rationalizing the denominator process will produce a fraction that needs to be reduced. This is the case with the example shown below.

$$\frac{14\sqrt{3}}{\sqrt{2}} = \frac{14\sqrt{3}}{\sqrt{2}} \cdot \frac{\sqrt{2}}{\sqrt{2}} = \frac{14\sqrt{6}}{2} = 7\sqrt{6}$$

Rationalize the denominator of $\frac{4\sqrt{11}}{\sqrt{10}}$ and then reduce the new fraction.

$\frac{2\sqrt{110}}{5}$

40

Rationalize the denominator of $\frac{6\sqrt{23}}{\sqrt{3}}$ and then reduce the new fraction.

$2\sqrt{69}$

41

As a first step in simplifying any fraction, try to reduce it. Then rationalize the denominator. Complete this process which is shown below.

$$\frac{\sqrt{15}}{\sqrt{10}} = \frac{\sqrt{3}}{\sqrt{2}} = \underline{}$$

$\frac{\sqrt{3}}{\sqrt{2}} \cdot \frac{\sqrt{2}}{\sqrt{2}} = \frac{\sqrt{6}}{2}$

42

Completely simplify the radical fraction shown below by first reducing it and then rationalizing the denominator.

$$\frac{\sqrt{6}}{\sqrt{21}} = \underline{}$$

$\frac{\sqrt{2}}{\sqrt{7}} \cdot \frac{\sqrt{7}}{\sqrt{7}} = \frac{\sqrt{14}}{7}$

43

Completely simplify. $\frac{\sqrt{15}}{\sqrt{10}} = \underline{}$

$\frac{\sqrt{3}}{\sqrt{2}} = \frac{\sqrt{6}}{2}$

44

Completely simplify. $\frac{\sqrt{21}}{\sqrt{33}} = \underline{}$

$\frac{\sqrt{7}}{\sqrt{11}} = \frac{\sqrt{77}}{11}$

45

Completely simplify. $\frac{\sqrt{6}}{\sqrt{10}} = \underline{}$

$\frac{\sqrt{3}}{\sqrt{5}} = \frac{\sqrt{15}}{5}$

46

Completely simplify. $\frac{\sqrt{6}}{\sqrt{21}} = \underline{}$

$\frac{\sqrt{2}}{\sqrt{7}} = \frac{\sqrt{14}}{7}$

SIMPLIFYING RADICAL FRACTIONS   109

47
Completely simplify. $\dfrac{\sqrt{22}}{\sqrt{77}} = $ _____ $\dfrac{\sqrt{14}}{7}$

48
Study the steps in the following simplification.

$$\dfrac{8\sqrt{7}}{\sqrt{6}} = \dfrac{8\sqrt{7}}{\sqrt{6}} \cdot \dfrac{\sqrt{6}}{\sqrt{6}} = \dfrac{8\sqrt{42}}{6} = \dfrac{4\sqrt{42}}{3}$$

Completely simplify. $\dfrac{15\sqrt{7}}{\sqrt{10}} = $ _____ $\dfrac{15\sqrt{70}}{10} = \dfrac{3\sqrt{70}}{2}$

49
Completely simplify. $\dfrac{4\sqrt{7}}{\sqrt{3}} = $ _____ $\dfrac{4\sqrt{21}}{3}$

50
Completely simplify. $\dfrac{7\sqrt{10}}{\sqrt{7}} = $ _____ $\dfrac{7\sqrt{70}}{7} = \sqrt{70}$

51
Completely simplify. $\dfrac{5\sqrt{3}}{\sqrt{2}} = $ _____ $\dfrac{5\sqrt{6}}{2}$

52
Completely simplify. $\dfrac{4\sqrt{3}}{\sqrt{11}} = $ _____ $\dfrac{4\sqrt{33}}{11}$

53
Study the steps in the following simplification.

$$\dfrac{\sqrt{6}}{2\sqrt{5}} = \dfrac{\sqrt{6}}{2\sqrt{5}} \cdot \dfrac{\sqrt{5}}{\sqrt{5}} = \dfrac{\sqrt{30}}{2 \cdot 5} = \dfrac{\sqrt{30}}{10}$$

Completely simplify. $\dfrac{\sqrt{10}}{7\sqrt{3}} = $ _____ $\dfrac{\sqrt{30}}{7 \cdot 3} = \dfrac{\sqrt{30}}{21}$

54
Completely simplify. $\dfrac{\sqrt{11}}{5\sqrt{3}} = $ _____ $\dfrac{\sqrt{33}}{5 \cdot 3} = \dfrac{\sqrt{33}}{15}$

55
A fraction should be reduced before rationalizing its denominator.
Completely simplify. $\dfrac{3\sqrt{10}}{6\sqrt{15}} = $ _____ $\dfrac{\sqrt{2}}{2\sqrt{3}} = \dfrac{\sqrt{6}}{6}$

56
Reduce a fraction whenever that is possible — before and/or after rationalizing the denominator.

Completely simplify. $\dfrac{6\sqrt{33}}{5\sqrt{22}} = $ _____

$\dfrac{6\sqrt{3}}{5\sqrt{2}} = \dfrac{6\sqrt{6}}{10} = \dfrac{3\sqrt{6}}{5}$

57
Some radical fractions contain radical expressions that should be simplified before rationalizing the denominator. The example below shows such a situation.

$$\dfrac{\sqrt{11}}{\sqrt{12}} = \dfrac{\sqrt{11}}{\sqrt{4}\cdot\sqrt{3}} = \dfrac{\sqrt{11}}{2\sqrt{3}} = \dfrac{\sqrt{33}}{6}$$

Completely simplify $\dfrac{\sqrt{3}}{\sqrt{50}}$ by first simplifying $\sqrt{50}$.

$\dfrac{\sqrt{3}}{5\sqrt{2}} = \dfrac{\sqrt{6}}{10}$

58
Completely simplify $\dfrac{\sqrt{5}}{\sqrt{18}}$ by first simplifying $\sqrt{18}$.

$\dfrac{\sqrt{5}}{3\sqrt{2}} = \dfrac{\sqrt{10}}{6}$

59
Completely simplify $\dfrac{2\sqrt{7}}{\sqrt{8}}$ by first simplifying $\sqrt{8}$.

$\dfrac{2\sqrt{7}}{2\sqrt{2}} = \dfrac{\sqrt{7}}{\sqrt{2}} = \dfrac{\sqrt{14}}{2}$

60
Study the following example involving reducing, simplifying $\sqrt{18}$, and rationalizing the denominator.

$$\dfrac{6\sqrt{5}}{7\sqrt{90}} = \dfrac{6\sqrt{1}}{7\sqrt{18}} = \dfrac{6}{7\cdot 3\sqrt{2}} = \dfrac{2}{7\sqrt{2}} = \dfrac{2\sqrt{2}}{7\cdot 2} = \dfrac{\sqrt{2}}{7}$$

Complete the following simplification.

$\dfrac{12\sqrt{5}}{2\sqrt{54}} = \dfrac{6\sqrt{5}}{\sqrt{54}} = \dfrac{6\sqrt{5}}{3\sqrt{6}} = $ _____

$\dfrac{2\sqrt{5}}{\sqrt{6}} = \dfrac{2\sqrt{30}}{6} = \dfrac{\sqrt{30}}{3}$

61
As the fractions of the preceding frame illustrate, completely simplifying can require:
    a.   Reducing more than once.
    b.   Simplifying individual radical expressions.
    c.   Rationalizing the denominator.
Complete the following simplification.

$\dfrac{14\sqrt{11}}{7\sqrt{32}} = \dfrac{2\sqrt{11}}{\sqrt{32}} = \dfrac{2\sqrt{11}}{4\sqrt{2}} = $ _____

$\dfrac{\sqrt{11}}{2\sqrt{2}} = \dfrac{\sqrt{22}}{4}$

62
Completely simplify. $\dfrac{14\sqrt{7}}{\sqrt{18}} = $ _____

$\dfrac{7\sqrt{14}}{3}$

63
Completely simplify. $\dfrac{18\sqrt{5}}{\sqrt{72}} = $ _____

$\dfrac{3\sqrt{10}}{2}$

64
For the fraction shown below, the radical expressions in both the numerator and the denominator need to be simplified first.

$$\dfrac{\sqrt{27}}{\sqrt{20}} = \dfrac{3\sqrt{3}}{2\sqrt{5}} = \dfrac{3\sqrt{15}}{10}$$

Completely simplify $\dfrac{\sqrt{32}}{\sqrt{75}}$ by first simplifying $\sqrt{32}$ and $\sqrt{75}$.

$\dfrac{4\sqrt{2}}{5\sqrt{3}} = \dfrac{4\sqrt{6}}{15}$

65
Completely simplify $\dfrac{\sqrt{8}}{\sqrt{27}}$ by first simplifying $\sqrt{8}$ and $\sqrt{27}$.

$\dfrac{2\sqrt{2}}{3\sqrt{3}} = \dfrac{2\sqrt{6}}{9}$

66
Completely simplify $\dfrac{\sqrt{18}}{\sqrt{125}}$ by first simplifying $\sqrt{18}$ and $\sqrt{125}$.

$\dfrac{3\sqrt{2}}{5\sqrt{5}} = \dfrac{3\sqrt{10}}{25}$

67
Completely simplify. $\dfrac{\sqrt{28}}{\sqrt{75}} = $ _____

$\dfrac{2\sqrt{21}}{15}$

68
Completely simplify. $\dfrac{\sqrt{20}}{\sqrt{63}} = $ _____

$\dfrac{2\sqrt{35}}{21}$

69
Completely simplifying a fraction may require:
    a. Reducing.
    b. Simplifying one or two radical expressions.
    c. Rationalizing the denominator.
    d. Reducing again.
Complete the following simplification.

$$\dfrac{8\sqrt{35}}{12\sqrt{90}} = \dfrac{2\sqrt{7}}{3\sqrt{18}} = \dfrac{2\sqrt{7}}{9\sqrt{2}} = $$ _____

$\dfrac{2\sqrt{14}}{18} = \dfrac{\sqrt{14}}{9}$

70
Completely simplify. $\dfrac{10\sqrt{3}}{7\sqrt{5}} = $ _____

$\dfrac{2\sqrt{15}}{7}$

71
Completely simplify. $\dfrac{5\sqrt{20}}{9\sqrt{4}} = $ _____

$\dfrac{5\sqrt{5}}{9}$

72
Completely simplify. $\dfrac{-3\sqrt{2}}{\sqrt{128}} = $ _____

$\dfrac{-3}{8}$

73
Completely simplify. $\dfrac{6\sqrt{96}}{-16\sqrt{3}} = $ _____

$\dfrac{-3\sqrt{2}}{2}$

74
Completely simplify. $\dfrac{6}{5\sqrt{18}} = $ _____

$\dfrac{\sqrt{2}}{5}$

75
The example below shows how to handle a rational number radicand.

$\sqrt{\dfrac{3}{5}}$ can be written as $\dfrac{\sqrt{3}}{\sqrt{5}}$

$\sqrt{\dfrac{15}{22}}$ can be written as _____

$\dfrac{\sqrt{15}}{\sqrt{22}}$

76
$\sqrt{\dfrac{2}{3}}$ can be written as $\dfrac{\sqrt{2}}{\sqrt{3}}$

$\sqrt{\dfrac{5}{7}}$ can be written as _____

$\dfrac{\sqrt{5}}{\sqrt{7}}$

77
$\dfrac{-7}{8}\sqrt{\dfrac{4}{5}} = \dfrac{-7\sqrt{4}}{8\sqrt{5}}$ and $\dfrac{-2}{5}\sqrt{\dfrac{7}{16}} = $ _____

$\dfrac{-2\sqrt{7}}{5\sqrt{16}}$

78
To simplify $\dfrac{-7}{8}\sqrt{\dfrac{4}{5}}$ the following steps are used.

$\dfrac{-7}{8}\sqrt{\dfrac{4}{5}} = \dfrac{-7\sqrt{4}}{8\sqrt{5}} = \dfrac{-7 \cdot 2}{8\sqrt{5}} = \dfrac{-7}{4\sqrt{5}} = \dfrac{-7\sqrt{5}}{20}$

Complete the following simplification.

$\dfrac{-2}{5}\sqrt{\dfrac{7}{16}} = \dfrac{-2\sqrt{7}}{5\sqrt{16}} = $ _____

$\dfrac{-2\sqrt{7}}{5 \cdot 4} = \dfrac{-\sqrt{7}}{5 \cdot 2} = \dfrac{-\sqrt{7}}{10}$

Simplifying Radical Fractions

79
Complete the following simplification.

$$\frac{6}{11}\sqrt{\frac{5}{2}} = \frac{6\sqrt{5}}{11\sqrt{2}} = \underline{\qquad}$$

$$\frac{6\sqrt{10}}{11 \cdot 2} = \frac{3\sqrt{10}}{11}$$

80
Completely simplify. $\frac{3}{5}\sqrt{\frac{3}{4}} = \underline{\qquad}$

$$\frac{3\sqrt{3}}{10}$$

81
Completely simplify. $\frac{4}{7}\sqrt{\frac{9}{10}} = \underline{\qquad}$

$$\frac{6\sqrt{10}}{35}$$

82
Completely simplify. $\frac{8}{9}\sqrt{\frac{7}{12}} = \underline{\qquad}$

$$\frac{4\sqrt{21}}{27}$$

Feedback Unit 1

This quiz reviews the preceding unit. Answers are at the back of the book.

Completely simplify each of the following.

1. $\dfrac{5\sqrt{2}}{\sqrt{7}} = \underline{\qquad}$

2. $\dfrac{2\sqrt{30}}{\sqrt{35}} = \underline{\qquad}$

3. $\dfrac{5\sqrt{12}}{2\sqrt{15}} = \underline{\qquad}$

4. $\dfrac{2\sqrt{50}}{5\sqrt{9}} = \underline{\qquad}$

5. $\dfrac{3}{8}\sqrt{\dfrac{4}{7}} = \underline{\qquad}$

6. $\dfrac{3}{2\sqrt{5}} = \underline{\qquad}$

7. $\dfrac{-6}{7}\sqrt{\dfrac{2}{3}} = \underline{\qquad}$

8. $\dfrac{7\sqrt{5}}{5\sqrt{7}} = \underline{\qquad}$

9. $\dfrac{3\sqrt{5}}{4\sqrt{3}} = \underline{\qquad}$

10. $\dfrac{5\sqrt{32}}{2\sqrt{24}} = \underline{\qquad}$

Unit 2: Reducing Radical Fractions with Indices Greater Than 2

1
Any radical fraction in which the numerator and denominator contain the same nonzero radical expression is equal to 1.

$$\frac{\sqrt[6]{5}}{\sqrt[6]{5}} = 1 \qquad \frac{\sqrt[8]{65}}{\sqrt[8]{65}} = 1 \qquad \frac{\sqrt[3]{17}}{\sqrt[3]{17}} = \text{\_\_\_\_\_}$$

1

2
When a nth root is divided by itself the quotient is 1.

$$\frac{\sqrt[4]{23}}{\sqrt[4]{23}} = 1 \qquad \frac{\sqrt[9]{41}}{\sqrt[9]{41}} = 1 \qquad \frac{\sqrt[6]{7^4}}{\sqrt[6]{7^4}} = \text{\_\_\_\_\_}$$

1

3
The fraction $\frac{\sqrt[3]{11}}{\sqrt[4]{11}}$ is not equal to 1 because the numerator is a cube root and the denominator is a _____ _____.

4th root

4
To **reduce** $\frac{\sqrt[3]{18}}{\sqrt[3]{28}}$, the following steps are used.

$$\frac{\sqrt[3]{18}}{\sqrt[3]{28}} = \frac{\sqrt[3]{2}}{\sqrt[3]{2}} \cdot \frac{\sqrt[3]{9}}{\sqrt[3]{14}} = \frac{\cancel{\sqrt[3]{2}}}{\cancel{\sqrt[3]{2}}} \cdot \frac{\sqrt[3]{9}}{\sqrt[3]{14}} = \frac{\sqrt[3]{9}}{\sqrt[3]{14}}$$

What was the common factor in this example?

$\sqrt[3]{2}$

5
Reduce $\frac{\sqrt[5]{28}}{\sqrt[5]{21}}$ by dividing out the common factor $\sqrt[5]{7}$.

$\frac{\sqrt[5]{4}}{\sqrt[5]{3}}$

6
Reduce $\frac{\sqrt[8]{65}}{\sqrt[8]{90}}$ by dividing out the common factor $\sqrt[8]{5}$.

$\frac{\sqrt[8]{13}}{\sqrt[8]{18}}$

SIMPLIFYING RADICAL FRACTIONS   115

7
The reducing example below shows a situation in which the common factor is the numerator. The final step of the process depends upon the fact that $\sqrt[7]{1} = 1$.

$$\frac{\sqrt[7]{9}}{\sqrt[7]{27}} = \frac{\sqrt[7]{9}}{\sqrt[7]{9}} \cdot \frac{\sqrt[7]{1}}{\sqrt[7]{3}} = \frac{\sqrt[7]{1}}{\sqrt[7]{3}} = \frac{1}{\sqrt[7]{3}}$$

Reduce. $\dfrac{\sqrt[6]{6}}{\sqrt[6]{42}}$ $\dfrac{1}{\sqrt[6]{7}}$

8
Reduce. $\dfrac{\sqrt[5]{3}}{\sqrt[3]{15}}$ Can't be reduced. Indices are different.

9
Reduce. $\dfrac{\sqrt[4]{10}}{\sqrt[4]{90}}$ $\dfrac{1}{\sqrt[4]{9}}$

10
Reduce. $\dfrac{\sqrt[7]{12}}{\sqrt[7]{30}}$ $\dfrac{\sqrt[7]{2}}{\sqrt[7]{5}}$

11
The reducing example below shows a situation in which the common factor is the denominator.

$$\frac{\sqrt[8]{28}}{\sqrt[8]{7}} = \frac{\sqrt[8]{7}}{\sqrt[8]{7}} \cdot \frac{\sqrt[8]{4}}{\sqrt[8]{1}} = \frac{\sqrt[8]{4}}{\sqrt[8]{1}} = \frac{\sqrt[8]{4}}{1} = \sqrt[8]{4}$$

Reduce. $\dfrac{\sqrt[7]{42}}{\sqrt[7]{14}}$ $\sqrt[7]{3}$

12
Reduce. $\dfrac{\sqrt[3]{36}}{\sqrt[3]{18}}$ $\sqrt[3]{2}$

13
Reduce. $\dfrac{\sqrt[9]{45}}{\sqrt[9]{27}}$ $\dfrac{\sqrt[9]{5}}{\sqrt[9]{3}}$

14
Reduce. $\dfrac{\sqrt[7]{52}}{\sqrt[7]{8}}$

$\dfrac{\sqrt[7]{13}}{\sqrt[7]{2}}$

15
Notice that $\dfrac{8\sqrt[4]{17}}{5\sqrt[4]{24}}$ cannot be reduced. Neither the coefficients nor the 4th roots have a common factor?

Should $\dfrac{4\sqrt[6]{21}}{7\sqrt[6]{4}}$ be reduced?

No

16
Study the following example which shows that the coefficients and radical expressions are separately reduced.

$$\dfrac{6\sqrt[8]{55}}{10\sqrt[8]{45}} = \dfrac{6}{10} \cdot \dfrac{\sqrt[8]{55}}{\sqrt[8]{45}} = \dfrac{3}{5} \cdot \dfrac{\sqrt[8]{11}}{\sqrt[8]{9}} = \dfrac{3\sqrt[8]{11}}{5\sqrt[8]{9}}$$

Use the same process to reduce $\dfrac{8\sqrt[5]{24}}{6\sqrt[5]{56}}$.

$\dfrac{4\sqrt[5]{3}}{3\sqrt[5]{7}}$

17
Reduce. $\dfrac{30\sqrt[3]{25}}{18\sqrt[3]{35}}$

$\dfrac{5\sqrt[3]{5}}{3\sqrt[3]{7}}$

18
Reduce. $\dfrac{12\sqrt[8]{20}}{20\sqrt[8]{16}}$

$\dfrac{3\sqrt[8]{5}}{5\sqrt[8]{4}}$

19
Reduce. $\dfrac{13\sqrt[9]{13}}{26\sqrt[9]{52}}$

$\dfrac{1}{2\sqrt[9]{4}}$

20
Reduce. $\dfrac{21\sqrt[4]{36}}{14\sqrt[4]{54}}$

$\dfrac{3\sqrt[4]{2}}{2\sqrt[4]{3}}$

SIMPLIFYING RADICAL FRACTIONS   117

21

Some radical fractions contain radical expressions that should be simplified before reducing. The example below shows such a situation.

$$\frac{2\sqrt[3]{21}}{3\sqrt[3]{8}} = \frac{2}{3} \cdot \frac{\sqrt[3]{21}}{\sqrt[3]{8}} = \frac{2}{3} \cdot \frac{\sqrt[3]{21}}{2} = \frac{\sqrt[3]{21}}{3}$$

The reducing was possible because $\sqrt[3]{8} = $ _____.

2

22

$2^3 = 8 \quad 3^3 = 27 \quad 4^3 = 64 \quad 5^3 = 125$

The fact that 125 is a perfect cube is used in the reducing example shown below.

$$\frac{7\sqrt[3]{375}}{10\sqrt[3]{31}} = \frac{7}{10} \cdot \frac{\sqrt[3]{125 \cdot 3}}{\sqrt[3]{31}} = \frac{7}{10} \cdot \frac{5\sqrt[3]{3}}{\sqrt[3]{31}} = \frac{7\sqrt[3]{3}}{2\sqrt[3]{31}}$$

Use a perfect cube factor in reducing $\dfrac{6\sqrt[3]{47}}{5\sqrt[3]{128}}$.

$\dfrac{6}{5} \cdot \dfrac{\sqrt[3]{47}}{4\sqrt[3]{2}} = \dfrac{3\sqrt[3]{47}}{10\sqrt[3]{2}}$

23

Reduce $\dfrac{8\sqrt[3]{81}}{12\sqrt[3]{50}}$. The process will include simplifying one of the cube roots.

$\dfrac{2}{3} \cdot \dfrac{3\sqrt[3]{3}}{\sqrt[3]{50}} = \dfrac{2\sqrt[3]{3}}{\sqrt[3]{50}}$

24

$2^4 = 16 \quad 3^4 = 81 \quad 4^4 = 256 \quad 5^4 = 625$

The fact that 256 is a perfect 4th power is used in the reducing example shown below.

$$\frac{3\sqrt[4]{768}}{8\sqrt[4]{97}} = \frac{3}{8} \cdot \frac{\sqrt[4]{256 \cdot 3}}{\sqrt[4]{97}} = \frac{3}{8} \cdot \frac{4\sqrt[4]{3}}{\sqrt[4]{97}} = \frac{3\sqrt[4]{3}}{2\sqrt[4]{97}}$$

Use a perfect 4th power in reducing $\dfrac{5\sqrt[4]{48}}{10\sqrt[4]{29}}$.

$\dfrac{1}{2} \cdot \dfrac{2\sqrt[4]{3}}{\sqrt[4]{29}} = \dfrac{\sqrt[4]{3}}{\sqrt[4]{29}}$

25

Reduce $\dfrac{20\sqrt[4]{81}}{21\sqrt[4]{625}}$. The process will include simplifying the 4th roots.

$\dfrac{20 \cdot 3}{21 \cdot 5} = \dfrac{4}{7}$

26

$2^5 = 32 \quad 3^5 = 243 \quad 4^5 = 1032 \quad 5^5 = 3125$

The fact that 243 is a perfect 5th power is used in the reducing example shown below.

$$\frac{12\sqrt[5]{3125}}{20\sqrt[5]{486}} = \frac{3}{5} \cdot \frac{\sqrt[5]{5^5}}{\sqrt[5]{3^5 \cdot 2}} = \frac{3}{5} \cdot \frac{5}{3\sqrt[5]{2}} = \frac{1}{\sqrt[5]{2}}$$

Use a perfect 5th power in reducing $\dfrac{10\sqrt[5]{2064}}{8\sqrt[5]{37}}$.

$\dfrac{5}{4} \cdot \dfrac{\sqrt[5]{2 \cdot 4^5}}{\sqrt[5]{37}} = \dfrac{5\sqrt[5]{2}}{\sqrt[5]{37}}$

27

Reduce $\dfrac{7\sqrt[5]{243}}{12\sqrt[5]{7}}$. The process will include simplifying at least one 5th root.

$\dfrac{7}{12} \cdot \dfrac{3}{\sqrt[5]{7}} = \dfrac{7}{4\sqrt[5]{7}}$

28

$\sqrt[n]{x^n} = x$ is an identity for any positive real number x.

Use this fact to reduce the fraction $\dfrac{14\sqrt[6]{3^4}}{11\sqrt[6]{7^6}}$.

$\dfrac{14}{11} \cdot \dfrac{\sqrt[6]{3^4}}{7} = \dfrac{2\sqrt[6]{3^4}}{11}$

29

$\sqrt[n]{x^n} = x$ is an identity for any positive real number x.

Reduce $\dfrac{7\sqrt[3]{5^4}}{10\sqrt[3]{18}}$ by first simplifying $\sqrt[3]{5^4}$.

$\dfrac{7}{10} \cdot \dfrac{5\sqrt[3]{5}}{\sqrt[3]{18}} = \dfrac{7\sqrt[3]{5}}{2\sqrt[3]{18}}$

30

Reduce $\dfrac{-8\sqrt[6]{10}}{10\sqrt[6]{25}}$. Look for opportunities to simplify any radical.

$\dfrac{-4\sqrt[6]{2}}{5\sqrt[6]{5}}$

31

Reduce $\dfrac{12\sqrt[3]{22}}{8\sqrt[3]{30}}$. Look for opportunities to simplify any radical.

$\dfrac{3\sqrt[3]{11}}{2\sqrt[3]{15}}$

32

Reduce $\dfrac{12\sqrt[3]{54}}{9\sqrt[3]{2}}$. Look for opportunities to simplify any radical.

$\dfrac{4}{3} \cdot \dfrac{3\sqrt[3]{2}}{\sqrt[3]{2}} = 4$

SIMPLIFYING RADICAL FRACTIONS   119

33
Reduce $\dfrac{20\sqrt[3]{34}}{14\sqrt[3]{32}}$. Simplify, if possible, any radical.

$\dfrac{10}{7} \cdot \dfrac{\sqrt[3]{34}}{2\sqrt[3]{4}} = \dfrac{5\sqrt[3]{17}}{7\sqrt[3]{2}}$

34
Reduce $\dfrac{-9\sqrt[3]{5000}}{30\sqrt[3]{27}}$. Simplify, if possible, any radical.

$\dfrac{-3}{10} \cdot \dfrac{10\sqrt[3]{5}}{3} = -\sqrt[3]{5}$

35
Reduce $\dfrac{6\sqrt[4]{32}}{20\sqrt[4]{21}}$. Simplify, if possible, any radical.

$\dfrac{3}{10} \cdot \dfrac{2\sqrt[4]{2}}{\sqrt[4]{21}} = \dfrac{3\sqrt[4]{2}}{5\sqrt[4]{21}}$

36
Reduce $\dfrac{-8\sqrt[7]{325}}{14\sqrt[7]{175}}$. Simplify, if possible, any radical.

$\dfrac{-4}{7} \cdot \dfrac{\sqrt[7]{13}}{\sqrt[7]{7}} = \dfrac{-4\sqrt[7]{13}}{7\sqrt[7]{7}}$

37
Reduce $\dfrac{10\sqrt[4]{162}}{15\sqrt[4]{32}}$. Simplify, if possible, any radical.

$\dfrac{2}{3} \cdot \dfrac{\sqrt[4]{81}}{\sqrt[4]{16}} = \dfrac{2}{3} \cdot \dfrac{3}{2} = 1$

38
Study the example below in which the radicands are written as power expressions.

$\dfrac{4\sqrt[3]{5^4}}{25\sqrt[3]{2^5}} = \dfrac{4}{25} \cdot \dfrac{\sqrt[3]{5^3 \cdot 5}}{\sqrt[3]{2^3 \cdot 2^2}} = \dfrac{4}{25} \cdot \dfrac{5\sqrt[3]{5}}{2\sqrt[3]{2^2}} = \dfrac{2\sqrt[3]{5}}{5\sqrt[3]{2^2}}$

Reduce $\dfrac{55\sqrt[4]{5^6}}{25\sqrt[4]{11^4}}$. Simplify, if possible, any radical.

$\dfrac{11}{5} \cdot \dfrac{5\sqrt[4]{5^2}}{11} = \sqrt[4]{25}$

39
Reduce. $\dfrac{40\sqrt[5]{3^9}}{30\sqrt[5]{2^6}}$

$\dfrac{4}{3} \cdot \dfrac{3\sqrt[5]{3^4}}{2\sqrt[5]{2}} = \dfrac{2\sqrt[5]{3^4}}{\sqrt[5]{2}}$

40
Reduce. $\dfrac{4\sqrt[3]{6x^7}}{12\sqrt[3]{10x^2}}$

$\dfrac{1}{3} \cdot \dfrac{3\sqrt[3]{3x^5}}{\sqrt[3]{5}} = \dfrac{x\sqrt[3]{3x^2}}{\sqrt[3]{5}}$

41
Reduce. $\dfrac{8\sqrt[7]{12x^5}}{10\sqrt[7]{15x^{12}}}$

$\dfrac{4}{5} \cdot \dfrac{\sqrt[7]{4}}{\sqrt[7]{5x^7}} = \dfrac{4\sqrt[7]{4}}{5x\sqrt[7]{5}}$

42

The nth root of a fraction is equal to the division of the nth roots as shown in the example below.

$$\sqrt[3]{\frac{16x^5y^2}{78xy^8}} = \frac{\sqrt[3]{16x^5y^2}}{\sqrt[3]{78xy^8}} = \frac{\sqrt[3]{8x^4}}{\sqrt[3]{39y^6}} = \frac{2x\sqrt[3]{x}}{y^2\sqrt[3]{39}}$$

Reduce $\sqrt[5]{\dfrac{7x^8y^9}{32xy^{12}}}$ by first writing it as the division of nth roots.

$$\frac{\sqrt[5]{7x^8y^9}}{\sqrt[5]{32xy^{12}}} = \frac{\sqrt[5]{7x^7}}{\sqrt[5]{2^5y^3}} = \frac{x\sqrt[5]{7x^2}}{2\sqrt[5]{y^3}}$$

43
Reduce. $\sqrt[4]{\dfrac{32x^8y^9}{1250y^{11}}}$

$$\frac{\sqrt[4]{16x^8y^9}}{\sqrt[4]{625y^{11}}} = \frac{\sqrt[4]{2^4x^8}}{\sqrt[4]{5^4y^2}} = \frac{2x^2}{5\sqrt[4]{y^2}}$$

44
Reduce. $\sqrt[5]{\dfrac{3 \cdot 5^6}{4 \cdot 3^9}}$

$$\frac{\sqrt[5]{5^6}}{\sqrt[5]{4 \cdot 3^8}} = \frac{5\sqrt[5]{5}}{3\sqrt[5]{4 \cdot 3^3}}$$

45
Reduce. $\sqrt[3]{\dfrac{432}{2 \cdot 5^9}}$

$$\frac{\sqrt[3]{216}}{\sqrt[3]{5^9}} = \frac{6}{5^3}$$

46
Reduce. $\sqrt[3]{\dfrac{48}{250}}$

$$\frac{\sqrt[3]{24}}{\sqrt[3]{125}} = \frac{2\sqrt[3]{3}}{5}$$

Simplifying Radical Fractions

Feedback Unit 2

This quiz reviews the preceding unit. Answers are at the back of the book.

Reduce each of the following. Simplify radicals when possible.

1. $\dfrac{6\sqrt[7]{15}}{12\sqrt[7]{25}}$

2. $\dfrac{39\sqrt[3]{5000}}{65\sqrt[3]{81}}$

3. $\dfrac{18\sqrt[5]{2^6 y^3}}{22\sqrt[5]{3^7 y^{10}}}$

4. $\dfrac{14\sqrt[7]{6x^{13}}}{21\sqrt[7]{20x^3}}$

5. $\dfrac{\sqrt[3]{250x^6}}{\sqrt[3]{432}}$

6. $\dfrac{\sqrt[5]{24x^6 y^4}}{\sqrt[5]{18x^8 y^9}}$

7. $\sqrt[3]{\dfrac{5}{135}}$

8. $\sqrt[4]{\dfrac{3^2 x^7}{3^3 x^2}}$

Unit 3: Simplifying Radical Fractions

The following mathematical term is crucial to an understanding of this unit.

Completely simplified

1

The fraction $\dfrac{7\sqrt[5]{11}}{3\sqrt[5]{6}}$ has been completely reduced.

However, it is not considered **completely simplified**

because its denominator contains the radical

expression $\sqrt[5]{6}$. Is $\dfrac{5\sqrt[7]{13 \cdot 3^5}}{19 \cdot 3}$ completely simplified? Yes, there is no radical in the denominator.

2
Which fraction shown below is completely simplified?

$$\frac{3\sqrt[4]{7}}{\sqrt[4]{5}} \qquad \frac{17\sqrt[3]{14 \cdot 11^2}}{55} \qquad \frac{4\sqrt[8]{3}}{5\sqrt[8]{17}}$$

$\dfrac{17\sqrt[3]{14 \cdot 11^2}}{55}$

3
The fraction $\dfrac{3\sqrt[4]{7}}{\sqrt[4]{5}}$ is not completely simplified because

it has the radical expression $\sqrt[4]{5}$ in its denominator. The

denominator's radical is eliminated in the steps shown below.

$$\frac{3\sqrt[4]{7}}{\sqrt[4]{5}} = \frac{3\sqrt[4]{7}}{\sqrt[4]{5}} \cdot 1 = \frac{3\sqrt[4]{7}}{\sqrt[4]{5}} \cdot \frac{\sqrt[4]{5^3}}{\sqrt[4]{5^3}} = \frac{3\sqrt[4]{7 \cdot 5^3}}{\sqrt[4]{5^4}} = \frac{3\sqrt[4]{7 \cdot 5^3}}{5}$$

The process above is dependent upon multiplying the original
fraction by 1 written as _____.

$\dfrac{\sqrt[4]{5^3}}{\sqrt[4]{5^3}}$

4
The fraction $\dfrac{5\sqrt[5]{13}}{4\sqrt[5]{3^3}}$ is completely simplified in the following steps.

$$\frac{5\sqrt[5]{13}}{4\sqrt[5]{3^3}} = \frac{5\sqrt[5]{13}}{4\sqrt[5]{3^3}} \cdot \frac{\sqrt[5]{3^2}}{\sqrt[5]{3^2}} = \frac{5\sqrt[5]{13 \cdot 3^2}}{4\sqrt[5]{3^5}} = \frac{5\sqrt[5]{13 \cdot 9}}{4 \cdot 3} = \frac{5\sqrt[5]{117}}{12}$$

The process of eliminating the radical from the denominator
depends on the multiplication $\sqrt[5]{3^3} \cdot \sqrt[5]{3^2} = \sqrt[5]{3^5} =$ _____. 3

5
The fraction $\dfrac{23}{\sqrt[9]{5^4 \cdot 7^6}}$ is completely simplified in the following steps.

$$\frac{23}{\sqrt[9]{5^4 \cdot 7^6}} = \frac{23}{\sqrt[9]{5^4 \cdot 7^6}} \cdot \frac{\sqrt[9]{5^5 \cdot 7^3}}{\sqrt[9]{5^5 \cdot 7^3}} = \frac{23\sqrt[9]{5^5 \cdot 7^3}}{\sqrt[9]{5^9 \cdot 7^9}} = \frac{23\sqrt[9]{5^5 \cdot 7^3}}{5 \cdot 7}$$

The process of eliminating the radical from the denominator
depends on raising each factor of its radicand to the _____ power.   9th

SIMPLIFYING RADICAL FRACTIONS   123

6 $\dfrac{\sqrt[3]{7}}{\sqrt[3]{10}} = \dfrac{\sqrt[3]{7}}{\sqrt[3]{10}} \cdot 1 = \dfrac{\sqrt[3]{7}}{\sqrt[3]{10}} \cdot$ _____

To completely simplify $\dfrac{\sqrt[3]{7}}{\sqrt[3]{10}}$, begin by multiplying it by _____.

$1 = \dfrac{\sqrt[3]{10^2}}{\sqrt[3]{10^2}}$

7 $\dfrac{\sqrt[3]{7}}{\sqrt[3]{10}} = \dfrac{\sqrt[3]{7}}{\sqrt[3]{10}} \cdot \dfrac{\sqrt[3]{10^2}}{\sqrt[3]{10^2}} = \dfrac{\sqrt[3]{7 \cdot 10^2}}{\sqrt[3]{10^3}} = \dfrac{\sqrt[3]{700}}{10}$

To completely simplify $\dfrac{\sqrt[4]{2}}{\sqrt[4]{7}}$, begin by multiplying it by _____.

$1 = \dfrac{\sqrt[4]{7^3}}{\sqrt[4]{7^3}}$

8 $\dfrac{\sqrt[4]{2}}{\sqrt[4]{7}} = \dfrac{\sqrt[4]{2}}{\sqrt[4]{7}} \cdot \dfrac{\sqrt[4]{7^3}}{\sqrt[4]{7^3}} = \dfrac{\sqrt[4]{2 \cdot 7^3}}{\sqrt[4]{7^4}} = \dfrac{\sqrt[4]{2 \cdot 7^3}}{7}$

To completely simplify $\dfrac{8}{\sqrt[5]{11}}$, begin by multiplying it by _____.

$1 = \dfrac{\sqrt[5]{11^4}}{\sqrt[5]{11^4}}$

9 $\dfrac{8}{\sqrt[5]{11}} = \dfrac{8}{\sqrt[5]{11}} \cdot \dfrac{\sqrt[5]{11^4}}{\sqrt[5]{11^4}} = \dfrac{8\sqrt[5]{11^4}}{\sqrt[5]{11^5}} = \dfrac{8\sqrt[5]{11^4}}{11}$

To completely simplify $\dfrac{\sqrt[6]{3}}{\sqrt[6]{2^4}}$, begin by multiplying it by _____.

$1 = \dfrac{\sqrt[6]{2^2}}{\sqrt[6]{2^2}}$

10 $\dfrac{\sqrt[6]{3}}{\sqrt[6]{2^4}} = \dfrac{\sqrt[6]{3}}{\sqrt[6]{2^4}} \cdot \dfrac{\sqrt[6]{2^2}}{\sqrt[6]{2^2}} = \dfrac{\sqrt[6]{3 \cdot 2^2}}{\sqrt[6]{2^6}} = \dfrac{\sqrt[6]{12}}{2}$

To completely simplify $\dfrac{\sqrt[8]{9}}{4\sqrt[8]{7^3}}$, begin by multiplying it by _____.

$1 = \dfrac{\sqrt[8]{7^5}}{\sqrt[8]{7^5}}$

124   CHAPTER 3

11

$$\frac{\sqrt[8]{9}}{4\sqrt[8]{7^3}} = \frac{\sqrt[8]{9}}{4\sqrt[8]{7^3}} \cdot \frac{\sqrt[8]{7^5}}{\sqrt[8]{7^5}} = \frac{\sqrt[8]{9 \cdot 7^5}}{4\sqrt[8]{7^8}} = \frac{\sqrt[8]{9 \cdot 7^5}}{28}$$

To completely simplify $\dfrac{2\sqrt[5]{11}}{\sqrt[5]{3^4}}$, begin by multiplying it by _____.

$1 = \dfrac{\sqrt[5]{3}}{\sqrt[5]{3}}$

12

Completely simplify. $\dfrac{2\sqrt[5]{11}}{\sqrt[5]{3^4}}$

$\dfrac{2\sqrt[5]{33}}{3}$

13

Completely simplify. $\dfrac{\sqrt[9]{4}}{\sqrt[9]{x^4}}$

$\dfrac{\sqrt[9]{4x^5}}{x}$

14

Completely simplify. $\dfrac{\sqrt[5]{y^3}}{\sqrt[5]{x^4}}$

$\dfrac{\sqrt[5]{xy^3}}{x}$

15

The process used in the preceding frames is **rationalizing the denominator**. Rationalize the denominator of $\dfrac{\sqrt[7]{x^4}}{\sqrt[7]{3^5}}$.

$\dfrac{\sqrt[7]{9x^4}}{3}$

16

Rationalize the denominator of $\dfrac{\sqrt[5]{5^3}}{\sqrt[5]{11^2}}$. Because of the size of the numbers, leave them written with exponents.

$\dfrac{\sqrt[5]{5^3 \cdot 11^3}}{11}$

In some situations it is necessary to find the prime factors of a composite number. This can be accomplished using tree-diagrams like the two shown at the right.

1. Begin with the composite number at the top.
2. Find any two factors of the number other than the number itself or 1.
3. If the branch ends in a prime number, circle it. Otherwise, repeat the process.

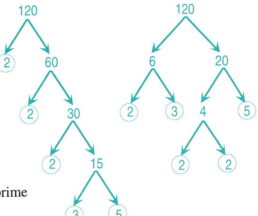

The numbers in the circles represent the prime factors of the original composite number.

$$120 = 2^3 \cdot 3 \cdot 5$$

17
To rationalize the denominator of $\dfrac{\sqrt[6]{5}}{\sqrt[6]{24}}$, the prime factors of 24 are used.

$$\dfrac{\sqrt[6]{5}}{\sqrt[6]{24}} = \dfrac{\sqrt[6]{5}}{\sqrt[6]{8 \cdot 3}} = \dfrac{\sqrt[6]{5}}{\sqrt[6]{2^3 \cdot 3}} = \dfrac{\sqrt[6]{5 \cdot 2^3 \cdot 3^5}}{2 \cdot 3}$$

Complete the first step for rationalizing the denominator of $\dfrac{\sqrt[4]{7}}{\sqrt[4]{72}}$ by writing 72 as the product of its prime factors.

$$72 = 8 \cdot 9 = 2^3 \cdot 3^2$$

18
$$\dfrac{\sqrt[4]{7}}{\sqrt[4]{72}} = \dfrac{\sqrt[4]{7}}{\sqrt[4]{2^3 \cdot 3^2}} = \dfrac{\sqrt[4]{7 \cdot 2 \cdot 3^2}}{\sqrt[4]{2^4 \cdot 3^4}} = \dfrac{\sqrt[4]{7 \cdot 2 \cdot 3^2}}{2 \cdot 3} = \dfrac{\sqrt[4]{7 \cdot 2 \cdot 3^2}}{6}$$

The example above shows how the prime factors of the denominator's radicand are used. Rationalize the denominator of $\dfrac{\sqrt[5]{3}}{\sqrt[5]{8}}$.

$$\dfrac{\sqrt[5]{3}}{\sqrt[5]{2^3}} = \dfrac{\sqrt[5]{3 \cdot 2^2}}{\sqrt[5]{2^5}} = \dfrac{\sqrt[5]{12}}{2}$$

126 CHAPTER 3

19 Simplify $\dfrac{\sqrt[3]{4}}{\sqrt[3]{25}}$ by first finding the prime factors of the denominator's radicand.

$$\dfrac{\sqrt[3]{4}}{\sqrt[3]{5^2}} = \dfrac{\sqrt[3]{4 \cdot 5}}{\sqrt[3]{5^3}} = \dfrac{\sqrt[3]{20}}{5}$$

20 Simplify $\dfrac{\sqrt[5]{y^3}}{\sqrt[5]{4x^3}}$ by first finding the prime factors of the denominator's radicand.

$$\dfrac{\sqrt[5]{y^3}}{\sqrt[5]{2^2 x^3}} = \dfrac{\sqrt[5]{2^3 x^2 y^3}}{\sqrt[5]{2^5 x^5}} = \dfrac{\sqrt[5]{8x^2 y^3}}{2x}$$

21 To completely simplify $\dfrac{\sqrt[3]{18}}{\sqrt[3]{14}}$ begin by reducing it.

$$\dfrac{\sqrt[3]{18}}{\sqrt[3]{14}} = \dfrac{\sqrt[3]{2}}{\sqrt[3]{2}} \cdot \dfrac{\sqrt[3]{9}}{\sqrt[3]{7}} = \dfrac{\sqrt[3]{9}}{\sqrt[3]{7}}$$

Finish the simplification by rationalizing the denominator of $\dfrac{\sqrt[3]{9}}{\sqrt[3]{7}}$.

$$\dfrac{\sqrt[3]{9 \cdot 7^2}}{\sqrt[3]{7^3}} = \dfrac{\sqrt[3]{9 \cdot 7^2}}{7}$$

22 Reduce $\dfrac{\sqrt[5]{28}}{\sqrt[5]{21}}$ and then complete its simplification.

$$\dfrac{\sqrt[5]{4}}{\sqrt[5]{3}} = \dfrac{\sqrt[5]{4 \cdot 3^4}}{3}$$

23 Reduce $\dfrac{\sqrt[6]{6}}{\sqrt[6]{42}}$ and then complete its simplification.

$$\dfrac{1}{\sqrt[6]{7}} = \dfrac{\sqrt[6]{7^5}}{7}$$

24 Reduce both the radicands and coefficients of $\dfrac{30\sqrt[3]{25}}{18\sqrt[3]{35}}$ and then complete its simplification.

$$\dfrac{5\sqrt[3]{5 \cdot 7^2}}{21}$$

25 Reduce both the radicands and coefficients of $\dfrac{13\sqrt[9]{13}}{26\sqrt[9]{52}}$ and then complete its simplification.

$$\dfrac{1}{2\sqrt[9]{2^2}} = \dfrac{\sqrt[9]{2^7}}{4}$$

SIMPLIFYING RADICAL FRACTIONS   127

26
Simplify $\dfrac{21\sqrt[4]{36}}{14\sqrt[4]{54}}$ and, after rationalizing the denominator, look for an opportunity to reduce again.

$\dfrac{3\sqrt[4]{2}}{2\sqrt[4]{3}} = \dfrac{3\sqrt[4]{2 \cdot 3^3}}{2 \cdot 3} = \dfrac{\sqrt[4]{2 \cdot 3^3}}{2}$

27
Simplify $\dfrac{\sqrt[3]{162}}{\sqrt[3]{50}}$ by reducing, simplifying one of the cube roots, and rationalizing the denominator.

$\dfrac{\sqrt[3]{81}}{\sqrt[3]{25}} = \dfrac{3\sqrt[3]{3}}{\sqrt[3]{5^2}} = \dfrac{3\sqrt[3]{15}}{5}$

28
Simplify completely. $\dfrac{4\sqrt[3]{6x^7}}{12\sqrt[3]{10x^2}}$

$\dfrac{x\sqrt[3]{3x^2}}{3\sqrt[3]{5}} = \dfrac{x\sqrt[3]{3 \cdot 5^2 x^2}}{15}$

29
Simplify completely. $\dfrac{40\sqrt[5]{3^9}}{30\sqrt[5]{2^6}}$

$\dfrac{2\sqrt[5]{3^4}}{\sqrt[5]{2}} = \sqrt[5]{3^4 2^4}$

30
Simplify completely. $\dfrac{8\sqrt[7]{12x^5}}{10\sqrt[7]{15x^{12}}}$

$\dfrac{4\sqrt[7]{4}}{5x\sqrt[7]{5}} = \dfrac{4\sqrt[7]{4 \cdot 5^6}}{25x}$

FEEDBACK UNIT 3

This quiz reviews the preceding unit. Answers are at the back of the book.

Completely simplify each of the following.

1. $\dfrac{\sqrt[6]{10}}{\sqrt[6]{3^2}}$

2. $\dfrac{\sqrt[3]{17}}{\sqrt[3]{9}}$

3. $\dfrac{11\sqrt[5]{12}}{22\sqrt[5]{16}}$

4. $\dfrac{8\sqrt[7]{10}}{14\sqrt[7]{25}}$

5. $\dfrac{30\sqrt[3]{6000}}{50\sqrt[3]{162}}$

6. $\dfrac{14\sqrt[5]{2^4}}{21\sqrt[5]{3^4 y^{10}}}$

7. $\dfrac{9\sqrt[4]{32}}{30\sqrt[4]{10}}$

8. $\dfrac{-14\sqrt[7]{75}}{6\sqrt[7]{175}}$

UNIT 4: RADICAL FRACTIONS WITH BINOMIAL DENOMINATORS

The following mathematical term is crucial to an understanding of this unit.

Conjugate

1
Multiply $(5 - \sqrt{3})(4 + 2\sqrt{3})$ by multiplying each term of $(5 - \sqrt{3})$ by $(4 + 2\sqrt{3})$.

$20 + 10\sqrt{3} - 4\sqrt{3} - 6$
$14 + 6\sqrt{3}$

2
Multiply. $(\sqrt{5} - 3\sqrt{2})(\sqrt{5} + 6\sqrt{2})$

$5 + 6\sqrt{10} - 3\sqrt{10} - 36$
$-31 + 3\sqrt{10}$

3
The **conjugate** of $(5 - 4\sqrt{2})$ is $(5 + 4\sqrt{2})$. The binomials are identical except that the second terms have opposite signs. Multiply. $(5 - 4\sqrt{2})(5 + 4\sqrt{2})$

$25 + 20\sqrt{2} - 20\sqrt{2} - 32$
-7

4
The conjugate of $(3 + 9\sqrt{5})$ is $(3 - 9\sqrt{5})$. The binomials have opposites as their second terms.
Multiply. $(3 + 9\sqrt{5})(3 - 9\sqrt{5})$

$9 - 27\sqrt{5} + 27\sqrt{5} - 405$
-396

5
Multiply $(7 - 2\sqrt{3})$ and its conjugate, $(7 + 2\sqrt{3})$.

$49 + 14\sqrt{3} - 14\sqrt{3} - 12$
37

6
Multiply $(2\sqrt{7} - \sqrt{5})$ and its conjugate, $(2\sqrt{7} + \sqrt{5})$.

$28 + 2\sqrt{35} - 2\sqrt{35} - 5$
23

Simplifying Radical Fractions

7
When $(4 + \sqrt{7})$ and its conjugate are multiplied, does the result contain a square root expression?

No, $(4 + \sqrt{7})(4 - \sqrt{7}) = 9$

8
When $(3 - \sqrt{10})$ and its conjugate are multiplied, does the result contain a square root expression?

No,
$(3 - \sqrt{10})(3 + \sqrt{10}) = -1$

9
When $(2\sqrt{5} - \sqrt{6})$ and its conjugate are multiplied, does the result contain a square root expression?

No,
$(2\sqrt{5} - \sqrt{6})(2\sqrt{5} + \sqrt{6}) = 14$

10
Both $(8 - \sqrt{5})$ and its conjugate $(8 + \sqrt{5})$ are irrational numbers. Is their product irrational?

No,
$(8 - \sqrt{5})(8 + \sqrt{5}) = 59$

11
Both $(-6 + \sqrt{2})$ and its conjugate $(-6 - \sqrt{2})$ are irrational numbers. Is their product irrational?

No,
$(-6 + \sqrt{2})(-6 - \sqrt{2}) = 34$

12
The product of $(2\sqrt{3} - \sqrt{5})$ and its conjugate is a(n) _____ (rational, irrational) number.

rational

13
To rationalize the denominator of $\frac{13}{4\sqrt{5} + 7}$ the conjugate of the denominator is used in the process shown below.

$$\frac{13}{4\sqrt{5} + 7} = \frac{13}{4\sqrt{5} + 7} \cdot \frac{4\sqrt{5} - 7}{4\sqrt{5} - 7} = \frac{52\sqrt{5} - 91}{31}$$

In the process shown above, the original denominator $4\sqrt{5} + 7$ was an irrational number and the final denominator 31 is a(n) _____ (rational, irrational) number.

rational

14

For the fraction $\dfrac{7}{8-\sqrt{3}}$, what binomial can be multiplied by both the numerator and denominator to produce a fraction with a rational number denominator?

$8 + \sqrt{3}$

15

$$\dfrac{7}{8-\sqrt{3}} = \dfrac{7}{8-\sqrt{3}} \cdot \dfrac{8+\sqrt{3}}{8+\sqrt{3}} = \dfrac{56+7\sqrt{3}}{61}$$

The process above shows how the denominator of $\dfrac{7}{8-\sqrt{3}}$ is rationalized. What should be multiplied by both the numerator and denominator of $\dfrac{4}{9+3\sqrt{2}}$ to produce a fraction with a rational number denominator?

$9 - 3\sqrt{2}$

16

$$\dfrac{4}{9+3\sqrt{2}} = \dfrac{4}{9+3\sqrt{2}} \cdot \dfrac{9-3\sqrt{2}}{9-3\sqrt{2}} = \dfrac{36-12\sqrt{2}}{63}$$

The process above shows how the denominator of $\dfrac{4}{9+3\sqrt{2}}$ is rationalized. What integer is the highest common factor (HCF) for each term of $\dfrac{36-12\sqrt{2}}{63}$?

3

17

The fraction $\dfrac{36-12\sqrt{2}}{63}$ can be reduced by dividing out the common factor 3. What is the result?

$\dfrac{12-4\sqrt{2}}{21}$

18

To rationalize the denominator of $\dfrac{8}{5\sqrt{7}-6}$, what should be multiplied by both the numerator and denominator?

$5\sqrt{7} + 6$

19

Complete the rationalizing of the denominator shown below.

$$\dfrac{8}{5\sqrt{7}-6} = \dfrac{8}{5\sqrt{7}-6} \cdot \dfrac{5\sqrt{7}+6}{5\sqrt{7}+6} = \underline{\hspace{2cm}}$$

$\dfrac{40\sqrt{7}+48}{139}$

20

Can the fraction $\dfrac{40\sqrt{7}+48}{139}$ be reduced?

No, terms have 1 as HCF.

21

To rationalize the denominator of $\frac{5}{3-2\sqrt{7}}$, what should be multiplied by both the numerator and denominator?

$3 + 2\sqrt{7}$

22

Complete the rationalizing of the denominator shown below.

$$\frac{5}{3-2\sqrt{7}} = \frac{5}{3-2\sqrt{7}} \cdot \frac{3+2\sqrt{7}}{3+2\sqrt{7}} = \underline{}$$

$\frac{15+10\sqrt{7}}{-19}$

23

The fraction $\frac{15+10\sqrt{7}}{-19}$ has a negative denominator and positive denominators are preferred. Multiply each term by -1. What is the result?

$\frac{-15-10\sqrt{7}}{19}$

24

To rationalize the denominator of $\frac{8}{4-\sqrt{6}}$, what should be multiplied by both the numerator and denominator?

$4 + \sqrt{6}$

25

Complete the rationalizing of the denominator shown below.

$$\frac{8}{4-\sqrt{6}} = \frac{8}{4-\sqrt{6}} \cdot \frac{4+\sqrt{6}}{4+\sqrt{6}} = \underline{}$$

$\frac{32+8\sqrt{6}}{10}$

26

Can the fraction $\frac{32+8\sqrt{6}}{10}$ be further simplified?

Yes, $\frac{16+4\sqrt{6}}{5}$

27

To rationalize the denominator of $\frac{5-\sqrt{3}}{7+2\sqrt{3}}$, what should be multiplied by both the numerator and denominator?

$7 - 2\sqrt{3}$

28

Complete the rationalizing of the denominator shown below.

$$\frac{5-\sqrt{3}}{7+2\sqrt{3}} = \frac{5-\sqrt{3}}{7+2\sqrt{3}} \cdot \frac{7-2\sqrt{3}}{7-2\sqrt{3}} = \frac{35-17\sqrt{3}+6}{37} = \underline{}$$

$\frac{41-17\sqrt{3}}{37}$

29
Can the fraction $\dfrac{41 - 17\sqrt{3}}{37}$ be further simplified?

No

30
To rationalize the denominator of $\dfrac{3\sqrt{2} - 4}{1 + \sqrt{5}}$, what should be multiplied by both the numerator and denominator?

$1 - \sqrt{5}$

31
Complete the rationalizing of the denominator shown below.
$$\dfrac{3\sqrt{2} - 4}{1 + \sqrt{5}} = \dfrac{3\sqrt{2} - 4}{1 + \sqrt{5}} \cdot \dfrac{1 - \sqrt{5}}{1 - \sqrt{5}} = \underline{\qquad}$$

$\dfrac{3\sqrt{2} - 3\sqrt{10} - 4 + 4\sqrt{5}}{-4}$

32
Can the fraction $\dfrac{3\sqrt{2} - 3\sqrt{10} - 4 + 4\sqrt{5}}{-4}$ be written with a positive denominator?

Yes, $\dfrac{-3\sqrt{2} + 3\sqrt{10} + 4 - 4\sqrt{5}}{4}$

33
Completely simplify. $\dfrac{\sqrt{6} + 5}{3\sqrt{6} - 2}$

$\dfrac{28 + 17\sqrt{6}}{50}$

34
Completely simplify. $\dfrac{5\sqrt{2} + 7}{\sqrt{2} - 3}$

$\dfrac{-31 - 22\sqrt{2}}{7}$

35
Completely simplify. $\dfrac{4\sqrt{2} - \sqrt{3}}{\sqrt{2} + 3\sqrt{3}}$

$\dfrac{-17 + 13\sqrt{6}}{25}$

36
Completely simplify. $\dfrac{\sqrt{5} - \sqrt{7}}{\sqrt{5} + \sqrt{7}}$

$-6 + \sqrt{35}$

37
Completely simplify. $\dfrac{\sqrt{6} - \sqrt{5}}{\sqrt{6} + \sqrt{5}}$

$11 - 2\sqrt{30}$

SIMPLIFYING RADICAL FRACTIONS

FEEDBACK UNIT 4

This quiz reviews the preceding unit. Answers are at the back of the book.

Completely simplify each of the following.

1. $\dfrac{5}{1-\sqrt{2}}$

2. $\dfrac{3+\sqrt{5}}{6+\sqrt{5}}$

3. $\dfrac{\sqrt{7}}{\sqrt{3}+2}$

4. $\dfrac{4-\sqrt{5}}{4+\sqrt{5}}$

5. $\dfrac{5-3\sqrt{2}}{1+\sqrt{2}}$

6. $\dfrac{\sqrt{6}-2}{2\sqrt{6}-1}$

7. $\dfrac{\sqrt{5}-2\sqrt{7}}{\sqrt{5}+\sqrt{7}}$

8. $\dfrac{\sqrt{7}-\sqrt{3}}{5\sqrt{7}-4\sqrt{3}}$

UNIT 5: APPLICATIONS

In this Applications Section, the format of the text has been altered. Answers for the problems appear beneath them rather than in the right-hand column. Your studying emphasis should be on learning the best procedures to follow with word problems.

1
Solving a word problem requires translating words, phrases, and sentences.
    Words are often translated as variables, like N.
    Phrases are often translated as expressions, like N + 8.
    Sentences are translated as _____, like N + 8 = 23.

Answer: Equations. The equal sign of an equation is often the translation of the verb "is" from a sentence.

134 CHAPTER 3

2

One method of organizing the translation of a word problem is a chart like that shown below. The sentence to be written as an equation is: A number increased by 8 is 23.

word/phrase/sentence	translation
number	N
a number increased by 8	N + 8
A number increased by 8 is 23.	

Complete the table by writing the needed equation in its space.

Answer: N + 8 = 23

3

Construct a table for translating all the necessary words, phases, and sentences of the following:

    17 less than a number is 31.

Answer:

word/phrase/sentence	translation
number	N
17 less than a number	N − 17
17 less than a number is 31.	N − 17 = 31

4

Construct a table for translating all the necessary words, phases, and sentences of the following:

    The product of 6 and a number is 42.

Answer:

word/phrase/sentence	translation
number	N
the product of 6 and a number	6N
The product of 6 and a number is 42.	6N = 42

5

A phrase like "15 plus 3 times a number" is ambiguous because it could be translated as either of the following:

$(15 + 3) \cdot N$ or $15 + 3N$

To eliminate any confusion over meaning, we will translate phrases where the order of operations is not explicitly stated by doing multiplication before addition/subtraction. For example,

17 more than 3 times a number   translates as $3N + 17$

Write an open expression for: 9 more than 5 times a number.

Answer: $5N + 9$ or $9 + 5N$ are correct translations of the phrase.

6

Construct a table for translating all the necessary words, phases, and sentences of the following:

The product of 5 and a number increased by 2 is 37.

Answer:

word/phrase/sentence	translation
number	N
the product of 5 and a number	5N
the product of 5 and a number increased by 2	5N + 2
The product of 5 and a number increased by 2 is 37.	5N + 2 = 37

7

Construct a table for translating all the necessary words, phases, and sentences of the following:

The product of 9 and a number decreased by 11 is 16.

Answer:

word/phrase/sentence	translation
number	N
the product of 9 and a number	9N
the product of 9 and a number decreased by 11	9N – 11
The product of 9 and a number decreased by 11 is 16.	9N – 11 = 16

8

Construct a table for translating all the necessary words, phases, and sentences of the following:

   5 less than a number multiplied by 11 is 72.

Answer:

word/phrase/sentence	translation
number	N
a number multiplied by 11	11N
5 less than a number multiplied by 11	11N – 5
5 less than a number multiplied by 11 is 72.	11N – 5 = 72

FEEDBACK UNIT 5 FOR APPLICATIONS

This quiz reviews the preceding unit. Answers are at the back of the book.

Construct a table for translating all the necessary words, phases, and sentences of each of the following:

1. A number increased by 14 is 34.

2. 15 less than a number is 53.

3. 7 multiplied by a number is 56.

4. 19 more than 3 times a number is 52.

5. 17 decreased by a number times 11 is -16.

Summary for Chapter 3

The following mathematical terms are crucial to an understanding of this chapter.

Reduce a fraction   Rationalizing the denominator
Completely simplified   Conjugate

The first three units of this chapter teach the simplification of radical fractions in which both the numerator and denominator contain a single term. The simplification of such fractions is accomplished by the following process:

1. Reduce the coefficients and/or the radicands. $\dfrac{15\sqrt[3]{48}}{3\sqrt[3]{10}} = \dfrac{5\sqrt[3]{24}}{\sqrt[3]{5}}$

2. Simplify each radical where possible. $\dfrac{5\sqrt[3]{24}}{\sqrt[3]{5}} = \dfrac{5\sqrt[3]{8 \cdot 3}}{\sqrt[3]{5}} = \dfrac{10\sqrt[3]{3}}{\sqrt[3]{5}}$

3. Rationalize the denominator. $\dfrac{10\sqrt[3]{3}}{\sqrt[3]{5}} = \dfrac{10\sqrt[3]{3 \cdot 5^2}}{\sqrt[3]{5 \cdot 5^2}} = \dfrac{10\sqrt[3]{75}}{5}$

4. Reduce again. $\dfrac{10\sqrt[3]{75}}{5} = 2\sqrt[3]{75}$

In simplifying a radical fraction that has a binomial as its denominator, the process is: Rationalize the denominator by using its conjugate and then simplify wherever possible.

$$\dfrac{5 - 2\sqrt{7}}{4 + \sqrt{7}} = \dfrac{5 - 2\sqrt{7}}{4 + \sqrt{7}} \cdot \dfrac{4 - \sqrt{7}}{4 - \sqrt{7}} = \dfrac{34 - 13\sqrt{7}}{9}$$

Chapter 3 Mastery Test

The following questions test the objectives of Chapter 3. Answers are at the back of the book. The number in parentheses which follows each problem indicates the unit in which it can be learned.

Simplify each of the following.

1. $\dfrac{\sqrt{98}}{\sqrt{32}}$ (1)

2. $\dfrac{\sqrt{22}}{\sqrt{110}}$ (1)

3. $\dfrac{8\sqrt{13}}{\sqrt{2}}$ (1)

4. $\dfrac{6\sqrt{6}}{18\sqrt{7}}$ (1)

5. $\dfrac{10\sqrt{45}}{6\sqrt{5}}$ (1)

6. $\sqrt{\dfrac{14}{5}}$ (1)

7. $\dfrac{5}{7}\sqrt{\dfrac{3}{5}}$ (1)

8. $\dfrac{\sqrt[7]{42}}{\sqrt[7]{6}}$ (3)

9. $\dfrac{14\sqrt[3]{17}}{\sqrt[3]{216}}$ (3)

10. $\dfrac{\sqrt[9]{3^6 \cdot 5^8}}{\sqrt[9]{3^4 \cdot 5^{17}}}$ (3)

11. $\dfrac{13\sqrt[6]{5^4}}{39\sqrt[6]{5^5}}$ (3)

12. $\dfrac{6\sqrt[4]{625}}{25\sqrt[4]{27}}$ (3)

13. $\dfrac{-12\sqrt[3]{6}}{\sqrt[3]{54}}$ (3)

14. $\dfrac{7\sqrt[3]{6000}}{20\sqrt[3]{2}}$ (3)

15. $\dfrac{4}{2\sqrt{6}-5}$ (4)

16. $\dfrac{\sqrt{5}}{2\sqrt{10}-3\sqrt{3}}$ (4)

17. $\dfrac{3-2\sqrt{5}}{7+\sqrt{5}}$ (4)

18. $\dfrac{6-5\sqrt{2}}{6+5\sqrt{2}}$ (4)

19. $\dfrac{\sqrt{3}+2\sqrt{7}}{\sqrt{3}-5\sqrt{7}}$ (4)

20. $\dfrac{\sqrt{6}+4\sqrt{2}}{\sqrt{6}-7\sqrt{2}}$ (4)

21. Construct a table for translating all the necessary words, phases, and sentences of this problem: 17 increased by 4 times a number is 77. (5)

22. Construct a table for translating all the necessary words, phases, and sentences of this problem: Twice a number decreased by 12 is 46. (5)

CHAPTER 4 OBJECTIVES

The following problems illustrate the objectives of this chapter. At this time you are not expected to know how to do these problems. However, if all these problems are thoroughly understood, proceed directly to the Chapter Mastery Test. The number in parentheses which follows each problem indicates the unit in which it can be learned.

Replacements for the variables in this test are restricted to positive real numbers.

1. Write the following using rational number exponents. (1)
   a. $\sqrt[3]{5^4}$
   b. $\sqrt[6]{7}$
   c. $\sqrt[4]{2^7}$
   d. $\sqrt{11^5}$

2. Write the following as radical expressions. (1)
   a. $17^{\frac{1}{2}}$
   b. $5^{\frac{3}{4}}$
   c. $(-2)^{\frac{3}{5}}$
   d. $6^{\frac{7}{3}}$

3. Reduce the exponent where possible or state that it should not be reduced. (1)
   a. $11^{\frac{15}{20}}$
   b. $(-5)^{\frac{6}{8}}$
   c. $3^{\frac{10}{8}}$
   d. $(-3)^{\frac{3}{9}}$

4. $6^{\frac{4}{5}} \cdot 6^{\frac{1}{3}} = $ _____ (2)

5. $y^{\frac{3}{8}} \cdot y^{\frac{5}{6}} = $ _____ (2)

6. $\sqrt[3]{x^4} \cdot \sqrt[5]{x^3} = $ _____ (2)

7. $\sqrt[7]{x^2} \cdot \sqrt[5]{x} = $ _____ (2)

8. $\dfrac{x^{\frac{5}{3}}}{x^{\frac{5}{7}}} = $ _____ (3)

9. $\dfrac{x^{\frac{4}{5}}}{x^{\frac{1}{2}}} = $ _____ (3)

10. $\dfrac{\sqrt[3]{x^2}}{\sqrt[7]{x^2}} = $ _____ (3)

11. $\dfrac{\sqrt[4]{x^7}}{\sqrt[8]{x^3}} = $ _____ (3)

12. $(x^{\frac{2}{3}})^{\frac{3}{8}} = $ _____ (4)

13. $(x^{\frac{1}{4}})^{\frac{6}{7}} = $ _____ (4)

14. $(\sqrt[4]{x^5})^8 = $ _____ (4)

15. $(\sqrt[3]{x^2})^{10} = $ _____ (4)

Chapter 4

Rational Number Exponents

Unit 1: The Meaning of Rational Number Exponents

1
When a counting number is used as an exponent, it indicates the number of times its base is used as a factor. For example, 7^3 means $7 \cdot 7 \cdot 7$. Similarly, 5^4 means _____

$5 \cdot 5 \cdot 5 \cdot 5$

2
Negative integers as exponents are defined by the identity $x^{-n} = \frac{1}{x^n}$. For example, $6^{-9} = \frac{1}{6^9}$.

Similarly, $2^{-7} =$ _____

$\frac{1}{2^7}$

3

Zero as an exponent is undefined when the base is itself 0, but for all other real number bases $x^0 = 1$.

$7^0 = 1 \qquad (-5)^0 = 1 \qquad (\sqrt{2})^0 = $ _____

1

4

In this chapter, the use of rational numbers as exponents is learned. It needs to be remembered that the radical expression $\sqrt[8]{5^3}$ has 8 as its index and the radicand is _____.

5^3

5

$\sqrt[8]{5^3}$ is equivalent to $5^{\frac{3}{8}}$ where the numerator of $\frac{3}{8}$ is the exponent of the radicand and the denominator of $\frac{3}{8}$ is the index of the radical. Similarly, $\sqrt[7]{4^5}$ is equivalent to _____.

$4^{\frac{5}{7}}$

6

Radical expressions can be written using rational exponents.

$\sqrt[5]{(-3)^4} = (-3)^{\frac{4}{5}} \qquad$ and $\qquad \sqrt[9]{(-2)^5} = $ _____.

$(-2)^{\frac{5}{9}}$

7

Rational number exponents indicate radical expressions.

$(-3)^{\frac{2}{7}} = \sqrt[7]{(-3)^2} \qquad$ and $\qquad (-5)^{\frac{4}{5}} = $ _____.

$\sqrt[5]{(-5)^4}$

8

The numerator of a rational exponent is the power for the radicand of the radical expression.

$7^{\frac{2}{3}} = \sqrt[3]{7^2} \qquad$ and $\qquad 6^{\frac{5}{6}} = $ _____.

$\sqrt[6]{6^5}$

9

The denominator of a rational exponent is the index of the radical expression.

$\sqrt[9]{2^7} = 2^{\frac{7}{9}} \qquad$ and $\qquad \sqrt[12]{11^5} = $ _____.

$11^{\frac{5}{12}}$

Rational Number Exponents

10
Is the exponent of $\sqrt[7]{5^{10}}$ greater than 1?   Yes

11
A rational number exponent may be greater than 1.
Is $\sqrt[3]{6^4}$ equivalent to $6^{\frac{4}{3}}$?   Yes

12
The absence of an index for a radical means the index is 2.
Is $\sqrt{7^9}$ equivalent to $7^{\frac{9}{2}}$?   Yes

13
The absence of an exponent on a radicand means the exponent is 1. Is $\sqrt[4]{13}$ equivalent to $13^{\frac{1}{4}}$?   Yes

14
Write $\sqrt[5]{31^2}$ using a rational exponent. $31^{\frac{2}{5}}$

15
Write $3^{\frac{7}{5}}$ as a radical expression. $\sqrt[5]{3^7}$

16
Write $\sqrt[7]{5^2}$ using a rational exponent. $5^{\frac{2}{7}}$

17
Write $3^{\frac{7}{5}}$ as a radical expression. $\sqrt[5]{3^7}$

18
Write $\sqrt[4]{11^9}$ using a rational exponent. $11^{\frac{9}{4}}$

19
Write $8^{\frac{7}{2}}$ as a radical expression. $\sqrt{8^7}$

20
Write $\sqrt[5]{7}$ using a rational exponent. $7^{\frac{1}{5}}$

21
Write $x^{\frac{4}{3}}$ as a radical expression. $\sqrt[3]{x^4}$

22
Write $\sqrt{13}$ using a rational exponent. $13^{\frac{1}{2}}$

23
Caution must be used whenever reducing a rational exponent with a negative base. Is caution necessary in writing $7^{\frac{6}{10}}$ as $7^{\frac{3}{5}}$?   No, base is positive

24
Reducing a rational number exponent is always safe when the base of the exponent is positive. Is $10^{\frac{4}{8}}$ equivalent to $10^{\frac{1}{2}}$?   Yes, base is positive

25
When the base of a rational number exponent is negative, it is possible to alter the meaning by reducing the exponent. Is $(-5)^{\frac{2}{4}}$ equivalent to $(-5)^{\frac{1}{2}}$?   No

26

$(-5)^{\frac{1}{2}} = \sqrt{-5}$ has a negative radicand.

$(-5)^{\frac{2}{4}} = \sqrt[4]{(-5)^2} = \sqrt[4]{25}$ has a positive radicand.

Be careful in reducing a rational number exponent. Reduce a rational number exponent only if it does not change the sign of the _____.   radicand

27
When the base of a rational exponent is positive, its exponent should be reduced whenever possible.

$37^{\frac{2}{4}} = 37^{\frac{1}{2}}$ $19^{\frac{8}{12}} = 19^{\frac{2}{3}}$ $47^{\frac{10}{18}} =$ _____ $47^{\frac{5}{9}}$

28
A rational number exponent can be reduced when its base is _____.   positive

Rational Number Exponents 145

29
Rational number exponents with negative bases may have the sign of their radicands changed when the exponent is reduced.

The base of $(-7)^{\frac{3}{9}}$ is negative and the radicand of $\sqrt[9]{(-7)^3}$ is _____ (negative, positive).

negative, -343

30
The base of $(-7)^{\frac{1}{3}}$ is negative and the radicand of $\sqrt[3]{-7}$ is _____ (negative, positive).

negative, -7

31
$(-7)^{\frac{3}{9}} = (-7)^{\frac{1}{3}}$ because the signs of both radicands are negative. Is $(-2)^{\frac{3}{5}}$ equivalent to $(-2)^{\frac{6}{10}}$?

No, $(-2)^3$ is negative but $(-2)^6$ is positive.

32
A rational number exponent can be reduced only when the signs of the radicands are the same. When the base is positive, the radicand is always positive. When the base is negative, the radicand is _____ (always, sometimes) negative.

sometimes

33
Reducing an exponent when the base is positive is always safe. Reducing an exponent when the base is negative is safe only when the radicands have _____ (the same, different) signs.

the same

34
Is $7^{\frac{4}{10}}$ equivalent to $7^{\frac{2}{5}}$?

Yes

35
Is $23^{\frac{9}{10}}$ equivalent to $23^{\frac{27}{30}}$?

Yes

36
Is $3^{\frac{5}{6}}$ equivalent to $3^{\frac{6}{5}}$?

No

37. Is $(-10)^{\frac{8}{10}}$ equivalent to $(-10)^{\frac{4}{5}}$?   Yes, both $(-10)^8$ and $(-10)^4$ are positive.

38. Is $(-6)^{\frac{6}{14}}$ equivalent to $(-6)^{\frac{3}{7}}$?   No, $(-6)^6$ is positive and $(-6)^3$ is negative.

39. Is $5^{\frac{12}{30}}$ equivalent to $5^{\frac{2}{5}}$?   Yes

40. Is $(-3)^{\frac{5}{15}}$ equivalent to $(-3)^{\frac{1}{3}}$?   Yes

---

FEEDBACK UNIT 1

This quiz reviews the preceding unit. Answers are at the back of the book.

1. Write the following using rational number exponents.
    a. $\sqrt[3]{5^2}$
    b. $\sqrt[7]{11^5}$
    c. $\sqrt{7^3}$
    d. $\sqrt[5]{3^4}$

2. Write the following as radical expressions.
    a. $2^{\frac{5}{7}}$
    b. $6^{\frac{7}{4}}$
    c. $(-7)^{\frac{1}{3}}$
    d. $3^{\frac{5}{2}}$

3. Reduce the exponent if that is possible or state that it should not be reduced.
    a. $3^{\frac{14}{18}}$
    b. $(-7)^{\frac{4}{12}}$
    c. $11^{\frac{8}{12}}$
    d. $(-5)^{\frac{4}{6}}$

RATIONAL NUMBER EXPONENTS 147

UNIT 2: MULTIPLYING LIKE BASES WITH RATIONAL NUMBER EXPONENTS

Because there are some instances where rational number exponents with negative bases should not be simplified, all variables in the remainder of this chapter are restricted to positive number replacements.

1
In the power expression $-7x^2$ the numerical coefficient is -7 and the base of the exponent 2 is _____.

x

2
In the power expression $9x^5$ the numerical coefficient is _____.

9

3
The base of the power expression x^6 is x. What is the base of the power expression y^9?

y

4
Do the two power expressions 2^3 and 3^2 have the same base?

No

5
A counting number exponent shows the number of times its base is to be used as a factor. x^7 means that x is to be used _____ times as a factor.

7

6
In multiplying two power expressions with the same base, the exponents can be used to determine the total number of times the base is to be used as a factor. In the multiplication $x^5 \cdot x^9$, how many times is x used as a factor?

14

7
When two power expressions with the same base are multiplied, the exponents are added.

$x^5 \cdot x^9 = x^{14}$ $y^7 \cdot y^3 =$ _____

y^{10}

8
Simplify. $x^6 \cdot x^8 =$ _____

x^{14}

9
Simplify. $7^3 \cdot 7^2 =$ _____

7^5

10
Simplify. $(-6)^4 \cdot (-6)^7 =$ _____

$(-6)^{11}$

11
The use of negative integers as exponents maintains the property that when like bases are multiplied, the exponents are added.

$x^5 \cdot x^{-7} = x^{-2}$ $y^{-4} \cdot y^9 =$ _____

y^5

12
When like bases are multiplied, integer exponents are added.

$7^{-3} \cdot 7^{-5} = 7^{-8}$ $5^{-2} \cdot 5^{-7} =$ _____

5^{-9}

13
In multiplying like bases, the exponents are _____.

added

14
To multiply $3^{\frac{2}{11}} \cdot 3^{\frac{5}{11}}$ the following steps are used.

$3^{\frac{2}{11}} \cdot 3^{\frac{5}{11}} = 3^{\frac{2}{11}+\frac{5}{11}} = 3^{\frac{7}{11}}$

In this example, exponents were _____ (added, multiplied).

added

15
When like bases are multiplied, the exponents are added whether the exponents are counting numbers, integers or rational numbers. Complete the following multiplication.

$x^{\frac{2}{3}} \cdot x^{\frac{1}{5}} = x^{\frac{2}{3}+\frac{1}{5}} =$ _____

$x^{\frac{10}{15}+\frac{3}{15}} = x^{\frac{13}{15}}$

RATIONAL NUMBER EXPONENTS   149

16
In multiplying like bases, the exponents are added.
$$x^5 \cdot x^{-2} = x^{5+(-2)} = x^3$$
$$x^{\frac{2}{7}} \cdot x^{\frac{3}{7}} = x^{\frac{2}{7}+\frac{3}{7}} = x^{\frac{5}{7}}$$

Multiply. $x^{\frac{4}{9}} \cdot x^{\frac{10}{9}}$

$x^{\frac{14}{9}}$

17
If two rational numbers are to be added, they must have a common denominator.
$$x^{\frac{2}{3}} \cdot x^{\frac{3}{5}} = x^{\frac{10}{15}+\frac{9}{15}} = x^{\frac{19}{15}}$$

Multiply. $x^{\frac{1}{2}} \cdot x^{\frac{3}{4}}$

$x^{\frac{2}{4}+\frac{3}{4}} = x^{\frac{5}{4}}$

18
Multiply. $x \cdot x^{\frac{5}{6}}$

$x^{\frac{1}{1}} \cdot x^{\frac{5}{6}} = x^{\frac{6}{6}+\frac{5}{6}} = x^{\frac{11}{6}}$

19
Multiply. $x^{\frac{2}{3}} \cdot x^{\frac{1}{4}}$

$x^{\frac{8}{12}+\frac{3}{12}} = x^{\frac{11}{12}}$

20
Multiply. $x^{\frac{4}{7}} \cdot x^{\frac{3}{2}}$

$x^{\frac{29}{14}}$

21
Multiply. $x^{\frac{1}{2}} \cdot x^{\frac{4}{7}}$

$x^{\frac{15}{14}}$

22
Multiply. $x^{\frac{7}{9}} \cdot x^{\frac{5}{6}}$

$x^{\frac{29}{18}}$

23

To multiply $\sqrt{x} \cdot \sqrt[3]{x^5}$,

1. Write each factor using a rational exponent.
$$\sqrt{x} \cdot \sqrt[3]{x^5} = x^{\frac{1}{2}} \cdot x^{\frac{5}{3}}$$

2. Since the bases are the same, add exponents.
$$\sqrt{x} \cdot \sqrt[3]{x^5} = x^{\frac{1}{2}} \cdot x^{\frac{5}{3}} = x^{\frac{3}{6}+\frac{10}{6}} = x^{\frac{13}{6}}$$

3. Write $x^{\frac{13}{6}}$ as a radical expression.
$$x^{\frac{13}{6}} = \sqrt[6]{x^{13}}$$

4. If possible, simplify the radical expression.
$$\sqrt[6]{x^{13}} = \sqrt[6]{x \cdot x^{12}} = \underline{}$$

$x^2 \sqrt[6]{x}$

24

To multiply $\sqrt[3]{x} \cdot \sqrt[4]{x^3}$, rational exponents are used as follows:

$$\sqrt[3]{x} \cdot \sqrt[4]{x^3} = x^{\frac{1}{3}} \cdot x^{\frac{3}{4}} = x^{\frac{4}{12}+\frac{9}{12}} = x^{\frac{13}{12}}$$

Write $x^{\frac{13}{12}}$ as a simplified radical expression.

$\sqrt[12]{x^{13}} = x\sqrt[12]{x}$

25

Complete the following multiplication, and write the answer as a simplified radical expression.

$$\sqrt[8]{x^3} \cdot \sqrt[4]{x} = x^{\frac{3}{8}} \cdot x^{\frac{1}{4}} = x^{\frac{3}{8}+\frac{2}{8}} = \underline{}$$

$\sqrt[8]{x^5}$

26

In general, it is good procedure to write the answer in the same form as the original problem. In this case, that means that problems given with rational exponents should have their answers given as rational exponents; problems given as radical expressions should have their answers given as _____ _____.

radical expressions

27

$\sqrt[4]{x^3} \cdot \sqrt[7]{x^2} = \underline{}$

$\sqrt[28]{x^{29}} = x\sqrt[28]{x}$

28

$\sqrt{x^5} \cdot \sqrt[4]{x^3} = \underline{}$

$\sqrt[4]{x^{13}} = x^3 \sqrt[4]{x}$

RATIONAL NUMBER EXPONENTS 151

29. $\sqrt[3]{x^2} \cdot \sqrt{x^3} =$ _____ $\sqrt[6]{x^{13}} = x^2 \sqrt[6]{x}$

30. $\sqrt[3]{x} \cdot \sqrt[5]{x^2} =$ _____ $\sqrt[15]{x^{11}}$

31. $\sqrt[3]{x^2} \cdot \sqrt[4]{x^3} =$ _____ $\sqrt[12]{x^{17}} = x\sqrt[12]{x^5}$

FEEDBACK UNIT 2

This quiz reviews the preceding unit. Answers are at the back of the book.

1. $3^4 \cdot 3^7 =$ _____

2. $5^{-6} \cdot 5^{-3} =$ _____

3. $x^5 \cdot x^{-8} =$ _____

4. $y^{-1} \cdot y^5 =$ _____

5. $2^{\frac{3}{5}} \cdot 2^{\frac{1}{5}} =$ _____

6. $7^{\frac{-5}{13}} \cdot 7^{\frac{8}{13}} =$ _____

7. $x^{\frac{2}{3}} \cdot x^{\frac{1}{6}} =$ _____

8. $y^{\frac{1}{2}} \cdot y^{\frac{2}{7}} =$ _____

9. $x^{\frac{4}{5}} \cdot x^{\frac{7}{10}} =$ _____

10. $y^{\frac{3}{5}} \cdot y =$ _____

11. $\sqrt[5]{x^2} \cdot \sqrt[3]{x} =$ _____

12. $\sqrt[7]{y^3} \cdot \sqrt{y} =$ _____

13. $\sqrt[4]{z^3} \cdot \sqrt[8]{z^5} =$ _____

14. $\sqrt[6]{x^5} \cdot \sqrt[9]{x^4} =$ _____

Unit 3: Division of Like Bases with Rational Number Exponents

1
In dividing power expressions with the same base, the exponents determine the number of times the base will remain as a factor. In the division $\frac{x^9}{x^3}$, 9 factors of x are divided by _____ factors of x.

3

2
When power expressions with the same base are divided, the exponents are subtracted.

$$\frac{x^9}{x^3} = x^{9-3} = x^6 \quad \text{and} \quad \frac{x^8}{x^4} = \underline{}.$$

$x^{8-4} = x^4$

3
Divide. $\frac{x^7}{x^5} = \underline{}$

$x^{7-5} = x^2$

4
The division of $\frac{y^3}{y^7}$ may be completed in two ways.

$$\frac{y^3}{y^7} = y^{3-7} = y^{-4} \quad \text{or} \quad \frac{y^3}{y^7} = \frac{1}{y^{7-3}} = \frac{1}{y^4}$$

Do the two problems above have equivalent answers?

Yes, $\frac{1}{y^4} = y^{-4}$

5
Divide. $\frac{7^3}{7^8} = \underline{}$

$7^{-5} = \frac{1}{7^5}$

6
Divide. $\frac{(-6)^7}{(-6)^4} = \underline{}$

$(-6)^3$

RATIONAL NUMBER EXPONENTS   153

7
The use of negative integers as exponents maintains the property that when like bases are divided, the exponents are subtracted.

$$\frac{x^5}{x^{-7}} = x^{5-(-7)} = x^{12} \quad \text{and} \quad \frac{y^{-4}}{y^{-9}} = \underline{\qquad}.$$

$y^{-4-(-9)} = y^5$

8
When like bases are divided, the exponents are subtracted.

$$\frac{7^{-8}}{7^{-5}} = 7^{-3} \quad \text{and} \quad \frac{5^{-7}}{5^{-2}} = \underline{\qquad}.$$

5^{-5}

9
In dividing like bases, the exponents are _____.

subtracted

10
To divide $3^{\frac{5}{11}}$ by $3^{\frac{2}{11}}$, the following steps are used.

$$\frac{3^{\frac{5}{11}}}{3^{\frac{2}{11}}} = 3^{\frac{5}{11}-\frac{2}{11}} = 3^{\frac{3}{11}}$$

In the example above, the exponents were _____.

subtracted

11
In dividing like bases, the exponents are subtracted. This is the case whether the exponents are counting numbers, integers, or _____ _____.

rational numbers

12
Complete the division of $x^{\frac{5}{6}}$ by $x^{\frac{3}{8}}$ shown below.

$$\frac{x^{\frac{5}{6}}}{x^{\frac{3}{8}}} = x^{\frac{5}{6}-\frac{3}{8}} = x^{\frac{20}{24}-\frac{9}{24}} = \underline{\qquad}$$

$x^{\frac{11}{24}}$

13
When dividing like bases, the exponents are subtracted.

$$\frac{x^{\frac{1}{2}}}{x^{\frac{3}{2}}} = \underline{\qquad}$$

x^{-1}

154 CHAPTER 4

14. $\dfrac{x^{\frac{3}{4}}}{x^{\frac{2}{7}}} = \underline{\hspace{1cm}}$ $x^{\frac{13}{28}}$

15. $\dfrac{x^{\frac{1}{2}}}{x^{\frac{1}{3}}} = \underline{\hspace{1cm}}$ $x^{\frac{1}{6}}$

16. $\dfrac{x^2}{x^{\frac{1}{2}}} = \underline{\hspace{1cm}}$ $x^{\frac{3}{2}}$

17. $\dfrac{x^{\frac{2}{3}}}{x^{\frac{1}{4}}} = \underline{\hspace{1cm}}$ $x^{\frac{5}{12}}$

18. $\dfrac{x^{\frac{5}{4}}}{x} = \underline{\hspace{1cm}}$ $x^{\frac{1}{4}}$

19.
To divide $\dfrac{\sqrt[3]{x^4}}{\sqrt[5]{x}}$,

1. Write each radical using a rational exponent.

$$\dfrac{\sqrt[3]{x^4}}{\sqrt[5]{x}} = \dfrac{x^{\frac{4}{3}}}{x^{\frac{1}{5}}}$$

2. Since the bases are the same, subtract exponents.

$$\dfrac{\sqrt[3]{x^4}}{\sqrt[5]{x}} = \dfrac{x^{\frac{4}{3}}}{x^{\frac{1}{5}}} = x^{\frac{4}{3}-\frac{1}{5}} = x^{\frac{20}{15}-\frac{3}{15}} = x^{\frac{17}{15}}$$

3. Write $x^{\frac{17}{15}}$ as a radical expression.

$$x^{\frac{17}{15}} = \sqrt[15]{x^{17}}$$

4. If possible, simplify the radical expression.

$$\sqrt[15]{x^{17}} = \sqrt[15]{x^{15} \cdot x^2} = \underline{\hspace{1cm}}$$ $x\sqrt[15]{x^2}$

RATIONAL NUMBER EXPONENTS   155

20. $\dfrac{\sqrt[3]{x^5}}{\sqrt{x}} = $ _____ $\qquad x\sqrt[6]{x}$

21. $\dfrac{\sqrt[4]{x^3}}{\sqrt[5]{x}} = $ _____ $\qquad \sqrt[20]{x^{11}}$

22. $\dfrac{\sqrt[4]{x^3}}{\sqrt[7]{x^2}} = $ _____ $\qquad \sqrt[28]{x^{13}}$

23. $\dfrac{\sqrt[10]{x^3}}{\sqrt[5]{x}} = $ _____ $\qquad \sqrt[10]{x}$

FEEDBACK UNIT 3

This quiz reviews the preceding unit. Answers are at the back of the book.

1. $\dfrac{x^{\frac{2}{3}}}{x^{\frac{1}{4}}} = $ _____

2. $\dfrac{x^{\frac{5}{6}}}{x^{\frac{3}{4}}} = $ _____

3. $\dfrac{x^{\frac{9}{10}}}{x^{\frac{3}{4}}} = $ _____

4. $\dfrac{x^{\frac{5}{2}}}{x^{\frac{3}{4}}} = $ _____

5. $\dfrac{x^{\frac{9}{4}}}{x^{\frac{3}{5}}} = $ _____

6. $\dfrac{x^{\frac{1}{6}}}{x^{\frac{5}{8}}} = $ _____

7. $\dfrac{\sqrt[5]{x^8}}{\sqrt[7]{x^2}} = $ _____

8. $\dfrac{\sqrt[9]{x^8}}{\sqrt[3]{x}} = $ _____

9. $\dfrac{\sqrt[5]{x^4}}{\sqrt[8]{x^5}} = $ _____

10. $\dfrac{\sqrt[3]{x^7}}{\sqrt[6]{x}} = $ _____

Unit 4: Raising a Power to a Power with Rational Exponents

1
In the power expression $(x^5)^9$, the exponents have different bases. In $(x^5)^9$, the base of the exponent 9 is (x^5). The base of the exponent 5 is _____.

x

2
To simplify $(x^5)^9$, the exponents are multiplied.
$$(x^5)^9 = x^{5 \cdot 9} = x^{45} \qquad (y^5)^8 = _____$$

$y^{5 \cdot 8} = y^{40}$

3
$$(x^9)^3 = x^{9 \cdot 3} = x^{27} \qquad (x^7)^6 = _____$$

$x^{7 \cdot 6} = x^{42}$

4
The exponent zero for any nonzero base gives 1.
$$(5^0)^3 = 1^3 \quad \text{and} \quad (5^0)^3 = 5^{0 \cdot 3} = 5^0$$
Do the two problems above have equivalent answers?

Yes, $1^3 = 1$ and $5^0 = 1$

5
Simplify. $(7^3)^6 = $ _____

7^{18}

6
Simplify. $[(-6)^7]^8 = $ _____

$(-6)^{56}$

7
The use of negative exponents maintains the property that when raising a power to a power the exponents are multiplied.
$$(x^5)^{-7} = x^{5(-7)} = x^{-35} \qquad (x^5)^{-7} = \left(\frac{1}{x^5}\right)^7 = \frac{1}{x^{35}}$$

Are the answers to these two problems equivalent?

Yes, $x^{-35} = \frac{1}{x^{35}}$

8
When raising a power to a power, the exponents are multiplied.
$$(7^{-8})^{-4} = 7^{32} \qquad (5^{-7})^{-9} = _____$$

5^{63}

9
In raising a power to a power, exponents are _____.

multiplied

10
To simplify $\left(3^{\frac{2}{3}}\right)^{\frac{5}{9}}$, the following steps are used.

$$\left(3^{\frac{2}{3}}\right)^{\frac{5}{9}} = 3^{\frac{2}{3} \cdot \frac{5}{9}} = 3^{\frac{10}{27}}$$

In the example above, the exponents were _____.

multiplied

11
In raising a power to a power, the exponents are multiplied. This is the case whether the exponents are counting numbers, integers, or _____ _____.

rational numbers

12
When the base is a positive number, cancellation can be used to simplify the multiplication of exponents.

$$\left(5^{\frac{1}{2}}\right)^{\frac{2}{3}} = 5^{\frac{1}{2} \cdot \frac{2}{3}} = 5^{\frac{1}{1} \cdot \frac{1}{3}} = 5^{\frac{1}{3}}$$

Simplify. $\left(7^{\frac{3}{4}}\right)^{\frac{2}{5}} =$ _____

$7^{\frac{3}{10}}$

13
$(x^a)^b = x^{ab}$ $(x^3)^{-5} = x^{-15}$ $\left(x^{\frac{3}{7}}\right)^{\frac{4}{5}} =$ _____

$x^{\frac{12}{35}}$

14
$\left(3^{\frac{1}{5}}\right)^2 = 3^{\frac{1}{5} \cdot \frac{2}{1}} = 3^{\frac{2}{5}}$ $\left(x^{\frac{2}{3}}\right)^4 =$ _____

$x^{\frac{8}{3}}$

15
$\left(x^{\frac{5}{4}}\right)^4 = x^{\frac{5}{4} \cdot \frac{4}{1}} = x^5$ $\left(x^{\frac{7}{4}}\right)^2 =$ _____

$x^{\frac{7}{2}}$

16
$\left(3^{\frac{2}{3}}\right)^3 =$ _____

$3^2 = 9$

17
$\left(y^{\frac{5}{2}}\right)^3 =$ _____

$y^{\frac{15}{2}}$

18

$$\left(11^{\frac{1}{2}}\right)^2 = \underline{}$$

$11^1 = 11$

19

$$\left(x^{\frac{2}{5}}\right)^{10} = \underline{}$$

x^4

20

To simplify $\left(\sqrt[3]{x^5}\right)^2$,

1. Write the radical expression using a rational exponent.

$$\left(\sqrt[3]{x^5}\right)^2 = \left(x^{\frac{5}{3}}\right)^2$$

2. Simplify by multiplying exponents.

$$\left(\sqrt[3]{x^5}\right)^2 = \left(x^{\frac{5}{3}}\right)^2 = x^{\frac{10}{3}}$$

3. Write $x^{\frac{10}{3}}$ as a radical expression.

$$x^{\frac{10}{3}} = \sqrt[3]{x^{10}}$$

4. If possible, simplify the radical expression.

$$\sqrt[3]{x^{10}} = \sqrt[3]{x \cdot x^9} = \underline{}$$

$x^3 \sqrt[3]{x}$

21

Complete the simplification of $\left(\sqrt[4]{x^3}\right)^2$ shown below.

$$\left(\sqrt[4]{x^3}\right)^2 = \left(x^{\frac{3}{4}}\right)^2 = x^{\frac{3}{2}} = \underline{}$$

$\sqrt{x^3} = x\sqrt{x}$

22

Complete the simplification of $\left(\sqrt{x^3}\right)^6$ shown below.

$$\left(\sqrt{x^3}\right)^6 = \left(x^{\frac{3}{2}}\right)^6 = \underline{}$$

x^9

23

Complete the simplification of the cube of the fourth root of x^7 that is shown below.

$$\left(\sqrt[4]{x^7}\right)^3 = \left(x^{\frac{7}{4}}\right)^3 = x^{\frac{21}{4}} = \underline{}$$

$\sqrt[4]{x^{21}} = x^5 \sqrt[4]{x}$

24. $\left(\sqrt[5]{2^3}\right)^2 = $ _____

$\sqrt[5]{2^6} = 2\sqrt[5]{2}$

25. $\left(\sqrt{5}\right)^4 = $ _____

$5^2 = 25$

Feedback Unit 4

This quiz reviews the preceding unit. Answers are at the back of the book.

1. $\left(x^{\frac{5}{7}}\right)^{\frac{3}{4}} = $ _____

2. $\left(x^{\frac{1}{3}}\right)^{\frac{3}{5}} = $ _____

3. $\left(x^{\frac{2}{9}}\right)^{\frac{3}{8}} = $ _____

4. $\left(x^6\right)^{\frac{3}{4}} = $ _____

5. $\left(x^{\frac{1}{2}}\right)^3 = $ _____

6. $\left(x^{\frac{1}{5}}\right)^5 = $ _____

7. $\left(y^{\frac{4}{3}}\right)^2 = $ _____

8. $\left(z^{\frac{3}{4}}\right)^2 = $ _____

9. $\left(\sqrt[4]{x^3}\right)^8 = $ _____

10. $\left(\sqrt[3]{y^4}\right)^5 = $ _____

11. $\left(\sqrt{x^5}\right)^2 = $ _____

12. $\left(\sqrt[4]{w^2}\right)^6 = $ _____

Unit 5: Applications

In this Applications Section, the format of the text has been altered. Answers for the problems appear beneath them rather than in the right-hand column. Your studying emphasis should be on learning the best procedures to follow with word problems.

1

There is a 3-step process to solving word problems.
- a. Construct a table that translates necessary words, phrases, and sentences.
- b. Solve the equation(s) obtained.
- c. Check the answer in the original wording of the problem.

Begin the 3-step process by constructing a table of translations for:

   A number plus 13 is 25.

Answer:

word/phrase/sentence	translation
number	N
number plus 13	$N + 13$
A number plus 13 is 25.	$N + 13 = 25$

2

The 3-step process for solving word problems requires:
- a. Constructing a table of translations.
- b. Solving the equation.
- c. Checking the answer in the original sentence.

The first step in the process is of crucial importance because it creates an equation that will provide the solution. Use the equation $N + 13 = 25$ to complete the last 2 steps in solving:

   A number plus 13 is 25. What is the number?

Answer:
- b. $N = 12$
- c. It is true that 12 plus 13 is 25.

3

There is a 3-step process to solving word problems.
  a. Construct a table that translates necessary words, phrases, and sentences.
  b. Solve the equation(s) obtained.
  c. Check the answer in the original wording of the problem.

Begin the 3-step process by constructing a table of translations for:

  A number decreased by 7 is 44.

Answer:

word/phrase/sentence	translation
number	N
a number decreased by 7	N – 7
A number decreased by 7 is 44.	N – 7 = 44

4

The 3-step process for solving word problems requires:
  a. Constructing a table of translations.
  b. Solving the equation.
  c. Checking the answer in the original sentence.

Use the equation N – 7 = 44 to complete the last 2 steps in solving:

  A number decreased by 7 is 44. What is the number?

Answer:
  b. N = 51
  c. It is true that 51 decreased by 7 is 44.

5

There is a 3-step process to solving word problems.

    a.   Construct a table that translates necessary words, phrases, and sentences.
    b.   Solve the equation(s) obtained.
    c.   Check the answer in the original wording of the problem.

Begin the 3-step process by constructing a table of translations for:

13 more than twice a number is 41.

Answer:

word/phrase/sentence	translation
number	N
twice a number	2N
13 more than twice a number	2N + 13
13 more than twice a number is 41.	2N + 13 = 41

6

The 3-step process for solving word problems requires:

    a.   Constructing a table of translations.
    b.   Solving the equation.
    c.   Checking the answer in the original sentence.

Use the equation $2N + 13 = 41$ to complete the last 2 steps in solving:

13 more than twice a number is 41. What is the number?

Answer:
    b.   N = 14
    c.   It is true that 13 more than twice 14 is 41.

7

Use the 3-step process to solve:

7 less than 5 times a number equals 58.

What is the number?

Answer:
a.

word/phrase/sentence	translation
number	N
5 times a number	5N
7 less than 5 times a number	5N – 7
7 less than 5 times a number equals 58.	5N – 7 = 58

b. N = 13
c. It is true that 7 less than 5 times 13 equals 58.

FEEDBACK UNIT 5 FOR APPLICATIONS

This quiz reviews the preceding unit. Answers are at the back of the book.

Show all 3 steps necessary for solving each of the following.

1. A number increased by 11 equals 48. What is the number?

2. 15 less than a number is 17. What is the number?

3. 7 multiplied by a number gives 56. What is the number?

4. 19 more than twice a number equals 27. What is the number?

5. 4 less than the product of a number and 9 is 59. What is the number?

Summary for Chapter 4

The meaning of a rational number exponent is shown by the identity: $x^{\frac{a}{b}} = \sqrt[b]{x^a}$

Once defined, the rational number exponents are found to follow all the laws earlier used with counting number and integer exponents.

For any rational number replacements of $\frac{a}{b}$ and $\frac{c}{d}$, the following are identities.

$$x^{\frac{a}{b}} \cdot x^{\frac{c}{d}} = x^{\frac{a}{b} + \frac{c}{d}}$$

$$\left(x^{\frac{a}{b}}\right)^{\frac{c}{d}} = x^{\frac{ac}{bd}}$$

$$\frac{x^{\frac{a}{b}}}{x^{\frac{c}{d}}} = x^{\frac{a}{b} - \frac{c}{d}}, \ x \neq 0$$

$$x^{-\frac{a}{b}} = \frac{1}{x^{\frac{a}{b}}}, \ x \neq 0$$

$$\frac{1}{x^{-\frac{a}{b}}} = x^{\frac{a}{b}}, \ x \neq 0$$

$$x^0 = 1, \ x \neq 0$$

Chapter 4 Mastery Test

The following questions test the objectives of Chapter 4. Answers are at the back of the book. The number in parentheses which follows each problem indicates the unit in which it can be learned.

Replacements for the variables in this test are restricted to positive real numbers.

1. Write the following using rational number exponents. (1)

   a. $\sqrt[6]{11}$ b. $\sqrt{7^5}$

   c. $\sqrt[7]{6^3}$ d. $\sqrt[9]{(-7)^4}$

2. Write the following as radical expressions. (1)

   a. $13^{\frac{7}{3}}$ b. $2^{\frac{1}{2}}$

   c. $(-3)^{\frac{2}{6}}$ d. $5^{\frac{6}{7}}$

3. Reduce the exponent where possible or state that it should not be reduced. (1)

   a. $17^{\frac{14}{21}}$ b. $(-7)^{\frac{4}{10}}$

   c. $2^{\frac{12}{9}}$ d. $(-11)^{\frac{6}{16}}$

4. $3^{\frac{7}{8}} \cdot 3^{\frac{3}{5}} =$ _____ (2)

5. $y^{\frac{1}{3}} \cdot y^{\frac{2}{3}} =$ _____ (2)

6. $\sqrt{x} \cdot \sqrt[5]{x} =$ _____ (2)

7. $\sqrt[3]{x^2} \cdot \sqrt[4]{x^3} =$ _____ (2)

8. $\dfrac{x^{\frac{3}{4}}}{x^{\frac{1}{5}}} =$ _____ (3)

9. $\dfrac{x^{\frac{5}{6}}}{x^{\frac{3}{8}}} =$ _____ (3)

10. $\dfrac{\sqrt[4]{x^5}}{\sqrt[8]{x}} =$ _____ (3)

11. $\dfrac{\sqrt[5]{x^8}}{\sqrt{x}} =$ _____ (3)

12. $\left(x^{\frac{2}{3}}\right)^{\frac{3}{5}} =$ _____ (4)

13. $\left(x^2\right)^{\frac{3}{4}} =$ _____ (4)

14. $\left(\sqrt[5]{x^4}\right)^{10} =$ _____ (4)

15. $\left(\sqrt[3]{x^5}\right)^4 =$ _____ (4)

16. Show all 3 steps necessary for solving the following problem: 13 more than 3 times a number equals 34. What is the number? (5)

17. Show all 3 steps necessary for solving the following problem: 7 less than the product of a number and 8 is 113. What is the number? (5)

Chapter 5 Objectives

The following problems illustrate the objectives of this chapter. At this time you are not expected to know how to do these problems. However, if all these problems are thoroughly understood, proceed directly to the Chapter Mastery Test. The number in parentheses which follows each problem indicates the unit in which it can be learned.

1. $(-4x^3 - x + 2) + (-x^3 - x^2 - x + 5) =$ _____ (1)

2. $(2x - 7x^2 + 3) - (5x^2 - 10x + 3) =$ _____ (1)

3. $(2x^2 + 3x - 13) + (-3x^3 - 7x^2 + 5x - 2) =$ _____ (1)

4. $(4x^2 + 7x - 2x^3 - 8) - (5x^3 + 5x^2 + 11x - 4) =$ _____ (1)

5. $(3x^2 + 2x + 7) + (-3x - 8 + 7x^2) =$ _____ (1)

6. Subtract $3x^2 - x - 1$ from $7x^2 + x - 2$. (1)

7. $(5x^3 - 8 + 3x - 7x^2) + (8x^2 + 5x^3 - 8 + 7x) =$ _____ (1)

8. Subtract $2x^2 - 7x + 4$ from $5x^3 - x + 3$. (1)

9. $5x(3x^2 - x + 1) =$ _____ (2)

10. $x(4x^2 - 25) =$ _____ (2)

11. $(x + 7)(x - 2) =$ _____ (2)

12. $(k - s)(r - v) =$ _____ (2)

13. $(2x - 1)(3x + 1) =$ _____ (2)

14. $(5x - 7)^2 =$ _____ (2)

15. $(3x + 2)(3x - 2) =$ _____ (2)

16. $(6x - 7)(3x + 4) =$ _____ (2)

17. $(x - 3)(2x^2 - 7x + 1) =$ _____ (2)

18. $(3x - 2)(9x^2 + 6x + 4) =$ _____ (2)

19. $(2x - 5)(3x^2 - 4x - 5) =$ _____ (2)

20. Divide $x^2 + 8x + 16$ by $x + 3$. (3)

21. Divide $6x^2 - 5x + 1$ by $2x - 1$. (3)

22. Divide $49x^2 - 14$ by $7x - 4$. (3)

23. Divide $2x^3 - 9x^2 - 7x + 10$ by $x - 5$. (3)

24. Divide $9x^2 - 6x + 4$ by $3x + 2$. (3)

CHAPTER 5
POLYNOMIALS

UNIT 1: ADDITION AND SUBTRACTION OF POLYNOMIALS

The following mathematical terms are crucial to an understanding of this unit.

    Addition expression   Monomial
    Binomial   Trinomial
    Constant   Polynomial
    Polynomial over the integers   Ascending order
    Descending order   Like terms
    Minuend   Subtrahend

1
The **addition expression** $7y - 10$ has two terms, $7y$ and -10.
The terms are separated by + and − signs. How many terms are there in $7x^2$?   One

2

    $7x^2$ which has only one term is a **monomial**.
Is $x^2 - 7$ a monomial?   No

3

    $x^2 - 7$ has two terms and is a **binomial**.
Is $x + 6$ a binomial?   Yes

4

    $3x$ is a monomial; $3x - 6$ is a binomial.
$2x^2 - x - 1$ has three terms and is a **trinomial**.
Is $x^4 - x^2 + 3x$ a trinomial?   Yes

5
Trinomials always have _____ terms.   three

6

    $4x^3y^2$ is an example of a _____.   monomial

7
Binomials always have _____ terms.   two

8

    $4x^3 - 2x^2 + 5$ is an example of a _____.   trinomial

9
In the binomial $5x^2 - 16$, 5 is the coefficient of x^2 and -16, which has no variable factor, is a **constant**. What is the constant in $4x^2 - 9x + 3$?   3

10
In the trinomial $x^2 - 4x + 7$, what is the constant?   7

11
In the trinomial $3x^2 + x + 24$, what is the coefficient of x?   1

12
In $7x^3 - 3x^2 + 5x - 9$, what numbers are coefficients?   7, -3, 5

13
An addition expression of two or more terms is a **polynomial** if it meets three conditions.
1. Every exponent is a counting number.
2. All coefficients and constants are integers.
3. No variable is in a denominator.

Which of these requirements is not met by the following addition expression? $x^{-2} + 3x^{-1} - 3$ (1)

14
A polynomial as defined is often called a **polynomial over the integers** because every coefficient and constant must be an integer. Which of the three requirements for a polynomial is not met by the addition expression shown below?

$$\frac{1}{2}x^2 + 0.7x + 5$$ (2)

15
Which of the three requirements for a polynomial is not met by the following expression? $\frac{2x^2 - 7x}{x - 3}$ (3)

16
$4x^2 - 4$ is a polynomial. Is $3c + 4$ a polynomial?   Yes

17
$2x^2 - 7$ is a polynomial. Is $4x^3 - 2x$ a polynomial?   Yes

18
Is $4x - 6x^3$ a polynomial?   Yes

19
Is $4x^3 - 2x^2 - 7x + 9$ a polynomial?   Yes

20
The exponents of $4x - 6x^3$ increase in size from left to right. In this instance, the terms are arranged in **ascending order**. Are the terms of $7x^8 - 5x^3 + 3x^7 - 9x^4$ in ascending order?   No

21
The terms of $2x^4 - 4x^3 + 9x^2 - 8x + 12$ are in **descending order**. The exponents decrease in size from left to right. Are the terms of $12x^8 - 25x^5 + 14x^4 + 5x^2$ in descending order?

Yes

22
The terms of $-x + 2x^3 - 9 + 5x^2$ are in neither ascending nor descending order. In descending order the polynomial is written $2x^3 + 5x^2 - x - 9$. Write the polynomial with the terms in ascending order.

$-9 - x + 5x^2 + 2x^3$

23
Write $4x - 5x^3 - 7 + x^4 - x^2$ with the terms in descending order.

$x^4 - 5x^3 - x^2 + 4x - 7$

24
Write $4x^3 - x^5 + 2x - 7 - x^2$ with the terms in ascending order.

$-7 + 2x - x^2 + 4x^3 - x^5$

25
Write $x^4 - 2 + x + 8x^2 + x^3$ with the terms in ascending order.

$-2 + x + 8x^2 + x^3 + x^4$

26
Write $y^2 - 7 + 2y$ in descending order.

$y^2 + 2y - 7$

27
Write $2y^3 - 2y^2 + y + 12$ in ascending order.

$12 + y - 2y^2 + 2y^3$

28
The addition of $(2x^2 - x + 7)$ and $(3x^2 + x - 9)$ requires the combining of **like terms**.

$$(2x^2 - x + 7) + (3x^2 + x - 9)$$
$$(2x^2 + 3x^2) + (-x + x) + (7 - 9)$$
$$5x^2 + 0 - 2$$
$$5x^2 - 2$$

Were the constants, 7 and -9, like terms?

Yes

POLYNOMIALS

29

$2x^5$ and $-6x^5$ are like terms. $-8x^3$ and x^3 are like terms. Complete the following.

$(2x^5 - 3x^2 - x + 6 - 8x^3) + (x^4 - 6x^5 + x^3 - x + 9x^2)$
$(2x^5 - 6x^5) + x^4 + (-8x^3 + x^3) + (-3x^2 + 9x^2) + (-x - x) + 6$
_____

$-4x^5 + x^4 - 7x^3 + 6x^2 - 2x + 6$

30

Polynomials can be added vertically by placing like terms in the same vertical column.
Arrange $(4x^3 - 8x^2 - x - 6) + (x^3 + 4x^2 + x - 5)$ vertically and complete the addition.

$$\begin{array}{r} 4x^3 - 8x^2 - x - 6 \\ x^3 + 4x^2 + x - 5 \\ \hline 5x^3 - 4x^2 - 11 \end{array}$$

31

Arrange $(3x^4 - 2x + x^3 - 7) + (4x^3 - 2x^2 + 7x - 3)$ vertically and complete the addition.

$$\begin{array}{r} 3x^4 + x^3 + 0x^2 - 2x - 7 \\ 4x^3 - 2x^2 + 7x - 3 \\ \hline 3x^4 + 5x^3 - 2x^2 + 5x - 10 \end{array}$$

32

$(3x^2 - x - 1) + (x^2 - x - 1) =$ _____

$4x^2 - 2x - 2$

33

$(4 - 3x^2 - x) + (2x^2 + 5 + 9x) =$ _____

$-x^2 + 8x + 9$

34

$(3x^4 - 8x^3 + 2x^2 + 5x + 5) + (7x^3 - x^2 + 2x - 5)$
_____

$3x^4 - x^3 + x^2 + 7x$

35

$(4x^2 - 7x^4 + 3 - 5x + x^3) + (5x^4 - 9x + x^3 + x^2 + 3)$
_____

$-2x^4 + 2x^3 + 5x^2 - 14x + 6$

36

$(8x - 3x^2 - 15) + (4 - 2x^2 + 9x) =$ _____

$-5x^2 + 17x - 11$

37

$(9x^2 - x - 1) + (7x^2 - x - 11) =$ _____

$16x^2 - 2x - 12$

38

$(9x^3 + 2x + 12) + (-7x^3 - 9x^2 - 9x + 1)$

_____

$2x^3 - 9x^2 - 7x + 13$

39

$(7x^3 - 10 + 9x^2 + x) + (12 - 9x^3 + 2x - 15x^2)$

_____

$-2x^3 - 6x^2 + 3x + 2$

40

$(4x - 13 + 8x^2) + (9x^2 - x + 2) =$ _____

$17x^2 + 3x - 11$

41

$(7x^3 + 2x + x^2 - 1) + (7x^4 - 2x^2 + 5 - 8x^3 + x)$

_____

$7x^4 - x^3 - x^2 + 3x + 4$

42

$(3x^3 - x + 5) + (4x^2 - 8x + 12) =$ _____

$3x^3 + 4x^2 - 9x + 17$

43

$(-8x^4 - 7x^2 + 8) + (-x^2 + 4x^3 - x) =$ _____

$-8x^4 + 4x^3 - 8x^2 - x + 8$

44

A polynomial is subtracted by adding its opposite. Notice in the example below that the signs of all the terms of the polynomial being subtracted were changed before the like terms were added.

$$(x^2 - 3x + 5) - (2x^2 - 3x + 8)$$
$$(x^2 - 3x + 5) + [-(2x^2 - 3x + 8)]$$
$$x^2 - 3x + 5 - 2x^2 + 3x - 8$$
$$-x^2 - 3$$

Complete the subtraction shown below.

$$(x^2 + 6x + 4) - (x^2 - 7x - 9)$$
$$(x^2 + 6x + 4) + [-(x^2 - 7x - 9)]$$

_____

_____

$x^2 + 6x + 4 - x^2 + 7x + 9$

$13x + 13$

45
Complete the subtraction shown below.

$$(3x^2 - 4x + 10) - (x^2 - 9x + 13)$$
$$(3x^2 - 4x + 10) + [-(x^2 - 9x + 13)]$$

_____

_____

$3x^2 - 4x + 10 - x^2 + 9x - 13$

$2x^2 + 5x - 3$

46
Subtraction always requires that the signs of the subtracted polynomial be changed.

$$(x^2 + 4x - 6) - (2x^2 + 7x - 2)$$
$$x^2 + 4x - 6 - 2x^2 - 7x + 2$$

_____

$-x^2 - 3x - 4$

47
$$(x^2 + 3x - 12) - (5x^2 - 3x - 14)$$

_____

$-4x^2 + 6x + 2$

48
$$(3x^4 - 2x^2 + x + 1) - (5x^3 - 4x^2 + x - 1)$$

_____

$3x^4 - 5x^3 + 2x^2 + 2$

49
Polynomials are subtracted vertically by changing the signs of the terms of the subtrahend.
To subtract $(3x^3 - 2x^2 + x - 5)$ from $(5x^3 + 8x^2 - x + 3)$, the following steps are used:

$5x^3 + 8x^2 - x + 3$ (**minuend**) $5x^3 + 8x^2 - x + 3$
$3x^3 - 2x^2 + x - 5$ (**subtrahend**) $\underline{-3x^3 + 2x^2 - x + 5}$
                                                               $2x^3 + 10x^2 - 2x + 8$

Show the vertical subtraction of $(2x^3 - 4x^2 + x - 5)$ from $(4x^3 - 4x^2 + 4x + 7)$.

$4x^3 - 4x^2 + 4x + 7$
$\underline{-2x^3 + 4x^2 - x + 5}$
$2x^3 + 3x + 12$

50
Subtract $2x^2 - 6x - 9$ from $5x^2 - 3x - 7$.

$$\begin{array}{rrr} 5x^2 & -3x & -7 \\ -2x^2 & +6x & +9 \\ \hline 3x^2 & +3x & +2 \end{array}$$

51
Complete the following vertical subtraction.
$$\begin{array}{rrrr} & 5x^2 & +6x & +2 \quad \text{(minuend)} \\ 7x^3 & -2x^2 & +7x & -7 \quad \text{(subtrahend)} \\ \hline \end{array}$$

$$\begin{array}{rrrr} & 5x^2 & +6x & +2 \\ -7x^3 & +2x^2 & -7x & +7 \\ \hline -7x^3 & +7x^2 & -x & +9 \end{array}$$

52
Subtract $-8x^2 + 2x - 10$ from $4x^3 - 7x^2 + x - 9$.

$$\begin{array}{rrrr} 4x^3 & -7x^2 & +x & -9 \\ & 8x^2 & -2x & +10 \\ \hline 4x^3 & +x^2 & -x & +1 \end{array}$$

53
Subtract $3x - 1$ from $6x^2 + 3x - 1$.

$$\begin{array}{rrr} 6x^2 & +3x & -1 \\ & -3x & +1 \\ \hline 6x^2 & & \end{array}$$

54
$(3x^2 - 7x + 5) - (8x^3 + 4x^2 + 9x + 4)$
_____

$-8x^3 - x^2 - 16x + 1$

Feedback Unit 1

This quiz reviews the preceding unit. Answers are at the back of the book.

1. Write $-3 + 7x^2 - x - 9x^4$ in ascending order.

2. $(8x^2 - 3x) + (-2x^2 - 9x) =$ _____

3. $(-x - 9) - (-2x - 5) =$ _____

4. $(7x^3 - 2x + 12) + (3x^2 - 9x^3 - 11) =$ _____

5. $(7x - 7x^3 + x^2 - 15) + (-3 + 9x^3 + x^2 + x) =$ _____

6. $(8 - 7x^2 + 2x^3) - (8x^3 + 2x + 10) =$ _____

7. $(-2x + 23) - (-6x + 35) =$ _____

8. $(7x^2 - x + 3) - (-2x^3 + 9x^2 + 10) =$ _____

9. Add $3x^2 + 2$ and $x^2 + x - 2$ using a vertical arrangement.

10. Subtract $3x^2 - 3x$ from $-8x^2 + x$ using a vertical arrangement.

11. Subtract $-2x^2 - 2x + 9$ from $-7x^2 + 3$ using a vertical arrangement.

Unit 2: Multiplication of Polynomials

The following mathematical terms are crucial to an understanding of this unit.

FOIL multiplication
Trinomial perfect square
Difference of 2 squares
FOIL polynomial
Sum and difference of 2 terms

1

$a(b + c) = ab + ac$ is an identity for all real number replacements of a, b, and c. To multiply $3(x + 5)$, multiply each term of $(x + 5)$ by 3.

$$3(x + 5)$$
$$3 \cdot x + 3 \cdot 5$$
$$3x + 15$$

$-2(x - 6) = $ _____ $\qquad\qquad -2x + 12$

2

$x(2x + 4) = x \cdot 2x + x \cdot 4 = 2x^2 + 4x$
$x(7x - 1) = $ _____ $\qquad\qquad 7x^2 - x$

3

$2x(3x - 5) = 2x \cdot 3x - 2x \cdot 5 = 6x^2 - 10x$
$8x(3x + 5) = $ _____ $\qquad\qquad 24x^2 + 40x$

4

$a(b + c + d) = ab + ac + ad$ is an identity for all real number replacements of a, b, c, and d. To multiply $3x(x^2 + 2x - 3)$, multiply each term of $(x^2 + 2x - 3)$ by $3x$.

$3x(x^2 + 2x - 3) = 3x^3 + 6x^2 - 9x$
$2x(x^2 + 2x - 4) = $ _____ $\qquad\qquad 2x^3 + 4x^2 - 8x$

5

$-2x(x^2 + x + 1) = $ _____ $\qquad\qquad -2x^3 - 2x^2 - 2x$

POLYNOMIALS 177

6

$(a + b)(c + d) = ac + ad + bc + bd$ is an identity for all real number replacements of a, b, c, and d. Notice that each term of $(a + b)$ is multiplied by each term of $(c + d)$. Complete the following identity:

$(w + x)(y + z) = $ _____   wy + wz + xy + xz

7

To multiply $(4x + 5)(x - 3)$ each term of $(4x + 5)$ is multiplied by each term of $(x - 3)$.

$$(4x + 5)(x - 3)$$
$$4x(x - 3) + 5(x - 3)$$
$$4x^2 - 12x + 5x - 15$$
$$4x^2 - 7x - 15$$

$(3x + 2)(x + 6) = $ _____ $3x^2 + 20x + 12$

8

To multiply $(3x - 2)(2x + 1)$ four separate monomial multiplications are required.

$$(3x - 2)(2x + 1)$$
$$3x(2x + 1) - 2(2x + 1)$$
$$6x^2 + 3x - 4x - 2$$
$$6x^2 - x - 2$$

$(5x - 3)(2x + 7) = $ _____ $10x^2 + 29x - 21$

9

$(a + b)(c + d + e) = ac + ad + ae + bc + bd + be$ is an identity for all real number replacements of a, b, c, d, and e. Each term of $(a + b)$ is multiplied by each term of $(c + d + e)$. Complete the following identity:

$(v + w)(x + y + z) = $ _____   vx + vy + vz + wx + wy + wz

10

To multiply $(2x - 1)(x^2 - 3x + 1)$, six separate monomial multiplications are required.

$$(2x - 1)(x^2 - 3x + 1)$$
$$2x(x^2 - 3x + 1) - 1(x^2 - 3x + 1)$$
$$2x^3 - 6x^2 + 2x - x^2 + 3x - 1$$
$$2x^3 - 7x^2 + 5x - 1$$

$(2x - 5)(x^2 - x + 4) = $ _____

$2x^3 - 2x^2 + 8x - 5x^2 + 5x - 20$
$2x^3 - 7x^2 + 13x - 20$

11

To organize the work better, multiplying polynomials is sometimes arranged vertically. In the example below, a trinomial is multiplied by a binomial.

$$\begin{array}{r} x^2 - 3x - 6 \\ x - 3 \\ \hline (1) \quad -3x^2 + 9x + 18 \\ (2) \quad x^3 - 3x^2 - 6x \\ \hline x^3 - 6x^2 + 3x + 18 \end{array}$$

The partial product marked (1) is obtained by multiplying -3 times each term of $(x^2 - 3x - 6)$. The partial product marked (2) is obtained by multiplying _____ by each term of _____.

x

$(x^2 - 3x - 6)$

12

Complete the vertical multiplication shown below by adding the partial products, (1) and (2).

$$\begin{array}{r} x^2 + 3x - 4 \\ x + 2 \\ \hline (1) \quad 2x^2 + 6x - 8 \\ (2) \quad x^3 + 3x^2 - 4x \\ \hline \end{array}$$

$x^3 + 5x^2 + 2x - 8$

13
Complete the vertical multiplication shown below.

$$\begin{array}{r} 2x^2 + 3x - 1 \\ x - 5 \\ \hline -10x^2 - 15x + 5 \\ 2x^3 + 3x^2 - x \\ \hline 2x^3 - 7x^2 - 16x + 5 \end{array}$$

14
The multiplication of $(x - 4)$ and $(x^2 + 4x + 16)$ requires 6 monomial multiplications because there are two terms in $(x - 4)$ and three terms in $(x^2 + 4x + 16)$.

$(x - 4)(x^2 + 4x + 16) = $ _____

$x^3 + 4x^2 + 16x - 4x^2 - 16x - 64$
$x^3 - 64$

15
$(x + 2)(x^2 - 2x + 4) = $ _____

$x^3 + 8$

16
$(2x - 3)(x^2 + x + 2) = $ _____

$2x^3 - x^2 + x - 6$

17
$(2x - 3)(4x^2 + 6x + 9) = $ _____

$8x^3 - 27$

18
In multiplying two polynomials it is necessary for each term of the first polynomial to be multiplied by each term of the second polynomial. How many separate monomial multiplications are necessary to multiply a polynomial with 8 terms by a polynomial with 5 terms?

$8 \cdot 5 = 40$

19
Since the multiplication of polynomials always requires a number of separate monomial multiplications it _____ (is, is not) necessary to organize the work carefully.

is

20

To help in the multiplication of two binomials, a standard process is desirable. One such standard process is described by the acronym **FOIL multiplication**.

$$(w + x)(y + z) = wy + wz + xy + xz$$

The first (F) terms of the binomials are w and y.
The last (L) terms of the binomials are ____ and ____.   x, z

21

$$(w + x)(y + z) = wy + wz + xy + xz$$

In the diagram shown above,
  a. The pair of first (F) terms is: w, y
  b. The pair of outer (O) terms is: w, z
  c. The pair of inner (I) terms is: ____   x, y
  d. The pair of last (L) terms is: x, z

22

The multiplication of two binomials like $(x^2 + 1)$ and $(x + 3)$ always produces a **FOIL polynomial** which has four terms that are the result of:
  1. Multiplying the first (F) terms.
  2. Multiplying the outer (O) terms.
  3. Multiplying the inner (I) terms.
  4. Multiplying the ____ ____ terms.   last, (L)

23

$$(x^2 + 1)(x + 3) = x^3 + 3x^2 + x + 3$$

$x^3 + 3x^2 + x + 3$ is the FOIL polynomial for $(x^2 + 1)(x + 3)$.
What is the FOIL polynomial for $(2x + 5)(y + 4)$? $2xy + 8x + 5y + 20$

24
Find the FOIL polynomial for $(3x - 7)(2y + 3)$.

$6xy + 9x - 14y - 21$

25
Find the FOIL polynomial for $(x - 2)(x + 7)$.

$x^2 + 7x - 2x - 14$
$x^2 + 5x - 14$

26
The FOIL polynomial $x^2 + 7x - 2x - 14$ can be simplified because its O and I terms are like terms. Find the FOIL polynomial for $(x + 5)(x - 3)$ and, if possible, simplify.

$x^2 - 3x + 5x - 15$
$x^2 + 2x - 15$

27
Find the FOIL polynomial for $(x - 3)(2x - 7)$ and, if possible, simplify.

$2x^2 - 7x - 6x + 21$
$2x^2 - 13x + 21$

28
Find the FOIL polynomial for $(2x - 1)(3x - 4)$ and, if possible, simplify.

$6x^2 - 8x - 3x + 4$
$6x^2 - 11x + 4$

29
$(2x - 1)(3x - 5) =$ _____

$6x^2 - 13x + 5$

30
$(3x - 2)(3x + 2) =$ _____

$9x^2 - 4$

31
$(4x + 3)(3x - 5) =$ _____

$12x^2 - 11x - 15$

32
$(2x - 5)(2x - 7) =$ _____

$4x^2 - 24x + 35$

33
$(4y - 3)(4y + 3) =$ _____

$16y^2 - 9$

34
$(3x - 5)(3x - 5) =$ _____

$9x^2 - 30x + 25$

35

$(x - 8)(x + 8) =$ _____ $x^2 - 64$

36

$(5x + 2)(5x + 2) =$ _____ $25x^2 + 20x + 4$

37

The multiplication of two binomials can be done in many ways, but it is necessary for each term of the first binomial to be multiplied by each term of the second binomial. The FOIL process provides consistency, improves accuracy, and makes it easier to compare answers.

$(a + b)(c + d) =$ _____ $ac + ad + bc + bd$

38

$(p + j)(k + m) =$ _____ $pk + pm + jk + jm$

39

$(r - f)(b + t) =$ _____ $rb + rt - fb - ft$

40

$(a + b)^2 = (a + b)(a + b) =$ _____ $a^2 + 2ab + b^2$

41

$(c - k)^2 =$ _____ $c^2 - 2ck + k^2$

42

$(d + f)^2 =$ _____ $d^2 + 2df + f^2$

43

When a binomial is squared, as in $(d + f)^2$, the O and I terms are always identical. This fact provides a formula that may be applied to the square of any binomial.

$(x + y)^2 = x^2 + 2xy + y^2$
$(x + 7)^2 = x^2 + 2x(7) + (7)^2 =$ _____ $x^2 + 14x + 49$

44

$(x + y)^2 = x^2 + 2xy + y^2$ and $(x - y)^2 = x^2 - 2xy + y^2$ are identities. The sign of the second term of this trinomial will be the _____ (same, opposite) as the sign of the second term of the binomial.

same

45
When a binomial is squared, the result is a trinomial.

$$\begin{pmatrix} \text{The square} \\ \text{of the first} \\ \text{term} \end{pmatrix} + \begin{pmatrix} \text{Two times} \\ \text{the first times} \\ \text{the second} \end{pmatrix} + \begin{pmatrix} \text{The square} \\ \text{of the second} \\ \text{term} \end{pmatrix}$$

Use the identity $(x + y)^2 = x^2 + 2xy + y^2$ to complete:
$(3x + 5)^2 = $ _____

$(3x)^2 + 2(3x)(5) + (5)^2$
$9x^2 + 30x + 25$

46
Use the identity $(x + y)^2 = x^2 + 2xy + y^2$ to complete:

$(7x + 4)^2 = $ _____

$(7x)^2 + 2(7x)(4) + (4)^2$
$49x^2 + 56x + 16$

47

$49x^2 + 56x + 16$ is a **perfect square trinomial** because it is equal to $(7x + 4)^2$. Find the perfect square trinomial that is equal to $(2x + 9)^2$.

$4x^2 + 36x + 81$

48
Find the perfect square trinomial equal to $(5x - 4)^2$ using the identity: $(x - y)^2 = x^2 - 2xy + y^2$

$25x^2 - 40x + 16$

49

$(7x - 3)^2 = $ _____

$49x^2 - 42x + 9$

50

$(8x - 5)^2 = $ _____

$64x^2 - 80x + 25$

51

The binomials $(3x + 7)$ and $(3x - 7)$ are the **sum and difference** of $3x$ and 7.

$(3x + 7)(3x - 7) =$ _____

$9x^2 - 21x + 21x - 49$
$9x^2 - 49$

52

The product of the sum and difference of $5x$ and 9 is:

$(5x + 9)(5x - 9) =$ _____

$25x^2 - 45x + 45x - 81$
$25x^2 - 81$

53

When a sum and difference are multiplied using the FOIL process, the O and I terms are always _____ (opposites, identical).

opposites

54

The product of a sum and difference is a binomial with a minus sign separating the terms.

$(6x + 7)(6x - 7) =$ _____

$36x^2 - 49$

55

The product of a sum and difference is a **difference of two squares**.

$(2x + 1)(2x - 1) =$ _____

$4x^2 - 1$

56

Find the difference of two squares that equals the product of $(10x + 7)(10x - 7)$.

$100x^2 - 49$

57

$(7x - 3)(4x + 1) =$ _____

$28x^2 - 5x - 3$

58

$(3x - 8)^2 =$ _____

$9x^2 - 48x + 64$

59

$(9x - 7)(9x + 7) =$ _____

$81x^2 - 49$

60

$(5x - 3)(2x^2 + 7) =$ _____　　　　　$15x^3 + 35x - 6x^2 - 21$

FEEDBACK UNIT 2

This quiz reviews the preceding unit. Answers are at the back of the book.

1. $2x(2x^2 + x - 5) =$ _____

2. $-x(x^2 - 9) =$ _____

3. $-3x(7x^2 + 6x - 5) =$ _____

4. $-2x(4x - 1) =$ _____

5. $(x - 4)(x + 9) =$ _____

6. $(2x - 1)(3x + 1) =$ _____

7. $(4x - 3)(x + 2) =$ _____

8. $(4x - 3)(4x + 3) =$ _____

9. $(x + 6)(3x - 4) =$ _____

10. $(5x - 3)(2x - 3) =$ _____

11. $x(4x^2 - 2x + 7) =$ _____

12. $(4x + 3)(7x + 1) =$ _____

13. $(x + 3)(x^2 - 6x + 2) =$ _____

14. $(x - 5)(2x^2 + 5x - 3) =$ _____

15. $(6x + 7)^2 =$ _____

16. $(5x^2 + 1)(2x - 3) =$ _____

Unit 3: Division of Polynomials

The following mathematical terms are crucial to an understanding of this unit.

    Cyclic process   Dividend
    Divisor   Quotient
    Remainder

1
Long division is a **cyclic process** which means that it goes through a cycle of steps. The cycle may be repeated several times in doing a problem. A cycle of steps for long division is listed below.

$$2 \overline{\smash{)}1374} \quad \begin{array}{l} 6 \\ 12 \\ \overline{17} \end{array}$$

a. Divide 13 by 2.
b. Multiply 6 times 2.
c. Subtract 12 from 13.
d. "Bring down" the 7.

Will the cycle containing these four steps be repeated as the division continues?   Yes

2
The long division of polynomials is also a cyclic process. The steps in one cycle of the process are listed below.

$$x \overline{\smash{)}5x^2 + 6x} \quad \begin{array}{l} 5x \\ 5x^2 \\ \overline{0 + 6x} \end{array}$$

a. Divide $5x^2$ by x.
b. Multiply 5x times x.
c. Subtract $5x^2$ from $5x^2$.
d. "Bring down" the 6x.

Will the cycle containing these four steps be repeated as the division continues?   Yes

3
To divide $2x^2 + 5x + 7$ by $x + 2$, begin the cyclic process by dividing $2x^2$ by x. Place the result directly above the $2x^2$.

$$x + 2 \overline{\smash{)}2x^2 + 5x + 7} \qquad\qquad x + 2 \overline{\smash{)}2x^2 + 5x + 7}^{\,2x}$$

4

To divide $2x^2 + 5x + 7$ by $x + 2$, the first step in the long division process was completed in the previous frame. Complete the second step by multiplying $2x$ times $x + 2$ and placing the result directly beneath $2x^2 + 5x$.

$$\begin{array}{r} 2x \\ x + 2 \overline{\smash{)}\, 2x^2 + 5x + 7} \end{array}$$

$$\begin{array}{r} 2x \\ x + 2 \overline{\smash{)}\, 2x^2 + 5x + 7} \\ 2x^2 + 4x \end{array}$$

5

Complete the third step of the long division process by subtracting $2x^2 + 4x$ from $2x^2 + 5x$.

$$\begin{array}{r} 2x \\ x + 2 \overline{\smash{)}\, 2x^2 + 5x + 7} \\ 2x^2 + 4x \end{array}$$

$$\begin{array}{r} 2x \\ x + 2 \overline{\smash{)}\, 2x^2 + 5x + 7} \\ -2x^2 - 4x \\ x \end{array}$$

6

Finish the first cycle of the process for long division by "bringing" down the next term of $2x^2 + 5x + 7$.

$$\begin{array}{r} 2x \\ x + 2 \overline{\smash{)}\, 2x^2 + 5x + 7} \\ -2x^2 - 4x \\ x \end{array}$$

$$\begin{array}{r} 2x \\ x + 2 \overline{\smash{)}\, 2x^2 + 5x + 7} \\ -2x^2 - 4x \\ x + 7 \end{array}$$

7

Repeat the four steps of frames 3-6 and divide $x + 7$ by $x + 2$.

$$\begin{array}{r} 2x \\ x + 2 \overline{\smash{)}\, 2x^2 + 5x + 7} \\ -2x^2 - 4x \\ x + 7 \end{array}$$

$$\begin{array}{r} 2x + 1 \quad \text{R5} \\ x + 2 \overline{\smash{)}\, 2x^2 + 5x + 7} \\ -2x^2 - 4x \\ x + 7 \\ -x - 2 \\ 5 \end{array}$$

8

In the division shown at the right, $2x^2 + 5x + 7$ is the **dividend**, $x + 2$ is the **divisor**. The problem has $2x + 1$ R 5 as its **quotient**; its **remainder** is _____.

$$\begin{array}{r} 2x + 1 \text{ R } 5 \\ x + 2 \overline{\smash{\big)}\, 2x^2 + 5x + 7} \\ \underline{-2x^2 - 4x} \\ x + 7 \\ \underline{-x - 2} \\ 5 \end{array}$$

5

9

Begin the cyclic process of long division by dividing x^2 by x and placing the result directly above x^2.

$$x - 3 \overline{\smash{\big)}\, x^2 - 5x + 7}$$

$$\begin{array}{r} x\phantom{{}-5x+7} \\ x - 3 \overline{\smash{\big)}\, x^2 - 5x + 7} \end{array}$$

10

Continue the cyclic process of long division by multiplying x by $x - 3$ and placing the result directly beneath $x^2 - 5x$.

$$\begin{array}{r} x\phantom{{}-5x+7} \\ x - 3 \overline{\smash{\big)}\, x^2 - 5x + 7} \end{array}$$

$$\begin{array}{r} x\phantom{{}-5x+7} \\ x - 3 \overline{\smash{\big)}\, x^2 - 5x + 7} \\ x^2 - 3x \end{array}$$

11

Subtract $x^2 - 3x$ from $x^2 - 5x$ as the next step in the cyclic process of long division.

$$\begin{array}{r} x\phantom{{}-5x+7} \\ x - 3 \overline{\smash{\big)}\, x^2 - 5x + 7} \\ x^2 - 3x \end{array}$$

$$\begin{array}{r} x\phantom{{}-5x+7} \\ x - 3 \overline{\smash{\big)}\, x^2 - 5x + 7} \\ \underline{-x^2 + 3x} \\ -2x \end{array}$$

12

Bring down the + 7
and repeat another
cycle of the process
for long division.
This will complete
the problem.

```
            x
x – 3  ) x² – 5x  + 7
         -x² + 3x
              – 2x
```

```
             x  – 2    R1
x – 3  ) x²  – 5x  + 7
         -x² + 3x
              -2x  + 7
              2x  – 6
                   + 1
```

13

The long division of $x^2 - 5x + 7$ (the dividend) by $x - 3$ (the divisor) gives the result $x - 2$ R 1 (the quotient). To check the division multiply $(x - 2)$ and $(x - 3)$ and then add 1.

$(x - 2)(x - 3) + 1 =$ _____

$x^2 - 5x + 7$, which is the dividend

14

Complete the division
shown at the right
by dividing 7x – 3
by x – 2.

```
             3x
x – 2  ) 3x² + x  – 3
         3x² + 6x
              7x  – 3
```

```
             3x + 7    R11
x – 2  ) 3x² + x  – 3
         3x² + 6x
              7x  – 3
              7x  – 14
                   + 11
```

15

Complete the division
shown at the right

```
             x
2x – 1 ) 2x² – 7x  + 4
         2x² – x
              -6x
```

```
             x – 3    R1
2x – 1 ) 2x² – 7x  + 4
         2x² – x
              -6x + 4
              -6x + 3
                   + 1
```

16
Complete the division shown at the right

$$4x + 3 \overline{\smash{)}\, 4x^2 - 5x - 4}$$

$$\begin{array}{r} x - 2 \quad R2 \\ 4x + 3 \overline{\smash{)}\, 4x^2 - 5x - 4} \\ \underline{4x^2 + 3x } \\ -8x - 4 \\ \underline{-8x - 6} \\ +2 \end{array}$$

17
Complete the division of $4x^2 - 1$ by $2x + 1$. Notice that a 0x term was inserted in the dividend to make the division easier.

$$2x + 1 \overline{\smash{)}\, 4x^2 + 0x - 1}$$

$$\begin{array}{r} 2x - 1 \quad R0 \\ 2x + 1 \overline{\smash{)}\, 4x^2 + 0x - 1} \\ \underline{4x^2 + 2x } \\ -2x - 1 \\ \underline{-2x - 1} \\ 0 \end{array}$$

18
$25x^2 - 11$ is equivalent to $25x^2 + 0x - 11$. Divide $25x^2 - 11$ by $5x - 3$.

$$5x - 3 \overline{\smash{)}\, 25x^2 - 11}$$

$$\begin{array}{r} 5x + 3 \quad R\text{-}2 \\ 5x - 3 \overline{\smash{)}\, 25x^2 + 0x - 11} \\ \underline{25x^2 - 15x } \\ 15x - 11 \\ \underline{15x - 9} \\ -2 \end{array}$$

19
Divide $(9x^2 + 50)$ by $(3x + 7)$. It will make the problem easier if a 0x term is inserted in the dividend.

$$\begin{array}{r} 3x - 7 \quad R99 \\ 3x + 7 \overline{\smash{)}\, 9x^2 + 0x + 50} \\ \underline{9x^2 + 21x } \\ -21x + 50 \\ \underline{-21x - 49} \\ 99 \end{array}$$

20
Divide $15x^2 - 26x - 2$ by $5x - 2$.

$$\begin{array}{r} 3x - 4 \quad R\text{-}10 \\ 5x - 2 \overline{\smash{)}\, 15x^2 - 26x - 2} \\ \underline{15x^2 - 6x } \\ -20x - 2 \\ \underline{-20x + 8} \\ -10 \end{array}$$

21
Divide $16x^2 + 25$ by $4x + 5$.

$$\begin{array}{r} 4x - 5 \text{ R}50 \\ 4x+5 \overline{)\, 16x^2 + 0x + 25} \\ \underline{16x^2 + 20x} \\ -20x + 25 \\ \underline{-20x - 25} \\ 50 \end{array}$$

22
Divide $x^3 + x^2 - 9x - 8$ by $x + 3$. This will require three cycles of the division process.

$$\begin{array}{r} x^2 - 2x - 3 \text{ R}1 \\ x+3 \overline{)\, x^3 + x^2 - 9x - 8} \\ \underline{x^3 + 3x^2} \\ -2x^2 - 9x \\ \underline{-2x^2 - 6x} \\ -3x - 8 \\ \underline{-3x - 9} \\ 1 \end{array}$$

23
Divide.

$$x+1 \,\overline{)\, x^3 + 3x^2 + 3x + 1}$$

$x^2 + 2x + 1 \text{ R }0$

24
Divide.

$$x+2 \,\overline{)\, 2x^3 + 7x^2 + 2x - 10}$$

$2x^2 + 3x - 4 \text{ R }-2$

25
Divide.

$$x+3 \,\overline{)\, 2x^3 - 17x + 3}$$

$2x^2 - 6x + 1 \text{ R }0$

26
Divide.

$$2x+3 \,\overline{)\, 4x^3 - x + 10}$$

$2x^2 - 3x + 4 \text{ R }-2$

FEEDBACK UNIT 3

This quiz reviews the preceding unit. Answers are at the back of the book.

Divide each of the following.

1. $x + 5 \overline{) x^2 + 10x + 25}$

2. $4x - 3 \overline{) 16x^2 - 24x - 11}$

3. $3x - 1 \overline{) 15x^2 + x - 1}$

4. $x - 4 \overline{) 2x^2 - 5x + 8}$

5. $x - 3 \overline{) 2x^3 - 7x^2 + 2x + 3}$

6. $2x - 5 \overline{) 4x^2 - 27}$

7. $3x - 1 \overline{) 9x^3 - 4x + 1}$

8. $2x - 5 \overline{) 8x^3 - 60x^2 + 150x - 125}$

Unit 4: Applications

In this Applications Section, the format of the text has been altered. Answers for the problems appear beneath them rather than in the right-hand column. Your studying emphasis should be on learning the best procedures to follow with word problems.

1
The set of integers is $\{\ldots,-3, -2, -1, 0, 1, 2, 3, \ldots\}$. Integers such as 35 and 36 are consecutive integers. What is the next consecutive integer to 87?

Answer: 88

2
The 3-step process learned earlier for solving a word problem is:

    a. Construct a table of the necessary translations of words, phrases, and sentences.
    b. Solve any equation(s) obtained in the table.
    c. Check answers in the original statement of the word problem.

To solve a word problem, if N represents the smaller of 2 consecutive integers what would be the translation for the larger integer?

Answer: N + 1
Consecutive integers differ by 1. If N is the smaller, N + 1 is the larger.

3

The 3-step process is used to solve consecutive integer problems.

Two consecutive integers have a sum of 93. What are the integers?

a. A table is constructed to show the necessary translations.

word/phrase/sentence	translation
(first) integer	N
next consecutive integer	N + 1
two consecutive integers have a sum	N + (N + 1)
Two consecutive integers have a sum of 93.	N + (N + 1) = 93

b. The equation is solved to give N = 46 which means N + 1 = 47.
c. The results 46 and 47 are checked. It is true that 46 and 47 are consecutive integers with a sum of 93.

Use the 3-step process to solve the following consecutive integer problem.

Two consecutive integers have a product of 12. What are the integers?

Answer:

a.

word/phrase/sentence	translation
(first) integer	N
next consecutive integer	N + 1
two consecutive integers have a product	N(N + 1)
Two consecutive integers have a product of 12.	N(N + 1) = 12

b. $N^2 + N = 12$ or $N^2 + N - 12 = 0$ is solved by factoring.
(N + 4)(N − 3) = 0 gives N = -4 or N = 3
If N = -4 then N + 1 = -3.   If N = 3 then N + 1 = 4.
-4 and -3 is one pair of consecutive integers. 3 and 4 is another pair.

c. Both pairs are checked. It is true that -4 • -3 = 12. It is also true that 3 • 4 = 12. The word problem has 2 pairs of solutions.

4

The 3-step process is used to solve the following problem.

> Two integers are consecutive integers.
> Twice the smaller increased by the larger is 43.
> What are the integers?

The process begins with the construction of a table of translations. This table is crucial to the solution of the problem.

a.

word/phrase/sentence	translation
(first) integer	N
next consecutive integer	$N + 1$
twice the smaller (integer)	$2N$
twice the smaller increased by the larger	$2N + (N + 1)$
Twice the smaller increased by the larger is 43.	$2N + (N + 1) = 43$

Complete the last 2 steps of the process.

Answer:
b. $3N + 1 = 43$ and $N = 14$. If $N = 14$ then $N + 1 = 15$.
c. 14 and 15 are consecutive integers. Twice 14 is 28 which increased by 15 is 43.

5

Use the 3-step process to solve the following.

> Two integers are consecutive integers. Twice the smaller decreased by the larger is 53. What are the integers?

Answer:
a.

word/phrase/sentence	translation
(first) integer	N
next consecutive integer	$N + 1$
twice the smaller (integer)	$2N$
twice the smaller decreased by the larger	$2N - (N + 1)$
Twice the smaller decreased by the larger is 53.	$2N - (N + 1) = 53$

b. $N - 1 = 53$ and $N = 54$. If $N = 54$, then $N + 1 = 55$.
c. 54 and 55 are consecutive integers. Twice 54 is 108 decreased by 55 is 53.

6

The 3-step process is used to solve the following problem.

> Two integers are consecutive integers.
> The square of the smaller decreased by the larger is 5.
> What are the integers?

The process begins with the construction of a table of translations. This table is crucial to the solution of the problem.

a.

word/phrase/sentence	translation
(first) integer	N
next consecutive integer	$N + 1$
the square of the smaller (integer)	N^2
the square of the smaller decreased by the larger	$N^2 - (N + 1)$
The square of the smaller decreased by the larger is 5.	$N^2 - (N + 1) = 5$

Complete the last 2 steps of the process.

Answer:
b. $N^2 - N - 1 = 5$ or $N^2 - N - 6 = 0$ is solved by factoring.
   $(N - 3)(N + 2) = 0$ gives $N = 3$ or $N = -2$.
   If $N = 3$ then $N + 1 = 4$.   If $N = -2$ then $N + 1 = -1$.
c. 3 and 4 is one pair of consecutive integers. Another pair is -2 and -1.
   3^2 decreased by 4 is $9 - 4$ or 5. $(-2)^2$ decreased by -1 is $4 - (-1)$ or 5.

7

Use the three-step process to solve the following.

> Two integers are consecutive integers.
> Three times the smaller decreased by the larger is 75.
> What are the integers?

Answer:

a.

word/phrase/sentence	translation
(first) integer	N
next consecutive integer	$N + 1$
3 times the smaller (integer)	$3N$
3 times the smaller decreased by the larger	$3N - (N + 1)$
3 times the smaller decreased by the larger is 75.	$3N - (N + 1) = 75$

b. $2N - 1 = 75$ and $N = 38$. If $N = 38$, then $N + 1 = 39$.

c. 38 and 39 are consecutive integers. 3 times 38 is 114 decreased by 39 is 75.

8

Consecutive even integers are always 2 apart. For example, 84 and 86 are consecutive even integers. This means that if N is the smaller integer, then the next consecutive even integer is N + 2. Use this information and the 3-step process to solve the following.

> The sum of 2 consecutive even integers is 66.
> What are the integers?

Answer:

a.

word/phrase/sentence	translation
(first) integer	N
next consecutive even integer	$N + 2$
the sum of 2 consecutive even integers	$N + (N + 2)$
The sum of 2 consecutive even integers is 66.	$N + (N + 2) = 66$

b. $2N + 2 = 66$ and $N = 32$. If $N = 32$, then $N + 2 = 34$.

c. 32 and 34 are consecutive even integers. The sum of 32 and 34 is 66.

9

Consecutive odd integers, like consecutive even integers, are always 2 apart. For example, 7 and 9 are consecutive odd integers. This means that if N represents an odd integer, then the next consecutive odd integer is N + 2. Use this information and the 3-step process to solve the following.

   Two integers are consecutive odd integers.
   Three times the smaller decreased by the larger is 64.
   What are the integers?

Answer:

a.

word/phrase/sentence	translation
(first) odd integer	N
next consecutive odd integer	N + 2
3 times the smaller	3N
3 times the smaller decreased by the larger	3N − (N + 2)
3 times the smaller decreased by the larger is 64.	3N − (N + 2) = 64

b. 2N − 2 = 64 and N = 33. If N = 33, then N + 2 = 35.
c. 33 and 35 are consecutive odd integers.
   3 times 33 or 99 decreased by 35 is 64.

FEEDBACK UNIT 4 FOR APPLICATIONS

This quiz reviews the preceding unit. Answers are at the back of the book.

Show all steps in the 3-step process for solving each of the following.

1. Two integers are consecutive integers. The sum of the consecutive integers is 35. What are the integers?

2. Two integers are consecutive integers. Twice the smaller decreased by the larger is 10. What are the integers?

3. Two integers are consecutive even integers. Twice the smaller increased by the larger is 44. What are the integers?

4. Two integers are consecutive odd integers. Three times the smaller increased by twice the larger is 39. What are the integers?

5. Two integers are consecutive integers. Twice the smaller decreased by the larger is 42. What are the integers?

6. The square of an integer when decreased by 3 times the next consecutive integer is 25. What are the integers?

Summary for Chapter 5

The following mathematical terms are crucial to an understanding of this chapter.

Addition expression	Monomial
Binomial	Trinomial
Constant	Polynomial
Polynomial over the integers	Ascending order
Descending order	Like terms
Minuend	Subtrahend
FOIL multiplication	FOIL polynomial
Trinomial perfect square	Sum and difference of 2 terms
Difference of 2 squares	Dividend
Cyclic process	Quotient
Divisor	Remainder

Polynomials are addition expressions of two or more terms which are separated by + or – signs.

Polynomials do not have any of the following:
1. Negative exponents
2. Non-integer coefficients or constants
3. Variables in the denominator

Polynomials are added or subtracted by combining like terms.

Polynomials are multiplied in one of the following ways:
1. $a(b + c) = ab + ac$
2. $(a + b)(c + d) = ac + ad + bc + bd$
3. $(a + b)^2 = a^2 + 2ab + b^2$
4. $(a + b)(a - b) = a^2 - b^2$
5. $(a + b)(c + d + e) = ac + ad + ae + bc + bd + be$

Polynomials are divided by using the same cyclic process as is used in the division of decimal numbers.

Chapter 5 Mastery Test

The following questions test the objectives of Chapter 5. Answers are at the back of the book. The number in parentheses which follows each problem indicates the unit in which it can be learned.

1. $(x^2 + 3x + 2) + (5x + x^2 - 2) =$ _____ (1)
2. $(2x^2 - 3x + 1) - (5x^2 + x + 1) =$ _____ (1)
3. $(7x^3 - 2x^2 + x - 12) + (x^3 + 2x^2 + 3x - 13) =$ _____ (1)
4. $(7x^2 + x^3 - 1 + 3x) - (12 - x - x^3 + 3x^2) =$ _____ (1)
5. $(4x^3 - 2x^2 + 5x - 8) + (-5x^3 - x^2 + 7x + 5) =$ _____ (1)
6. Subtract $3x^3 + 2x - 6$ from $2x^3 + x - 8$. (1)
7. $(2x^5 + 3x^4 - x^2 + 1) + (5x^5 - 7x^4 - x^3 - 7x^2 - 5) =$ _____ (1)
8. Subtract $5x^3 - 2x^2 - 5x + 1$ from $7x^4 - 2x^3 + 3x - 1$. (1)

9. $2x(3x + 8) =$ _____ (2)
10. $5x^2(2x^2 - x + 3) =$ _____ (2)
11. $(x + 3)(x + 2) =$ _____ (2)
12. $(d + k)(m - n) =$ _____ (2)
13. $(5x - 3)(2x - 3) =$ _____ (2)
14. $(3x - 8)^2 =$ _____ (2)
15. $(2x + 3)(2x - 3) =$ _____ (2)
16. $(3x - 1)(2x + 5) =$ _____ (2)
17. $(x - 2)(x^2 + 3x + 5) =$ _____ (2)
18. $(4x + 7)(2x^2 - 3x + 2) =$ _____ (2)
19. $(2x - 3)(4x^2 + 6x + 9) =$ _____ (2)

20. Divide $x^2 + 8x + 16$ by $x + 4$. (3)
21. Divide $3x^2 - 22x + 7$ by $3x - 1$. (3)
22. Divide $4x^2 - 26$ by $2x - 5$. (3)
23. Divide $2x^3 - 7x^2 - 5x + 4$ by $x - 4$. (3)
24. Divide $8x^3 - 125$ by $2x - 5$. (3)

25. Show all steps necessary for solving the following problem: The sum of 2 consecutive integers is 85. What are the integers? (4)

26. Show all steps necessary for solving the following problem: Two integers are consecutive odd integers. Twice the larger integer increased by the smaller is 145. What are the integers? (4)

Chapter 6 Objectives

The following problems illustrate the objectives of this chapter. At this time you are not expected to know how to do these problems. However, if all these problems are thoroughly understood, proceed directly to the Chapter Mastery Test. The number in parentheses which follows each problem indicates the unit in which it can be learned.

Factor completely or state that the polynomial is prime.

1. $-8x + 9$ (1)
2. $7x - 14$ (1)
3. $6x^2 + 3x - 15$ (1)
4. $4x^2 + 2x - 14$ (1)
5. $x(3x - 7) - (3x - 7)$ (1)
6. $6x^3 - 18x^2 - x + 3$ (1)
7. $6x^3 - 2x^2 - 15x + 5$ (1)
8. $x^2 - x - 56$ (2)
9. $x^2 + 14xy + 48y^2$ (2)
10. $x^2 - 12xy + 36y^2$ (2)
11. $x^2 - 18xy - 40y^2$ (2)
12. $x^2 + 18xy - 63y^2$ (2)
13. $x^6 - 9y^2$ (2)
14. $7x^2 + 3xy - 10y^2$ (3)
15. $9x^2 - 16y^2$ (3)
16. $6x^2 + 29xy - 5y^2$ (3)
17. $4x^2 + 12xy + 9y^2$ (3)
18. $81x^4 - 25y^8$ (3)
19. $4x^4 - y^6$ (3)
20. $x^3 + 216$ (4)
21. $x^3 - 125$ (4)
22. $27x^3 + 64y^{12}$ (4)
23. $x^9 - 64y^3$ (4)
24. $(a - b)^2 - 11(a - b) - 26$ (5)
25. $x^2 - 9x(y + w) + 8(y + w)^2$ (5)
26. $x^2 - (y + 7)^2$ (5)
27. $9x^2 + 6xy + y^2 - z^2$ (5)
28. $x^2 - y^2 - 20yz - 100z^2$ (5)
29. $2x^2 - 22xy + 20y^2$ (6)
30. $4x^2 - 22xy + 10y^2$ (6)
31. $k^6 - r^6$ (6)
32. $m^9 - 1$ (6)
33. $20x^2 - 45y^2$ (6)
34. $7x^2 - 21xy - 28y^2$ (6)
35. $10 - 40(x - y)^2$ (6)

CHAPTER 6
FACTORING POLYNOMIALS

UNIT 1: FACTORING BY THE COMMON FACTOR METHOD

The following mathematical terms are crucial to an understanding of this unit.

- Integer factors
- Variable factors
- Common factor
- Factored form
- Prime polynomial
- Positive integer factors
- Factor
- Highest common factor (HCF)
- Factored completely

1
The **integer factors** of 5 are 1, -1, 5, and -5.
What are the integer factors of 7?   1, -1, 7, -7

2
The **positive integer factors** of 10 are 1, 2, 5, and 10.
What are the positive integer factors of 21?   1, 3, 7, 21

3
What are the positive integer factors of 12?   1, 2, 3, 4, 6, 12

4
What are the positive integer factors of 20?   1, 2, 4, 5, 10, 20

5
The **variable factors** of x^4 are x, x^2, x^3, and x^4.
What are the variable factors of y^5? y, y^2, y^3, y^4, y^5

6
To **factor** $-14x^2y$, it must be written as a multiplication.
Can $-14x^2y$ be factored as $-2xy \cdot 7x$?   Yes

7
Can $-14x^2y$ be factored as $-14xy \cdot x$?   Yes

8
Can $-14x^2y$ be factored as $7x^2 \cdot -2y$?   Yes

9
The monomial $10x^3y^4$ can be factored in many different ways. Three possible ways are:

$$10x^3y^4 = 2xy \cdot 5x^2y^3$$
$$10x^3y^4 = x^3y \cdot 10y^3$$
$$10x^3y^4 = 5x^2y^2 \cdot 2xy^2$$

Can $10x^3y^4$ be factored as $-2x^2y \cdot -5xy^3$?   Yes

10
   $4x^2y^3$ is a factor of $12x^5y^3$ because:

   a. 4 is a factor of 12,
   b. x^2 is a factor of x^5, and
   c. y^3 is a factor of y^3.

Is $8xy$ a factor of $12x^5y^3$?   No, 8 is not a factor of 12

FACTORING POLYNOMIALS   205

11
Two numbers, 1 and -1, are factors of every monomial. For example, $14x^2z^4$ can be factored as $1 \cdot 14x^2z^4$ or $-1 \cdot -14x^2z^4$. Is -1 a factor of $9x^4y$?

Yes, $-1 \cdot -9x^4y$

12
Is $-x^4y$ a factor of $-8x^6y^5$?

Yes, $-x^4y \cdot 8x^2y^4$

13
Is $6x^3y^4$ a factor of $12x^5y^2$?

No, y^4 is not a factor of y^2

14
Is $7x^2y^3$ a factor of $7x^2y^3$?

Yes, $7x^2y^3 \cdot 1$

15
Is $-2x$ a factor of $6x$?

Yes, $-2x \cdot -3$

16
The monomials $10x^2y^3$ and $25xy^4$ have many **common factors** (factors of both $10x^2y^3$ and $25xy^4$). Five of these common factors are: 5, xy, xy^3, 5y, $5xy^3$. Is $-x^2y^3$ a common factor of $10x^2y^3$ and $25xy^4$?

No, x^2 is not a common factor

17
The **highest common factor (HCF)** of $10x^2y^3$ and $25xy^4$ is $5xy^3$ because:

    a.   5 is the HCF of 10 and 25,
    b.   x is the HCF of x^2 and x, and
    c. y^3 is the HCF of y^3 and y^4.

What is the HCF for $3x^5$ and $8x^2$?

x^2

18
What is the HCF of $6x^2$ and $12x$?

$6x$

19
What is the HCF of $8x$ and $12y$?

4

20
What is the HCF of $5x^3$ and $7y^2$?

1

21

To factor $5x - 10$, it is necessary to find the HCF of $5x$ and 10. What is the HCF of $5x$ and 10?

5

22

Because 5 is the HCF of $5x$ and 10, the binomial $5x - 10$ is factored as follows:

$$5x - 10$$
$$5 \cdot x - 5 \cdot 2$$
$$5(x - 2)$$

The **factored form** of $5x - 10$ is $5(x - 2)$.
Is $5(x - 2)$ equal to $5x - 10$?

Yes

23

Find the factored form of $4x - 10$ by using the HCF of $4x$ and 10.

$2 \cdot 2x - 2 \cdot 5$
$2(2x - 5)$

24

As the first step in factoring $4xy^2 - 2xy$, find the HCF of $4xy^2$ and $2xy$.

$2xy$

25

The factored form of $4xy^2 - 2xy$ is found below.

$$4xy^2 - 2xy$$
$$2xy \cdot 2y - 2xy \cdot 1$$
$$2xy(2y - 1)$$

Find the factored form of $3x^2y - 12xy^3$ by using the HCF of the two terms of the binomial.

$3xy \cdot x - 3xy \cdot 4y^2$
$3xy(x - 4y^2)$

26

Find the HCF of $15x^2$ and $20x$ and factor $15x^2 + 20x$.

$5x(3x + 4)$

27

Find the factored form of $x^2 - 3x$.

$x(x - 3)$

28

Find the factored form of $x^2y + 5yx$.

$xy(x + 5)$

Factoring Polynomials

29
The polynomial $x^2y + 5yx$ is **factored completely** as $xy(x + 5)$ because 1 is the HCF of the terms of $(x + 5)$. Factor $6x^2 - 12xy$ completely.

$6x(x - 2y)$, 1 is HCF of x and 2y

30
Factor $8x^3y - 10z$ completely.

$2(4x^3y - 5z)$, 1 is HCF of $4x^3y$ and -5z

31
$-7x + 14z^2$ could be factored as $7(-x + 2z^2)$ or $-7(x - 2z^2)$. The factorization $-7(x - z^2)$ is preferred because the coefficient of the first term of the polynomial factor is positive. Factor $-4xy - 3yz$ in its preferred way.

$-y(4x + 3z)$

32
Factor $-15x - 3$ completely.

$-3(5x + 1)$

33
Factor $-9xy^3 + 15xy^2$ completely.

$-3xy^2(3y - 5)$

34
Factor $8x^3y^2z - 18xy^2z^2$ completely.

$2xy^2z(4x^2 - 9z)$

35
To factor $8x^3 - 10x^2 + 24x$, find the HCF of its 3 terms.

$$8x^3 - 10x^2 + 24x \qquad \text{HCF is } 2x$$
$$2x \cdot 4x^2 - 2x \cdot 5x + 2x \cdot 12$$
$$2x(4x^2 - 5x + 12)$$

Factor $x^3 - 3x^2 + 5x$ by finding the HCF of its 3 terms.

$x(x^2 - 3x + 5)$

36
Factor $2x^3 + 14x^2 - 6x$ by finding the HCF of its 3 terms.

$2x(x^2 + 7x - 3)$

37
Factor $x^3 - 2x^2 - x$ completely.

$x(x^2 - 2x - 1)$

38
Factor $4x^3 - 10x^2 - 16x$ completely.

$2x(2x^2 - 5x - 8)$

39
Factor $2x^2 - 10x + 14$ completely. $\qquad$ $2(x^2 - 5x + 7)$

40
Factor $6x^2y - 3xy + 9xy^2$ completely. $\qquad$ $3xy(2x - 1 + 3y)$

41
For some polynomials the HCF can be a binomial.
This is the case with the factoring shown below.

$$2x(x-3) + 5(x-3) \qquad \text{HCF is } (x-3)$$
$$2x \cdot (x-3) + 5 \cdot (x-3)$$
$$(x-3)(2x+5)$$

What is the HCF for $x^2(4x-1) - 6(4x-1)$? $\qquad$ $(4x-1)$

42
Factor $x^2(4x-1) - 6(4x-1)$ using $(4x-1)$ as the HCF. $\qquad$ $(4x-1)(x^2-6)$

43
Factor $4x(2x-1) + (2x-1)$. Notice that $(2x-1)$ is
equivalent to $1(2x-1)$. $\qquad$ $(2x-1)(4x+1)$

44
Factor $7x(x+4) - 5(x+4)$ completely. $\qquad$ $(x+4)(7x-5)$

45
Factor $3x(x-7) + (x-7)$ completely. $\qquad$ $(x-7)(3x+1)$

46
Factor $x^2(x+12) - 3(x+12)$ completely. $\qquad$ $(x+12)(x^2-3)$

47
Factor $(4x+5) - 5x(4x+5)$ completely. $\qquad$ $(4x+5)(1-5x)$

48

The four-term polynomial $8x^3 - 2x^2 - 12x + 3$ is factored by grouping its terms in pairs.

$$8x^3 - 2x^2 - 12x + 3$$
$$8x^3 - 2x^2 \quad - 12x + 3$$
$$2x^2(4x - 1) \quad - 3(4x - 1)$$
$$(4x - 1)(2x^2 - 3)$$

Check this factoring by multiplying $(4x - 1)(2x^2 - 3)$. $\qquad$ $8x^3 - 2x^2 - 12x + 3$

49

Complete the factoring of $x^3 - 3x^2 - 5x + 15$ shown below.

$$x^3 - 3x^2 - 5x + 15$$
$$x^3 - 3x^2 \quad - 5x + 15$$
$$x^2(x - 3) \quad - 5(x - 3)$$
$$\underline{\hspace{2in}} \qquad (x - 3)(x^2 - 5)$$

50

Factor $x^3 - 5x^2 + 4x - 20$ by grouping its terms. $\qquad$ $(x - 5)(x^2 + 4)$

51

Factor $2x^3 + 14x^2 - 3x - 21$ completely. $\qquad$ $(x + 7)(2x^2 - 3)$

52

Factor $5x^3 - 20x^2 - 3x + 12$ completely. $\qquad$ $(x - 4)(5x^2 - 3)$

53

Factor $xy + xz + ay + az$ completely. $\qquad$ $(y + z)(x + a)$

54

$5x - 9$ is a **prime polynomial** because 1 and -1 are the only common factors for $5x$ and 9. Is $-6x + 7$ prime? $\qquad$ Yes

55

Is $6x + 10$ prime? $\qquad$ No, it factors as $2(3x + 5)$

Feedback Unit 1

This quiz reviews the preceding unit. Answers are at the back of the book.

Factor each polynomial completely or state that it is prime.

1. $6x - 10$
2. $15x + 20$
3. $7x^2 - 21x$
4. $12x^2y + 18xy^2$
5. $-18x - 9$
6. $35x^3yz^2 - 21x^2y^2z$
7. $-x^3w^4 + 5xw^3$
8. $3x^3 - 9x^2 + 12x$
9. $x^4 - 3x^3 - 7x^2$
10. $3x^2 - 15x - 12$
11. $14x^3 - 7x^2 + 35x$
12. $9x^2 + 6x + 21$
13. $8x^2y + 28xy - 4xy^2$
14. $5x(2x + 1) - 3(2x + 1)$
15. $7x(x + 3) + 4(x + 3)$
16. $4x(3x - 2) - (3x - 2)$
17. $3x^3 - 6x^2 + 4x - 8$
18. $5x^3 - 15x^2 - x + 3$
19. $rs - ts + rx - tx$
20. $tx + tb - wx - wb$

Unit 2: Factoring Trinomials of the Form $x^2 + bx + c$

The following mathematical terms are crucial to an understanding of this unit.

FOIL multiplication of binomials
Prime trinomial
Difference of two squares
FOIL polynomial
Trinomial perfect square
Common factor

FACTORING POLYNOMIALS

1

To find the factored form of $x^2 - 7x - 8$, reverse the **FOIL multiplication of binomials**.

$$x^2 - 7x - 8$$
$$\begin{array}{cccc} F & O & I & L \\ x^2 & ?x & ?x & -8 \end{array}$$
$$(x + ?)(x + ?)$$

To replace the question marks in the problem above, it is necessary to find two integers with a product of -8 and a sum of -7. What are the two integers?

-8, 1

2

$x^2 - 7x - 8$ is factored in the following steps.

$$x^2 - 7x - 8$$
$$\begin{array}{cccc} F & O & I & L \\ x^2 & -8x & +1x & -8 \end{array}$$
$$(x - 8)(x + 1)$$

Check the factored form by multiplying $(x - 8)(x + 1)$.

$x^2 - 7x - 8$

3

To factor $x^2 - 5x + 4$, reverse the process for multiplying two binomials.

$$x^2 - 5x + 4$$
$$\begin{array}{cccc} F & O & I & L \\ x^2 & ?x & ?x & +4 \end{array}$$
$$(x + ?)(x + ?)$$

To replace the question marks in the problem above, it is necessary to find two integers with a product of 4 and a sum of -5. What are the two integers?

-4, -1

4

$x^2 - 5x + 4$ is factored in the following steps.

$$x^2 - 5x + 4$$
$$\begin{array}{cccc} F & O & I & L \\ x^2 & -4x & -1x & +4 \end{array}$$
$$(x - 4)(x - 1)$$

Check the factored form by multiplying $(x - 4)(x - 1)$.

$x^2 - 5x + 4$

5

Factor $x^2 + 5x + 6$ by first finding a **FOIL polynomial** in which the coefficients of the O and I terms have a product of 6 and a sum of 5.

$x^2 + 3x + 2x + 6$
$(x + 3)(x + 2)$

6

Factor $x^2 + 2x - 15$ by first finding a FOIL polynomial in which the coefficients of the O and I terms have a product of -15 and a sum of 2.

$x^2 + 5x - 3x - 15$
$(x + 5)(x - 3)$

7

Find the factored form of $x^2 + 10x + 24$. Look for two numbers which have a product of 24 and a sum of 10.

$x^2 + 6x + 4x + 24$
$(x + 6)(x + 4)$

8

Factor $x^2 - 5x - 24$ by first writing its FOIL polynomial.

$x^2 - 8x + 3x - 24$
$(x - 8)(x + 3)$

9

Factor. $x^2 + 10x + 16$

$(x + 8)(x + 2)$

10

$x^2 - 8x - 21$ is a **prime trinomial**; it has no associated FOIL polynomial. There are no factors of -21 with a sum of -8. Is $x^2 - 4x - 21$ prime?

No, it factors as $(x - 7)(x + 3)$

11

Is $x^2 + 7x - 10$ prime?

Yes, there are no factors of -10 with sum of 7.

12

Factor. $x^2 - 7x + 6$

$(x - 6)(x - 1)$

13
To factor $x^2 - 10xy + 21y^2$, its FOIL polynomial is used.

$$x^2 - 10xy + 21y^2$$
$$\begin{array}{cccc} F & O & I & L \\ x^2 & ?xy & ?xy & +21y^2 \end{array}$$
$$(x + ?y)(x + ?y)$$

To replace the question marks, find the two integers that have a product of 21 and a sum of -10.

-7, -3

14
$x^2 - 10xy + 21y^2$ is factored in the following steps.

$$x^2 - 10xy + 21y^2$$
$$\begin{array}{cccc} F & O & I & L \\ x^2 & -7xy & -3xy & +21y^2 \end{array}$$
$$(x - 7y)(x - 3y)$$

Check the factored form by multiplying $(x - 7y)(x - 3y)$.

$x^2 - 10xy + 21y^2$

15
Factor $x^2 + 13xy + 42y^2$ by first finding two integers with a product of 42 and a sum of 13.

$x^2 + 7xy + 6xy + 42y^2$
$(x + 7y)(x + 6y)$

16
Factor $x^2 - 7xy + 12y^2$ by writing its FOIL polynomial.

$x^2 - 3xy - 4xy + 12y^2$
$(x - 3y)(x - 4y)$

17
Factor. $x^2 - 9xy + 18y^2$

$(x - 6y)(x - 3y)$

18
Factor. $x^2 + 5xy - 7y^2$

prime

19
Factor. $x^2 + 15xy + 50y^2$

$(x + 5y)(x + 10y)$

20
Factor. $x^2 + 6xy + 9y^2$

$(x + 3y)(x + 3y)$
$(x + 3y)^2$

21

$x^2 + 6xy + 9y^2$ is a **trinomial perfect square**. Its FOIL polynomial is $x^2 + 3xy + 3xy + 9y^2$. Notice that its O and I terms are identical; both are $3xy$.

Is $x^2 - 18xy + 80y^2$ a trinomial perfect square?

No, it factors as $(x - 8y)(x - 10y)$

22

Is $x^2 - 18xy + 81y^2$ a trinomial perfect square?

Yes, it factors as $(x - 9y)^2$

23

Factor. $x^2 + 7xy - 30y^2$

$(x + 10y)(x - 3y)$

24

Factor. $x^2 - 8xy - 33y^2$

$(x - 11y)(x + 3y)$

25

Factor. $x^2 - 16xy + 64y^2$

$(x - 8y)^2$

26

Factor. $x^2 + 9xy - 10y^2$

$(x + 10y)(x - y)$

27

Factor. $x^2 + xy - 20y^2$

$(x + 5y)(x - 4y)$

28

Factor. $x^2 + 10xy + 25y^2$

$(x + 5y)^2$

29

To factor $x^2 - 25y^2$, its FOIL polynomial may be used.

$$\begin{array}{cccc} & x^2 & - 25y^2 & \\ F & O & I & L \\ x^2 & ?xy & ?xy & -25y^2 \end{array}$$
$$(x + ?y)(x + ?y)$$

To replace the question marks, notice that the sum of the O and I terms must be _____.

0

30

$x^2 - 25y^2$ is factored in the following steps.
$$x^2 - 25y^2$$
$$\begin{array}{cccc} F & O & I & L \\ x^2 & -5xy & +5xy & -25y^2 \end{array}$$
$$(x-5y)(x+5y)$$

Check the factored form by multiplying $(x-5y)(x+5y)$.

$x^2 - 25y^2$

31

$x^2 - 25y^2$ is a **difference of two squares** and is factored as the sum and difference of x and 5y.
Is $x^2 - 81y^4$ a difference of two squares?

Yes

32

Factor $x^2 - 81y^4$ as a difference of two squares by writing the two binomials that are the sum and difference of x and $9y^2$.

$(x + 9y^2)(x - 9y^2)$

33

Is $x^2 - 100y^2$ a difference of two squares?

Yes

34

Factor $x^2 - 100y^2$ as a difference of two squares.

$(x + 10y)(x - 10y)$

35

Factor. $x^2 - 9$

$(x + 3)(x - 3)$

36

Is $x^2 - 49y^4$ a difference of two squares?

Yes

37

Factor. $x^2 - 49y^4$

$(x + 7y^2)(x - 7y^2)$

38

Is $x^2 - 30y^2$ a difference of two squares?

No, $30y^2$ is not a perfect square.

39

Is $x^4 - 100y^6$ a difference of two squares?

Yes, $x^2 \cdot x^2$ and $10y^3 \cdot 10y^3$

40
Factor. $x^4 - 100y^6$

$(x^2 + 10y^3)(x^2 - 10y^3)$

41
Is $25x^2 + y^2$ the difference of two squares?

No, it is a sum and is a prime polynomial.

42
Factor. $x^6 - 4y^2$

$(x^3 + 2y)(x^3 - 2y)$

43
In factoring any polynomial always look for a **common factor** as is shown in the example below.

$$2x^2 + 10x + 8$$
$$2(x^2 + 5x + 4)$$
$$2(x + 4)(x + 1)$$

$2x^2 + 10x + 8$ has 3 factors which are 2, $(x + 4)$, and _____.

$(x + 1)$

44
What is the common factor for $6x^2 - 9x + 12$?

3

45
What is the common factor for $15x^3 - 35x^2 + 20x$?

$5x$

46
Complete the factoring shown below.

$$3x^2 - 21x + 30$$
$$3(x^2 - 7x + 10)$$
_____

$3(x - 5)(x - 2)$

47
Complete the factoring shown below.

$$x^3y - 9x^2y^2 + 20xy^3$$
$$xy(x^2 - 9xy + 20y^2)$$
_____

$xy(x - 5y)(x - 4y)$

48
What is the common factor for $5x^3y - 35x^2y^2 + 60xy^3$?

$5xy$

49
Find three factors of $5x^3y - 35x^2y^2 + 60xy^3$ by first using the common factor method.

$5xy(x - 3y)(x - 4y)$

50
Factor $4x^2y - 12xy - 40y$ completely.

$4y(x - 5)(x + 2)$

51
To factor any trinomial of the form $x^2 + bx + c$ where the coefficient of x^2 is 1, factors of c with a sum of b are needed. For example, $x^2 - 9x - 36$ is factored by finding factors of -36 with a sum of -9. Factor $x^2 - 9x - 36$.

$(x + 3)(x - 12)$

FEEDBACK UNIT 2

This quiz reviews the preceding unit. Answers are at the back of the book.

Factor each polynomial completely or state that it is prime.

1. $x^2 - 2x - 3$
2. $x^2 + 3x - 18$
3. $x^2 - 13x + 30$
4. $x^2 - x - 72$
5. $x^2 + 11x + 28$
6. $x^2 + 4x - 21$
7. $x^2 - 3xy + 2y^2$
8. $x^2 + 11xy + 24y^2$
9. $x^2 - xy - 42y^2$
10. $x^2 + 5xy + 4y^2$
11. $x^2 - 5xy - 14y^2$
12. $x^2 - 25y^2$
13. $x^2 - 36y^4$
14. $x^4 - 81y^6$
15. $x^8 - 4y^2$
16. $2x^2 - 12x - 14$
17. $5x^2 + 15x - 50$
18. $3x^3 - 18x^2 + 24x$
19. $2x^3y - 6x^2y - 8xy$
20. $2x^2y - 10xy - 72y$

Unit 3: Factoring Quadratics of the Form $ax^2 + bx + c$

The following mathematical terms are crucial to an understanding of this unit.

    FOIL polynomial   Prime polynomial
    Trinomial perfect square   Difference of two squares

1
Multiply $(3x - 8)(2x + 1)$ using the FOIL process.

$$\begin{array}{cccc} F & O & I & L \\ 6x^2 & +3x & -16x & -8 \\ 6x^2 & -13x & & -8 \\ F & O+I & & L \end{array}$$

2
$6x^2 + 3x - 16x - 8$ is a **FOIL polynomial** which is factored by grouping the FO and IL terms.

$$6x^2 + 3x - 16x - 8$$
$$6x^2 + 3x \qquad -16x - 8$$
$$3x(2x + 1) \qquad -8(2x + 1)$$

_____

Complete the factoring using $(2x + 1)$ as a common binomial factor.

$(2x + 1)(3x - 8)$

3
Every FOIL polynomial can be factored by grouping its FO and IL terms and using the common factor method. Complete the factoring shown below.

$$35x^2 + 20x - 21x - 12$$
$$35x^2 + 20x \qquad -21x - 12$$
$$5x(7x + 4) \qquad -3(7x + 4)$$

_____

$(7x + 4)(5x - 3)$

FACTORING POLYNOMIALS 219

4
Complete the factoring of the FOIL polynomial below.

$$4x^2 - 7x + 4x - 7$$
$$4x^2 - 7x \quad + 4x - 7$$
$$x(4x - 7) \quad + 1(4x - 7)$$
_____

$(4x - 7)(x + 1)$

5
Complete the factoring of the FOIL polynomial below.

$$10x^2 + 18xy + 15xy + 27y^2$$
$$10x^2 + 18xy \quad + 15xy + 27y^2$$
$$2x(5x + 9y) \quad + 3y(5x + 9y)$$
_____

$(5x + 9y)(2x + 3y)$

6
Complete the factoring of the FOIL polynomial below.

$$6x^2 + 18x - 7x - 21$$
$$6x^2 + 18x \quad - 7x - 21$$
$$6x(x + 3) \quad - 7(x + 3)$$
_____

$(x + 3)(6x - 7)$

7
Complete the factoring of the FOIL polynomial below.

$$12x^2 - 3xy + 32xy - 8y^2$$
$$12x^2 - 3xy \quad + 32xy - 8y^2$$
$$3x(4x - y) \quad + 8y(4x - y)$$
_____

$(4x - y)(3x + 8y)$

8
Which numbered step in the process shown below has not been previously explained?

$$8x^2 - 46x - 25$$
(1) $\quad 8x^2 - 50x + 4x - 25$
(2) $\quad 8x^2 - 50x \quad + 4x - 25$
(3) $\quad 2x(4x - 25) \quad + 1(4x - 25)$
(4) $\quad (4x - 25)(2x + 1)$

(1)

220   CHAPTER 6

9
The first step in factoring $8x^2 - 46x - 25$ is to find its FOIL polynomial.

$$8x^2 - 46x - 25$$

F	O	I	L
$8x^2$	?x	?x	-25

To replace the question marks, integers with a sum of -46 and a product of $8 \cdot -25 = -200$ must be found. Are -50 and 4 two such integers?

Yes, $-50 + 4 = -46$
and $-50 \cdot 4 = -200$

10
To find O + I terms for $8x^2 - 46x - 25$,

1. Multiply 8 times -25 which gives -200.
2. Find factors of -200 with a sum of -46.

A table like the one shown below will help in trying to find such a pair of integers.

Factors of -200	Sum of the factors
1 • -200	-199
2 • -100	-98
4 • -50	-46
5 • -40	-35
8 • -25	-17
10 • -20	-10

Which pair of integers will be the coefficients for the O + I terms?

4, -50

11
In the table shown in frame 10, the entries for the column labeled "Factors of -200" are found by trying each counting number, 1, 2, 3, 4, etc., as a factor of 200. Entries in the column labeled "Sum of the factors" are found by _____ the integers in the first column.

adding

12
To find the FOIL polynomial for $2x^2 - 3x - 20$, begin by multiplying the coefficients/constants of the first and the third terms. What is their product?

$2 \cdot -20 = -40$

FACTORING POLYNOMIALS 221

13
To write the FOIL polynomial for $2x^2 - 3x - 20$, complete the first column of the table shown below.

Factors of -40	Sum of the factors
_____	
_____	
_____	
_____	

$1 \cdot -40$
$2 \cdot -20$
$4 \cdot -10$
$5 \cdot -8$

14
To write the FOIL polynomial for $2x^2 - 3x - 20$, complete the second column of the table shown below.

Factors of -40	Sum of the factors
$1 \cdot -40$	_____
$2 \cdot -20$	_____
$4 \cdot -10$	_____
$5 \cdot -8$	_____

-39
-18
-6
-3

15
The completed table of frame 14 provides the information needed to write a FOIL polynomial for $2x^2 - 3x - 20$.

$$2x^2 - 3x - 20$$
$$\begin{array}{cccc} F & O & I & L \\ 2x^2 & \underline{} & \underline{} & -20 \end{array}$$

5x, -8x

16
Complete the factoring of $2x^2 - 3x - 20$ shown below.

$$2x^2 - 3x - 20$$
$$2x^2 + 5x - 8x - 20$$
_____
_____

$x(2x + 5) - 4(2x + 5)$
$(2x + 5)(x - 4)$

17
To write the FOIL polynomial for $4x^2 - 11x + 6$, it is necessary to find the factors of 24 ($4 \cdot 6$) with a sum of _____.

-11

18

To factor $4x^2 - 11x + 6$, construct a table that will show each pair of factors of 24 and the sum of each pair.

Factors	Sum
-1 • -24	-25
-2 • -12	-14
-3 • -8	-11
-4 • -6	-10

19

Use the information from the table constructed for frame 18 to write a FOIL polynomial for $4x^2 - 11x + 6$.

$4x^2 - 3x - 8x + 6$

20

Complete the factoring of $4x^2 - 11x + 6$ shown below.

$$4x^2 - 11x + 6$$
$$4x^2 - 3x - 8x + 6$$
_____
_____

$x(4x - 3) - 2(4x - 3)$

$(4x - 3)(x - 2)$

21

If necessary construct a table and then write a FOIL polynomial for $8x^2 - 18x - 5$.

$8x^2 + 2x - 20x - 5$

22

Factor. $8x^2 - 18x - 5$

$(4x + 1)(2x - 5)$

23

If necessary construct a table and then write a FOIL polynomial for $12x^2 - 5x - 2$.

$12x^2 + 3x - 8x - 2$

24

Factor. $12x^2 - 5x - 2$

$(4x + 1)(3x - 2)$

25

If necessary construct a table and then write a FOIL polynomial for $18x^2 + 9x - 2$.

$18x^2 - 3x + 12x - 2$

26

Factor. $18x^2 + 9x - 2$

$(6x - 1)(3x + 2)$

FACTORING POLYNOMIALS

27
Construct a table for $6x^2 - 5x - 2$. Does the trinomial have an associated FOIL polynomial?

Factors	Sum
1 • -12	-11
2 • -6	-4
3 • -4	-1

No, there are no factors of -12 with sum of -5

28
$6x^2 - 5x - 2$ is not factorable because it has no associated FOIL polynomial. Is $6x^2 - x - 12$ factorable?

Yes, its FOIL polynomial is $6x^2 + 8x - 9x - 12$

29
$6x^2 - 5x - 2$ is a **prime polynomial**; it has no associated FOIL polynomial. $6x^2 - x - 12$ is factorable; it has an associated FOIL polynomial. Is $5x^2 - 4x - 2$ prime?

Yes, there are no factors of -10 with sum of -4

30
Is $2x^2 + 7x + 6$ prime?

No, its FOIL polynomial is $2x^2 + 3x + 4x + 6$

31
Factor. $2x^2 + 7x + 6$

$(2x + 3)(x + 2)$

32
Is $6x^2 - 25x - 9$ prime?

No

33
Factor. $6x^2 - 25x - 9$

$6x^2 + 2x - 27x - 9$
$(3x + 1)(2x - 9)$

34
Is $4x^2 - 9x + 6$ prime?

Yes, 24 has no factors with a sum of -9

35
Is $8x^2 + 5x - 3$ prime?

No

36
Factor. $8x^2 + 5x - 3$ $\hfill (x + 1)(8x - 3)$

37
Factor $3x^2 - 5x + 7$ or state that it is prime. $\hfill \text{prime}$

38
Factor $6x^2 + 13x - 5$ or state that it is prime. $\hfill (2x + 5)(3x - 1)$

39
Factor $9x^2 + 38x + 8$ or state that it is prime. $\hfill (x + 4)(9x + 2)$

40
Factor $4x^2 - 7x + 9$ or state that it is prime. $\hfill \text{prime}$

41
Factor $4x^2 + 12x + 9$ or state that it is prime. $\hfill (2x + 3)^2$

42
$4x^2 + 12x + 9$ is a **trinomial perfect square**. Its FOIL polynomial is $4x^2 + 6x + 6x + 9$ and it has two important distinguishing characteristics:

1. The F and L terms are perfect squares.
   $4x^2 = (2x)^2$ and $9 = 3^2$

2. The O and I terms are identical.
   O term is $6x$ and I term is _____ $\hfill 6x$

43
To determine if $49x^2 + 14x + 1$ is a trinomial perfect square begin by finding its FOIL polynomial. What is it? $\hfill 49x^2 + 7x + 7x + 1$

44
The FOIL polynomial $49x^2 + 7x + 7x + 1$ is a trinomial perfect square. Its F and L terms are perfect squares and its O and I terms are identical. Factor. $49x^2 + 7x + 7x + 1$ $\hfill (7x + 1)^2$

45

The FOIL polynomial of a trinomial perfect square has two distinguishing characteristics.

1. The F and L terms are perfect squares.
2. The O and I terms are identical.

Is $9x^2 + 30x + 25$ a trinomial perfect square?

Yes, $(3x)^2 + 15x + 15x + 5^2$

46

Factor $9x^2 + 30x + 25$ as a trinomial perfect square.

$(3x + 5)^2$

47

The FOIL polynomial of a trinomial perfect square has:

1. F and L terms which are perfect squares.
2. Identical O and I terms.

Is $16x^2 - 40x + 25$ a trinomial perfect square?

Yes, $(4x)^2 - 20x - 20x + 5^2$

48

Factor $16x^2 - 40x + 25$ as a trinomial perfect square.

$(4x - 5)^2$

49

Is $49x^2 - 84x + 36$ a trinomial perfect square?

Yes, $(7x)^2 - 42x - 42x + 6^2$

50

Factor $49x^2 - 84x + 36$.

$(7x - 6)^2$

51

The middle term of a trinomial perfect square can have either a positive or negative coefficient. Can the last term be either positive or negative?

No, a perfect square cannot be negative

52

Determine if $16x^2 - 24xy + 9y^2$ is a trinomial perfect square, and if so, factor it.

Yes, $(4x - 3y)^2$

53

Determine if $4x^2 - 4x + 1$ is a trinomial perfect square, and if so, factor it.

Yes, $(2x - 1)^2$

54
Determine if $25x^2 + 15x + 9$ is a trinomial perfect square, and if so, factor it.

No, middle term would have to be 30x or -30x

55
Determine if $4x^2 - 12x - 9$ is a trinomial perfect square, and if so, factor it.

No, last term is -9 which is not a perfect square

56
$36x^2 - 49y^4$ is a **difference of two squares**.

$$36x^2 = (6x)^2 \quad \text{and} \quad 49y^4 = (7y^2)^2$$

$36x^2 - 49y^4$ factors as the sum and difference of 6x and _____.

$7y^2$

57
Factor $36x^2 - 49y^4$ by writing two binomials which are the sum and difference of 6x and $7y^2$.

$(6x + 7y^2)(6x - 7y^2)$

58
Is $9x^4 - 100y^2$ a difference of two squares?

Yes, $(3x^2)^2$ and $(10y)^2$

59
Factor $9x^4 - 100y^2$ by writing two binomials which are the sum and difference of $3x^2$ and 10y.

$(3x^2 + 10y)(3x^2 - 10y)$

60
Is $25x^8 + 64y^6$ a difference of two squares?

No, it is a sum and is prime.

61
Factor. $25x^8 - 64y^6$

$(5x^4 + 8y^3)(5x^4 - 8y^3)$

62
Factor. $100x^2 - 49$

$(10x + 7)(10x - 7)$

63

Any difference of two squares can be factored as a FOIL polynomial in which the O and I terms have opposite signs. For example,

$$64x^2 - 49y^2$$
$$64x^2 + (8x)(7y) - (8x)(7y) - 49y^2$$
$$64x^2 + 56xy - 56xy - 49y^2$$
$$8x(8x + 7y) - 7y(8x + 7y)$$
$$(8x + 7y)(8x - 7y)$$

Factor. $25x^2 - 64y^2$ $(5x + 8y)(5x - 8y)$

64

To factor any trinomial of the form $ax^2 + bx + c$, a reversal of FOIL multiplication may be used. The process requires writing a FOIL polynomial and separately factoring the FO and IL terms. For example, factor $8x^2 - 10x - 3$ by first finding its FOIL polynomial.

$8x^2 + 2x - 12x - 3$

$(4x + 1)(2x - 3)$

FEEDBACK UNIT 3

This quiz reviews the preceding unit. Answers are at the back of the book.

Factor each polynomial or state that it is prime.

1. $15x^2 + 7x - 2$
2. $6x^2 + 2x - 3$
3. $4x^2 + 9x - 9$
4. $b^2 - 2bc + c^2$
5. $3x^2 - x + 3$
6. $3x^2 + 10x + 3$
7. $7x^2 + 33x - 10$
8. $8x^2 - 19x + 5$
9. $100x^2 - 9z^2$
10. $2x^2 + x - 3$
11. $9x^2 - 2x - 3$
12. $6x^2 + 5x - 21$
13. $5x^2 + 18x - 8$
14. $25x^2 + 60xz + 36z^2$
15. $81 - d^2$
16. $12x^2 + 31x + 20$

Unit 4: Factoring Sums/Differences of Two Cubes

The following mathematical terms are crucial to an understanding of this unit.

Perfect cubes
Sum of two cubes
Difference of two cubes

1
Recall the multiplication of $(x - y)(x^2 + xy + y^2)$.

$$(x - y)(x^2 + xy + y^2)$$
$$x(x^2 + xy + y^2) - y(x^2 + xy + y^2)$$
$$x^3 + x^2y + xy^2 - x^2y - xy^2 - y^3$$

Simplify the 6-term polynomial shown above. $\qquad x^3 - y^3$

2
Multiply and simplify. $(2x - 3)(4x^2 + 6x + 9)$ $\qquad 8x^3 - 27$

3
Multiply and simplify. $(x - 4)(x^2 + 4x + 16)$ $\qquad x^3 - 64$

4
The multiplication of $(x - 5)(x^2 + 5x + 25)$, like those in frames 1-3, can be greatly simplified. Find the product. $\qquad x^3 - 125$

5
$$(x - 2)(x^2 + 2x + 4) = x^3 - 8$$
$$(x - 3)(x^2 + 3x + 9) = x^3 - 27$$
$$(x - 4)(x^2 + 4x + 16) = x^3 - 64$$
$$(x - 5)(x^2 + 5x + 25) = x^3 - 125$$

Is there a pattern shown by these four multiplications that would apply to multiplying $(x - 6)(x^2 + 6x + 36)$? $\qquad$ Yes, $x^3 - 216$

6
Using exponents, the pattern found in frame 5 would explain each of the following.

$$(x - 2)(x^2 + 2x + 2^2) = x^3 - 2^3$$
$$(x - 3)(x^2 + 3x + 3^2) = x^3 - 3^3$$
$$(x - 4)(x^2 + 4x + 4^2) = x^3 - 4^3$$

Use this pattern to multiply $(x - 7)(x^2 + 7x + 7^2)$.

$x^3 - 7^3$

7
Multiply. $(x - 13)(x^2 + 13x + 13^2)$

$x^3 - 13^3$

8
Multiply. $(x - y)(x^2 + xy + y^2)$

$x^3 - y^3$

9
To factor $x^3 - 2^3$ the pattern needs to be applied in reverse.

$(x - 3)(x^2 + 3x + 3^2)$ $\quad = x^3 - 3^3$
_____ $\quad = x^3 - 2^3$

$(x - 2)(x^2 + 2x + 2^2)$

10
To factor $x^3 - 5^3$ the pattern needs to be applied in reverse.

$x^3 - y^3 = (x - y)(x^2 + xy + y^2)$
$x^3 - 5^3 =$ _____

$(x - 5)(x^2 + 5x + 5^2)$

11
The terms of $x^3 - y^3$ are **perfect cubes**. x^3 is a perfect cube. y^3 is a perfect cube. Are the terms of $x^3 - 8$ perfect cubes?

Yes, $x^3 - 2^3$

12
The binomial $x^3 - y^3$ is a **difference of two cubes**.
To factor any difference of two cubes, use this pattern.

$$x^3 - y^3 = (x - y)(x^2 + xy + y^2)$$

Notice that one factor is a binomial, $(x - y)$.
The other factor is a trinomial, $(x^2 + xy + y^2)$.
Factor $w^3 - z^3$ using the pattern shown above.

$(w - z)(w^2 + wz + z^2)$

13

To factor any difference of two cubes, use this pattern.

$$x^3 - y^3 = (x - y)(x^2 + xy + y^2)$$

Factor. $a^3 - b^3$

$(a - b)(a^2 + ab + b^2)$

14

Factor. $k^3 - m^3$

$(k - m)(k^2 + km + m^2)$

15

To factor $x^3 - 125$, first write it as $x^3 - 5^3$.

Factor. $x^3 - 125$

$(x - 5)(x^2 + x \cdot 5 + 5^2)$
$(x - 5)(x^2 + 5x + 25)$

16

To factor $x^3 - 8y^3$, first write it as $x^3 - (2y)^3$.

Factor. $x^3 - 8y^3$

$(x - 2y)[x^2 + x(2y) + (2y)^2]$
$(x - 2y)(x^2 + 2xy + 4y^2)$

17

To factor $8x^3 - 125$, first write it as $(2x)^3 - 5^3$.

Factor. $8x^3 - 125$

$(2x - 5)[(2x)^2 + (2x)5 + 5^2]$
$(2x - 5)(4x^2 + 10x + 25)$

18

Factor. $x^3 - 216$

[Note: $x^3 - 216 = (x)^3 - (6)^3$]

$(x - 6)(x^2 + 6x + 36)$

19

Factor. $x^3 - 1000$

$(x - 10)(x^2 + 10x + 100)$

20

Factor. $27x^3 - 8$

$(3x - 2)(9x^2 + 6x + 4)$

21

Factor. $a^3 - 64$

$(a - 4)(a^2 + 4a + 16)$

22

Factor. $1 - 27x^6$

$(1 - 3x^2)(1 + 3x^2 + 9x^4)$

23
The binomial $8x^3 + 125$ is a **sum of two cubes**. The sum of two cubes is always factorable.
Is $8x^3 + 125$ factorable?

Yes

24
The pattern for the sum of two cubes is similar to that for the difference of two cubes.

$$x^3 + y^3 = (x + y)(x^2 - xy + y^2)$$

Use the pattern to factor $a^3 + b^3$.

$(a + b)(a^2 - ab + b^2)$

25
$x^3 + 1$ is the sum of two cubes. $x^3 + 1 = (x)^3 + (1)^3$
Use the pattern, $x^3 + y^3 = (x + y)(x^2 - xy + y^2)$, to factor $x^3 + 1$.

$(x + 1)(x^2 - x \cdot 1 + 1^2)$
$(x + 1)(x^2 - x + 1)$

26
Use the pattern, $x^3 + y^3 = (x + y)(x^2 - xy + y^2)$, to factor $x^3 + 8$.

$(x + 2)(x^2 - x \cdot 2 + 2^2)$
$(x + 2)(x^2 - 2x + 4)$

27
The binomial factor of $x^6 + 27y^3$ is $(x^2 + 3y)$.
What is the binomial factor of $27x^3 + 8$?

$(3x + 2)$

28
$x^3 + 27$ is the sum of two cubes. $x^3 - 27$ is the difference of two cubes. $x^3 - 8$ is a _____ (sum, difference) of two cubes.

difference

29
The binomial factor of $x^3 - 8$ is $(x - 2)$. Which of the following is the binomial factor of $x^3 - 125$?

$(x - 5)$ or $(x + 5)$?

$(x - 5)$

30
What is the binomial factor of $8x^3 - y^3$?

$(2x - y)$

31

The general form for factoring the sum of two cubes is

$$x^3 + y^3 = (x + y)(x^2 - xy + y^2)$$

Factor. $a^3 + b^3$ \hfill $(a + b)(a^2 - ab + b^2)$

32
Factor. $r^3 + s^3$ \hfill $(r + s)(r^2 - rs + s^2)$

33
Factor. $a^3 + 8$ \hfill $(a + 2)(a^2 - 2a + 4)$

34
Factor. $8x^3 + 27b^3$ \hfill $(2x + 3b)(4x^2 - 6xb + 9b^2)$

35
Factor. $125r^3 + w^3$ \hfill $(5r + w)(25r^2 - 5rw + w^2)$

36
Factor. $a^3b^3 + 27$
[Note: $a^3b^3 = (ab)^3$] \hfill $(ab + 3)(a^2b^2 - 3ab + 9)$

FEEDBACK UNIT 4

This quiz reviews the preceding unit. Answers are at the back of the book.

Factor each of the following polynomials.

1. $x^3 + 1$
2. $a^6 - 8$
3. $27z^3 - 1$
4. $8a^3 + 27b^3$
5. $64x^6 - 27y^3$
6. $b^3 + a^3$
7. $1000z^3 - 27x^6$
8. $x^3y^3 + z^3$
9. $8a^3b^6 - 1$
10. $m^6 - 64r^3s^6$
11. $x^3 + y^3$
12. $x^3 - y^3$

Unit 5: Reviewing and Extending Factoring Skills

The following mathematical terms are crucial to an understanding of this unit.

Common factor method
Trinomial perfect squares
Difference of two cubes
Substitution

Differences of two squares
FOIL polynomial
Sum of two cubes

1
Four types of factoring situations have been presented in this chapter. The **common factor method** is illustrated by the example below.

$$6x^3y - 9x^2y^5 = 3x^2y(2x - 3y^4)$$

Factor $12x^4y^2 - 10xy^3$ using the common factor method.

$2xy^2(6x^3 - 5y)$

2
If a polynomial is to be factored, the first attempt should be to apply the common factor method. For example, the binomial $4x^2 - 64$ has 4 as a common factor, but it is also a difference of two squares. How should the factoring of $4x^2 - 64$ begin?

$4(x^2 - 16)$

3
The second type of factoring situation presented in this chapter was for polynomials of the form $x^2 + bx + c$, where the coefficient of x^2 is 1. This method of factoring is illustrated by the example below.

$$x^2 - 13x + 30 = (x - 10)(x - 3)$$

Factor $x^2 - 9x - 90$ by finding factors of -90 with a sum of -9.

$(x - 15)(x + 6)$

4

Some **differences of two squares** are of the form $x^2 - a^2y^2$. This is the case with the example shown below.

$$x^2 - 9y^6 = (x + 3y^3)(x - 3y^3)$$

Factor $x^2 - 16$ as a difference of two squares. $\qquad (x + 4)(x - 4)$

5

Some **trinomial perfect squares** are of the form $x^2 - 2bx + b^2$. This is the case with the example shown below.

$$x^2 - 20x + 100 = (x - 10)^2$$

Factor $x^2 - 14x + 49$ as a trinomial perfect square. $\qquad (x - 7)^2$

6

The third type of factoring situation presented in this chapter was for polynomials of the form $ax^2 + bx + c$, where the coefficient of x^2 is not 1. This method of factoring is illustrated by the example below.

$$10x^2 - 9x - 9$$
$$10x^2 + 6x - 15x - 9$$
$$2x(5x + 3) - 3(5x + 3)$$
$$(5x + 3)(2x - 3)$$

Factor $12x^2 - 13x + 3$ by first writing its **FOIL polynomial**. $\qquad (4x - 3)(3x - 1)$

7

Some differences of two squares are of the form $a^2x^2 - b^2c^2$. This is the case with the example shown below.

$$36x^2 - 49y^4 = (6x + 7y^2)(6x - 7y^2)$$

Factor. $25x^8 - 64y^6$ $\qquad (5x^4 + 8y^3)(5x^4 - 8y^3)$

8

Some trinomial perfect squares are of the form $a^2x^2 - 2abx + b^2$. This is the case with the example shown below.

$$9x^2 - 48x + 64 = (3x - 8)^2$$

Factor $4x^2 - 28x + 49$ as a trinomial perfect square. $(2x - 7)^2$

9

The fourth, and last, type of factoring learned in this chapter is for the sum/difference of two cubes. This type of factoring is illustrated below.

$$f^3 - g^3 = (f - g)(f^2 + fg + g^2)$$

Factor $y^3 - 27$ as a **difference of two cubes**. $(y - 3)(y^2 + 3y + 9)$

10

A **sum of two cubes** is factored in the example below.

$$t^3 + u^3 = (t + u)(t^2 - tu + u^2)$$

Factor $r^3 + 64$ as a sum of two cubes. $(r + 4)(r^2 - 4r + 16)$

11
Factor. $x^2 - 36y^{10}$ $(x + 6y^5)(x - 6y^5)$

12
Factor. $18x^2y - 9xy^2 - 18y^4$ $9y(2x^2 - xy - 2y^3)$

13
Factor. $x^2 - 64y^4$ $(x + 8y^2)(x - 8y^2)$

14
Factor. $x^2 - 5x - 14$ $(x - 7)(x + 2)$

15
Factor. $4x^2 - 3x - 10$ $(x - 2)(4x + 5)$

16
Factor. $x^2 - 6x + 8$ $(x - 4)(x - 2)$

17
Factor. $7x^3 - 21x$

$7x(x^2 - 3)$

18
Factor. $x^2 + 2x - 15$

$(x + 5)(x - 3)$

19
Factor. $x^3 - 125$

$(x - 5)(x^2 + 5x + 25)$

20
Factor. $x^2 - 25y^2$

$(x + 5y)(x - 5y)$

21
Factor. $y^3 + 8$

$(y + 2)(y^2 - 2x + 4)$

22
Factor. $100x^2 - 1$

$(10x + 1)(10x - 1)$

23
Some factoring situations look far more difficult than necessary. Study the two examples below.

$$5xy - 7y \qquad 5x\left(\frac{7-w^3}{12}\right) - 7\left(\frac{7-w^3}{12}\right)$$

$$y(5x - 7) \qquad \left(\frac{7-w^3}{12}\right)(5x - 7)$$

The fraction $\left(\frac{7-w^3}{12}\right)$ in one of the examples above makes its factoring seem more difficult than the other example. Actually, the examples are identical when $\left(\frac{7-w^3}{12}\right) =$ _____.

y

24

To factor $(x-y)^2 - 5(x-y) + 6$, the binomial $(x-y)$ may be replaced by a single variable.

$$(x-y)^2 - 5(x-y) + 6$$
Let $(x-y) = k$
$$k^2 - 5k + 6$$
$$(k-3)(k-2)$$
Since $(x-y) = k$
$$[(x-y)-3][(x-y)-2]$$
$$(x-y-3)(x-y-2)$$

This example shows the use of **substitution** in a factoring situation. First k was substituted for $(x+y)$. After the factoring was completed $(x+y)$ was substituted for ____.

k

25

To factor $(a-3)^2 - 5(a-3)y + 4y^2$ using substitution, begin by selecting a variable to replace the binomial $(a-3)$. If k is substituted for $(a-3)$,

$(a-3)^2 - 5(a-3)y + 4y^2$ becomes _____

$k^2 - 5ky + 4y^2$

26

Factor. $k^2 - 5ky + 4y^2$

$(k-4y)(k-y)$

27

Since $k = (a-3)$, substitute $(a-3)$ for k in $(k-4y)(k-y)$. Use parentheses in making this substitution.

$[(a-3)-4y][(a-3)-y]$
$(a-3-4y)(a-3-y)$

28

The example below illustrates factoring by substitution.

$$(x-7)^2 + 10(x-7)y + 16y^2$$

Substitute k for $(x-7)$ $k^2 + 10ky + 16y^2$
Factor $(k+2y)(k+8y)$
Substitute $(x-7)$ for k $[(x-7)+2y][(x-7)+8y]$
Remove parentheses   _____

$(x-7+2y)(x-7+8y)$

29

The example below illustrates factoring by substitution.

$$x^2 + 8x(y-3) + 15(y-3)^2$$

Substitute k for (y – 3) $\quad x^2 + 8xk + 15k^2$
Factor $\quad (x + 5k)(x + 3k)$
Substitute (y – 3) for k $\quad [x + 5(y-3)][x + 3(y-3)]$
Remove parentheses $\quad\underline{\hspace{3cm}}$ $\quad (x + 5y - 15)(x + 3y - 9)$

30

The example below illustrates factoring by substitution.

$$x^2 - x(y-6) - 20(y-6)^2$$

Substitute k for (y – 6) $\quad x^2 - xk - 20k^2$
Factor $\quad (x - 5k)(x + 4k)$
Substitute (y – 6) for k $\quad [x - 5(y-6)][x + 4(y-6)]$
Remove parentheses $\quad\underline{\hspace{3cm}}$ $\quad (x - 5y + 30)(x + 4y - 24)$

31

Complete the factoring of $x^2 + 5x(y-2) - 24(y-2)^2$ using substitution,

$$x^2 + 5x(y-2) - 24(y-2)^2$$
$$x^2 + 5xk - 24k^2$$
$$(x + 8k)(x - 3k)$$
$$[x + 8(y-2)][x - 3(y-2)]$$
$$\underline{\hspace{3cm}}$$

$(x + 8y - 16)(x - 3y + 6)$

32

Complete the factoring shown below.

$$x^2 - 9x(y-7) + 20(y-7)^2$$
$$x^2 - 9xk + 20k^2$$
$$(x - 5k)(x - 4k)$$
$$\underline{\hspace{3cm}}$$
$$\underline{\hspace{3cm}}$$

$[x - 5(y-7)][x - 4(y-7)]$
$(x - 5y + 35)(x - 4y + 28)$

33

Factor. $x^2 - 2x(a+b) - 15(a+b)^2$
Note: Use parentheses in substituting k for (a + b).

$[x - 5(a+b)][x + 3(a+b)]$
$(x - 5a - 5b)(x + 3a + 3b)$

34

Factor. $x^2 + 7x(r+s) + 12(r+s)^2$

$(x + 4r + 4s)(x + 3r + 3s)$

FACTORING POLYNOMIALS 239

35
Factor. $r^2 - 7r(x-y) + 10(x-y)^2$

$(r - 5x + 5y)(r - 2x + 2y)$

36
Factor. $(x+a)^2 - 11(x+a)y + 10$

$(x + a - 10)(x + a - 1)$

37
Factor. $(x-y)^2 + 7(x-y)z - 8z^2$

$(x - y + 8z)(x - y - z)$

38
$4x^2 - (z+w)^2$ is a difference of two squares.
Use substitution to factor it.

$4x^2 - k^2$
$[2x + (z+w)][2x - (z+w)]$
$(2x + z + w)(2x - z - w)$

39
Factor. $(r-s)^2 - 9z^2$

$(r - s + 3z)(r - s - 3z)$

40
The clue to factoring $x^2 - 10x + 25 - y^2$ is that the first three terms are a trinomial perfect square. Complete the factoring shown below.

$$x^2 - 10x + 25 - y^2$$
$$(x^2 - 10x + 25) - y^2$$
$$(x-5)^2 - y^2$$
Let $k = (x-5)$ $\qquad k^2 - y^2$
_____
_____

$(k + y)(k - y)$
$(x - 5 + y)(x - 5 - y)$

41
Factor $x^2 + 16x + 64 - y^2$ by first grouping the first three terms as a trinomial perfect square.

$(x + 8)^2 - y^2$
$(x + 8 + y)(x + 8 - y)$

42
Factor $4x^2 + 12x + 9 - y^2$ by first grouping the first three terms as a trinomial perfect square.

$(2x + 3)^2 - y^2$
$(2x + 3 + y)(2x + 3 - y)$

43

The clue to factoring $x^2 - y^2 + 6yz - 9z^2$ is that the last three terms can be made into a trinomial perfect square. Complete the factoring shown below.

$$x^2 - y^2 + 6yz - 9z^2$$
$$x^2 - (y^2 - 6yz + 9z^2)$$
$$x^2 - (y - 3z)^2$$

Let $k = (y - 3z)$ $\quad x^2 - k^2$

_____ $[x + (y - 3z)][x - (y - 3z)]$
_____ $(x + y - 3z)(x - y + 3z)$

44

Begin the factoring of $x^2 - y^2 + 14y - 49$ by first using -1 as a common factor for the last three terms.

$x^2 - (y^2 - 14y + 49)$

45

Complete the factoring of $x^2 - y^2 + 14y - 49$ shown below.

$$x^2 - y^2 + 14y - 49$$
$$x^2 - (y^2 - 14y + 49)$$

_____ $x^2 - (y - 7)^2$
_____ $(x + y - 7)(x - y + 7)$

46

Factor $x^2 - y^2 - 10y - 25$ by grouping the last three terms.

$x^2 - (y^2 + 10y + 25)$
$x^2 - (y + 5)^2$
$(x + y + 5)(x - y - 5)$

47

Factor $4a^2 - x^2 + 6x - 9$ by grouping the last three terms.

$4a^2 - (x^2 - 6x + 9)$
$4a^2 - (x - 3)^2$
$(2a + x - 3)(2a - x + 3)$

48

Factor. $x^2 - y^2 - 2y - 1$

$(x + y + 1)(x - y - 1)$

49

Factor $x^2 + 10x + 25 - 81y^2$ by first deciding whether to group the first three terms or the last three terms.

$(x + 5)^2 - (9y)^2$
$(x + 5 + 9y)(x + 5 - 9y)$

50
Factor. $x^2 + 6x + 9 - 25y^2$ $(x + 3 + 5y)(x + 3 - 5y)$

51
Factor. $r^2 + 4r + 4 - 16s^2$ $(r + 2 + 4s)(r + 2 - 4s)$

52
Factor. $r^2 - x^2 + 4x - 4$ $(r + x - 2)(r - x + 2)$

53
Factor. $x^2 - w^2 - 12w - 36$ $(x + w + 6)(x - w - 6)$

54
Factor. $9x^2 - 12x + 4 - 25z^2$ $(3x - 2 + 5z)(3x - 2 - 5z)$

Feedback Unit 5

This quiz reviews the preceding unit. Answers are at the back of the book.

Factor each polynomial completely or state that it is prime.

1. $10x^2y^3 - 15xy^2$
2. $x^2 + 4x - 21$
3. $3x^2 + 13x - 10$
4. $x^2 - x - 20$
5. $6x^2 + x - 2$
6. $49a^4b^3 + 28a^2b^2$
7. $4x^2 - 9y^4$
8. $(x + y)^2 - 2(x + y) - 15$
9. $3 - 2(x - y) - (x - y)^2$
10. $10x^2 + 33x - 7$
11. $x^2 + x - 72$
12. $x^2 + 2xy + y^2$
13. $4a^4 - 25b^2$
14. $a^2 - 2ab + b^2$
15. $9x^2w^5 + 6x^2w^3$
16. $x^2 + 16x - 17$
17. $(a - b)^2 + 5(a - b) - 6$
18. $4y^2 + 5yw - 6w^2$
19. $4x^2 + 4x + 1 - z^2$
20. $1 - a^2 + 10ab - 25b^2$

Unit 6: Factoring Polynomials with More Than Two Factors

1
The polynomial $5x - 10$ has two factors: 5 and $(x - 2)$.
The polynomial $2x^2 - 6x - 8$ has three factors which are found in the following steps.

$$2x^2 - 6x - 8$$
$$2(x^2 - 3x - 4)$$
$$2(x - 4)(x + 1)$$

The three factors of $2x^2 - 6x - 8$ are:
_____, _____, _____. $2, (x - 4), (x + 1)$

2
For any attempt at factoring a polynomial, the first step should always be to remove a common factor if one exists. Factor. $2x^2 - 14x + 20$ $2(x - 5)(x - 2)$

3
Factor $6x^2 - 6$ completely. $6(x + 1)(x - 1)$

4
Factor $8x^2 - 32x + 32$ completely. $8(x - 2)^2$

5
Factor $10x^2 - 35x + 15$ completely. $5(2x - 1)(x - 3)$

6
Factor $12x^2 - 27$ completely. $3(2x + 3)(2x - 3)$

7
Factor $20x^2 + 140x + 245$ completely. $5(2x + 7)^2$

8
Factor $3x^2 + 6x - 6$ completely. $3(x^2 + 2x - 2)$

9
Factor $7x^3 + 56$ completely. $7(x + 2)(x^2 - 2x + 4)$

10

Some polynomials, like $x^2 + 7$, are prime.
Other polynomials, like $7x^3 + 56$, are factorable and their factoring is not complete until each polynomial is prime.
Is $7x^6 - 7y^6$ prime?

No

11

The first steps for factoring $7x^6 - 7y^6$ are shown below.

$$7x^6 - 7y^6$$
7 is a common factor. $\quad 7(x^6 - y^6)$
$x^6 - y^6$ is a difference of 2 squares. $\quad 7([x^3]^2 - [y^3]^2)$
$$7(x^3 + y^3)(x^3 - y^3)$$

Finish the factoring by factoring each binomial.

$7(x + y)(x^2 - xy + y^2) \cdot$
$(x - y)(x^2 + xy + y^2)$

12

The first steps for factoring $x^4 - 13x^2y^2 + 36y^4$ are shown below.

$$x^4 - 13x^2y^2 + 36y^4$$
$$[x^2]^2 - 13x^2y^2 + 36[y^2]^2$$
$$(x^2 - 4y^2)(x^2 - 9y^2)$$

Finish the factoring by factoring each binomial.

$(x + 2y)(x - 2y) \cdot$
$(x + 3y)(x - 3y)$

13
Factor $4x^2 - 8x - 60$ completely.

$4(x - 5)(x + 3)$

14
Factor $6x^2 - 24$ completely.

$6(x + 2)(x - 2)$

15
Factor $2x^2 - 50$ completely.

$2(x + 5)(x - 5)$

16
Factor $7rx^2 - 3rx + 19r$ completely.

$r(7x^2 - 3x + 19)$

17
Factor $x^4 - y^4$ completely.

$(x^2 + y^2)(x + y)(x - y)$

Feedback Unit 6

This quiz reviews the preceding unit. Answers are at the back of the book.

Factor each polynomial completely or state that it is prime.

1. $x^2 + 2x - 15$
2. $3x^2 - 12x - 15$
3. $5a^2 - 5b^2$
4. $8x^3 - y^3$
5. $54x^3 - 16y^6$
6. $5x^3y^2 + 15x^2y^2$
7. $ax^2 - 3ay^2$
8. $a^6 + b^6$
9. $4xr^6 + 4xs^3$
10. $5x^2 - 20x - 25$
11. $a^2 - 25b^2$
12. $13a^4b^2c^3 + 52a^2b^2c^2$
13. $a^4 - b^4$
14. $3a^6 - 3b^6$
15. $18x^2 + 15x - 12$
16. $(x + y)^2 - 4(x + y) + 3$
17. $9x^2 + 29x + 6$
18. $5x^3 - 40y^3$
19. $x^4 - 5x^2 + 4$
20. $x^4 - 16$

UNIT 7: APPLICATIONS

In this Applications Section, the format of the text has been altered. Answers for the problems appear beneath them rather than in the right-hand column. Your studying emphasis should be on learning the best procedures to follow with word problems.

1

The figure at the right is a rectangle. Its dimensions are: length which is designated by L and width designated by W. The perimeter of the rectangle is found using the formula P = 2L + 2W. The area of the rectangle is found using the formula A = LW. Find the perimeter and area of a rectangle with width 7 feet and length 10 feet.

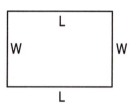

Answer:
P = 2L + 2W = 14 + 20 = 34   The perimeter is 34 feet.
A = LW = 7 • 10 = 70   The area is 70 square feet.

2

If the length (L) of a rectangle is 8 feet longer than the width (W), then the length can be translated as W + 8. What is the translation of the length of a rectangle when it is 4 inches more than the width?

Answer: The length (L) can be translated as W + 4.

3

The 3-step process for solving a word problem is:

a. Construct a table of the necessary translations of words, phrases, and sentences.
b. Solve any equation(s) obtained in the table.
c. Check answers in the original statement of the word problem.

Begin the process for solving the following problem by constructing a table of translations.

> The perimeter of a rectangle is 48 feet. The length of the rectangle is 6 feet more than its width. What are the rectangle's dimensions?

Answer:

word/phrase/sentence	translation
width (of the rectangle)	W
length	$W + 6$
perimeter ($P = 2L + 2W$)	$2(W + 6) + 2W$
The perimeter of a rectangle is 48 feet.	$2(W + 6) + 2W = 48$

4

Use the equation found in the table of frame 3. Complete the last 2 steps in solving the following problem.

> The perimeter of a rectangle is 48 feet. The length of the rectangle is 6 feet more than its width. What are the rectangle's dimensions?

Answer:
b. $2(W + 6) + 2W = 48$ or $4W = 36$ solves as $W = 9$.
If length is $W + 6$ (see the table) and $W = 9$, then length is 15.
c. When a rectangle has 9 feet as its width and 15 feet as its length, then its length is 6 more than its width and its perimeter is 48 feet.

5

Use the 3-step process when solving word problems like the following.

> The width of a rectangle is 9 less than the length. The perimeter of the rectangle is 90. What are the rectangle's dimensions?

Begin the solution process by constructing a table of solutions in which the first line is the length of the rectangle.

Answer:

word/phrase/sentence	translation
length (of the rectangle)	L
width	$L - 9$
perimeter ($P = 2L + 2W$)	$2L + 2(L - 9)$
The perimeter of a rectangle is 90.	$2L + 2(L - 9) = 90$

6

Use the equation found in the table of frame 5. Complete the last 2 steps in solving the following problem.

> The width of a rectangle is 9 less than the length. The perimeter of the rectangle is 90. What are the rectangle's dimensions?

Answer:
b. $2L + 2(L - 9) = 90$ or $4L = 108$ solves as $L = 27$.
    If width is $L - 9$ (see the table) and $L = 27$, then width is 18.
c. When a rectangle has 18 as its width and 27 as its length, then its width is 9 less than its length and its perimeter is 90.

7

Use the 3-step process when solving word problems like the following.

> The length of a rectangle is twice the width. The perimeter of the rectangle is 72. What are the rectangle's dimensions?

Begin the solution process by constructing a table of solutions in which the first line is the width of the rectangle.

Answer:

word/phrase/sentence	translation
width (of the rectangle)	W
length	2W
perimeter (P = 2L + 2W)	2(2W) + 2W
The perimeter of a rectangle is 72.	2(2W) + 2W = 72

8

Use the equation found in the table of frame 7. Complete the last 2 steps in solving the following problem.

> The length of a rectangle is twice the width. The perimeter of the rectangle is 72. What are the rectangle's dimensions?

Answer:
b. 2(2W) + 2W = 72 or 6W = 72 solves as W = 12.
    If length is 2W (see the table) and W = 12, then length is 24.
c. When a rectangle has 12 as its width and 24 as its length, then its length is twice its width and its perimeter is 72.

9

Use the 3-step process when solving word problems like the following.

> The length of a rectangle is 1 more than the width. The area of the rectangle is 30 square units. What are the rectangle's dimensions?

Begin the solution process by constructing a table of solutions in which the first line is the width of the rectangle.

Answer:

word/phrase/sentence	translation
width (of the rectangle)	W
1 more than the width	$W + 1$
length (1 more than the width)	$W + 1$
area ($A = LW$)	$(W + 1)W$
The area of a rectangle is 30.	$(W + 1)W = 30$

10

Use the equation found in the table of frame 9. It can be solved by factoring. Complete the last 2 steps in solving the following problem.

> The length of a rectangle is 1 more than the width. The area of the rectangle is 30 square units. What are the rectangle's dimensions?

Answer:
b. $(W + 1)W = 30$ or $W^2 + W - 30 = 0$ factors to $(W - 5)(W + 6) = 0$. Solutions of the equation are 5 and -6, but -6 cannot be the width of a rectangle so $W = 5$ is the only solution for this situation. If $W = 5$ then the length is $W + 1$ or 6.
c. When a rectangle has 5 as its width and 6 as its length, then its length is 1 more than the width and its area is 30.

Feedback Unit 7 for Applications

This quiz reviews the preceding unit. Answers are at the back of the book.

Use the 3-step process of this unit to solve each of the following problems.

1. The perimeter of a rectangle is 28. The length of a rectangle is 4 more than the width. What are the rectangle's dimensions?

2. The width of a rectangle is 2 less than the length. The perimeter of a rectangle is 60. What are the rectangle's dimensions?

3. The perimeter of a rectangle is 42. The length of a rectangle is twice the width. What are the rectangle's dimensions?

4. The length of a rectangle is 1 more than twice the width. The perimeter of a rectangle is 44. What are the rectangle's dimensions?

5. The perimeter of a rectangle is 22. The length of a rectangle is 1 less than 3 times the width. What are the rectangle's dimensions?

6. The area of a rectangle is 63 square units. The length of a rectangle is 2 more than the width. What are the rectangle's dimensions?

Summary for Chapter 6

The following mathematical terms are crucial to an understanding of this chapter.

Integer factors	Positive integer factors
Variable factors	Factor
Common factor	Highest common factor (HCF)
Factored form	Factored completely
Prime polynomial	FOIL multiplication of binomials
FOIL polynomial	Prime trinomial
Trinomial perfect square	Difference of two squares
Perfect cubes	Difference of two cubes
Sum of two cubes	Common factor method
Substitution	

The factoring methods studied in this chapter included:

1. Polynomials with common factors for all terms that are not 1 or -1.

    $$ax + ay = a(x + y)$$

2. Factoring polynomials of the form $x^2 + bx + c$ where the coefficient of x^2 is 1.

    $$x^2 + (a + b)x + ab = (x + a)(x + b)$$

    a. Some of these polynomials are differences of two squares.

    $$x^2 - y^2 = (x + y)(x - y)$$

    b. Some of these polynomials are trinomial perfect squares.

    $$x^2 + 2xy + y^2 = (x + y)^2 \quad \text{or} \quad x^2 - 2xy + y^2 = (x - y)^2$$

3. Factoring polynomials of the form $ax^2 + bx + c$ where the coefficient of x^2 is not 1.

    $$acx^2 + (ad + bc)x + bd = (ax + b)(cx + d)$$

    a. Some of these polynomials are differences of two squares.

    $$a^2x^2 - b^2y^2 = (ax + by)(ax - by)$$

    b. Some of these polynomials are trinomial perfect squares.

    $$a^2x^2 + 2abx + b^2 = (ax + b)^2 \quad \text{or} \quad a^2x^2 - 2abx + b^2 = (ax - b)^2$$

4. Factoring the sum or difference of two cubes.

    $$x^3 + y^3 = (x + y)(x^2 - xy + y^2) \quad \text{and} \quad x^3 - y^3 = (x - y)(x^2 + xy + y^2)$$

Chapter 6 Mastery Test

The following questions test the objectives of Chapter 6. Answers are at the back of the book. The number in parentheses which follows each problem indicates the unit in which it can be learned.

Factor completely or state that the polynomial is prime.

1. $-35x^3 - 20x$ (1)
2. $-27x + 9$ (1)
3. $19x^4y^2 - 3x^3y^2 + 7x^2y^2$ (1)
4. $6x(2x - 1) - 5(2x - 1)$ (1)
5. $12a(5a + 2) - (5a + 2)$ (1)
6. $xy - xz + by - bz$ (1)
7. $x^2 - 5x + 7x - 35$ (1)
8. $a^2 - 8a - 20$ (2)
9. $x^2 - 5x + 4$ (2)
10. $b^2 - 19bc + 34c^2$ (2)
11. $x^2 + 4xy - 21y^2$ (2)
12. $c^2 - 25d^2$ (2)
13. $x^2 + 18xy + 81y^2$ (2)
14. $5x^2 + 7x - 6$ (3)
15. $4x^2 + 15x + 9$ (3)
16. $12x^2 + xy - 6y^2$ (3)
17. $10a^2 - 11ab - 6b^2$ (3)
18. $100x^2 - 49$ (3)
19. $64x^2 - 48x + 9$ (3)
20. $y^3 + 1$ (4)
21. $x^3 - 8$ (4)
22. $x^3 - 27y^6$ (4)
23. $125x^3 + 64$ (4)
24. $(x - 5)^2 - 6(x - 5) - 7$ (5)
25. $x^2 - x(a + b) - 12(a + b)^2$ (5)
26. $25x^6 - (r - s)^2$ (5)
27. $r^2 - s^2 - 6s - 9$ (5)
28. $b^2 + 2bc + c^2 - w^2$ (5)
29. $7x^2 - 21x - 70$ (6)
30. $4x^3 - 32y^3$ (6)
31. $13x^2w^5 - 13y^2w^5$ (6)
32. $x^4 - 81$ (6)
33. $6x^2 + 22x + 4$ (6)
34. $x^4 - 10x^2 + 9$ (6)
35. $5x^2 + 35x + 60$ (6)
36. The perimeter of a rectangle is 88. The length of a rectangle is 6 more than the width. What are the rectangle's dimensions? (7)
37. The length of a rectangle is 1 more than 3 times the width. The perimeter of a rectangle is 90. What are the rectangle's dimensions? (7)

Chapter 7 Objectives

The following problems illustrate the objectives of this chapter. At this time you are not expected to know how to do these problems. However, if all these problems are thoroughly understood, proceed directly to the Chapter Mastery Test. The number in parentheses which follows each problem indicates the unit in which it can be learned.

1. $\dfrac{-12x^2y^8z}{-14x^2y^3z^2} = \underline{\hspace{2cm}}$ (1)

2. $\dfrac{2x^3yz^3}{-4x^6y^2z^2} = \underline{\hspace{2cm}}$ (1)

3. $\dfrac{12-18x}{3x-2} = \underline{\hspace{2cm}}$ (1)

4. $\dfrac{6x-x^2}{x^2-12x+36} = \underline{\hspace{2cm}}$ (1)

5. $\dfrac{42x^2-xy-y^2}{36x^2-12xy+y^2} = \underline{\hspace{2cm}}$ (1)

6. $\dfrac{4x^2-25y^2}{4x^2-20xy+25y^2} = \underline{\hspace{2cm}}$ (1)

7. $\dfrac{x^3-8y^3}{3x^2-13xy+14y^2} \cdot \dfrac{9x^2-49y^2}{x^2+2xy+4y^2} = \underline{\hspace{2cm}}$ (2)

8. $\dfrac{x^2-4y^2}{x^2+3xy+9y^2} \cdot \dfrac{x^3-27y^3}{x^2-5xy+6y^2} = \underline{\hspace{2cm}}$ (2)

9. $\dfrac{x^2-y^2}{6x^2+6xy} \div \dfrac{5x^2-6xy+y^2}{2} = \underline{\hspace{2cm}}$ (2)

10. $\dfrac{\dfrac{xy-5y^2}{x^2y-xy^2-6y^3}}{\dfrac{5y-x}{x^2-2xy-8y^2}} = \underline{\hspace{2cm}}$ (2)

11. $\dfrac{5x-9}{2x-17} - \dfrac{2-x}{2x-17} = \underline{\hspace{2cm}}$ (3)

12. $\dfrac{2x^2+3}{2x-1} - \dfrac{5x+4}{2x-1} = \underline{\hspace{2cm}}$ (3)

13. $\dfrac{x+1}{x^2-5x+6} + \dfrac{x+7}{x^2-4} = \underline{\hspace{2cm}}$ (4)

14. $\dfrac{x+8}{4x^2-1} + \dfrac{x-1}{2x^2+13x-7} = \underline{\hspace{2cm}}$ (4)

15. $\dfrac{x+3}{x^2+4x+16} - \dfrac{x+2}{x^3-64} = \underline{\hspace{2cm}}$ (4)

16. $\dfrac{x-8}{4x^2-6x+9} - \dfrac{x+1}{8x^3+27} = \underline{\hspace{2cm}}$ (4)

17. $\dfrac{\dfrac{x-4}{5} - \dfrac{1}{x}}{\dfrac{x^2-25}{10x^2}} = \underline{\hspace{2cm}}$ (4)

18. $\dfrac{\dfrac{3}{x} - \dfrac{2}{y}}{\dfrac{2x^2-xy-3y^2}{x^2y^2}} = \underline{\hspace{2cm}}$ (4)

Chapter 7
Polynomial Fractions

Unit 1: Simplifying Fractions

The following mathematical terms are crucial to an understanding of this unit.

    Reduce (fraction)   Simplify (polynomial fraction)
    Opposite (of a polynomial)

1

The fraction $\frac{21x^4}{28x}$ can be **reduced** because 7x is a common factor of both $21x^4$ and 28x.

$$\frac{21x^4}{28x} = \frac{7x \cdot 3x^3}{7x \cdot 4} = \frac{7x}{7x} \cdot \frac{3x^3}{4} = 1 \cdot \frac{3x^3}{4} = \frac{3x^3}{4}$$

The reduction is possible because $\frac{7x}{7x} = $ _____ .

2
To reduce (**simplify**) any fraction find a common factor for its numerator and denominator. What is the common factor for $\dfrac{15x^2}{20x^5}$?

$5x^2$

3
Complete the simplification below.

$$\dfrac{15x^2}{20x^5} = \dfrac{5x^2 \cdot 3}{5x^2 \cdot 4x^3} = \underline{\qquad}$$

$\dfrac{3}{4x^3}$

4

Begin the simplification of $\dfrac{-12x^2y^2}{18x^4y}$ by first finding a common factor for $-12x^2y^2$ and $18x^4y$.

$6x^2y$

5

$\dfrac{5x^3y^2}{-15xy^5}$ can be simplified to either $\dfrac{x^2}{-3y^3}$ or $\dfrac{-x^2}{3y^3}$. The preferred simplification is $\dfrac{-x^2}{3y^3}$ because the denominator has a positive coefficient. Simplify. $\dfrac{8xyz^2}{-6x^2yz}$

$\dfrac{-4z}{3x}$

6
Simplify. $\dfrac{25x^6y^2z^3}{-10x^3y^3z^3}$

$\dfrac{-5x^3}{2y}$

7
Simplify. $\dfrac{-16x^2y^5z}{-24xy^4z^2}$

$\dfrac{2xy}{3z}$

8
Simplify. $\dfrac{xyz}{x^2y^5z}$

$\dfrac{1}{xy^4}$

9
Simplify. $\dfrac{xy^5z^3}{x^4yz^5}$

$\dfrac{y^4}{x^3z^2}$

10
Simplify. $\dfrac{6x^5y^2}{8xy^8}$

$\dfrac{3x^4}{4y^6}$

POLYNOMIAL FRACTIONS   257

11
The division of any polynomial by itself is 1.

$$\frac{4x^2 - 17x - 6}{4x^2 - 17x - 6} = 1 \quad \text{and} \quad \frac{5 - x^2}{5 - x^2} = \underline{}$$

1

12
The division of any polynomial by its **opposite** (the signs of all terms are opposites) is equal to -1.

$$\frac{-3x^2 + 5x + 9}{3x^2 - 5x - 9} = -1 \quad \text{and} \quad \frac{7 - 4x^2}{4x^2 - 7} = \underline{}$$

-1

13
The fraction $\frac{(x-3)(x+5)}{5x(x-3)}$ can be simplified

because $\frac{(x-3)}{(x-3)} = \underline{}$.

1

14
$$\frac{(x-3)(x+5)}{5x(x-3)} = \frac{\cancel{(x-3)}(x+5)}{5x\cancel{(x-3)}} = \frac{x+5}{5x}$$

Can $\frac{x+5}{5x}$ be simplified?

No, there is no common factor for x + 5 and 5x.

15
Simplify. $\frac{(4x-7)(3x+5)}{3x(4x-7)}$

$\frac{3x+5}{3x}$

16
Simplify. $\frac{(x+7)(2x+5)}{8x(x+7)}$

$\frac{(2x+5)}{8x}$

17
The fraction $\frac{(x-2)(x+7)}{(3x+7)(2-x)}$ can be simplified

because $\frac{(x-2)}{(2-x)} = \underline{}$.

-1

18
$$\frac{(x-2)(x+7)}{(3x+7)(2-x)} = \frac{\overset{-1}{\cancel{(x-2)}}(x+7)}{(3x+7)\underset{1}{\cancel{(2-x)}}} = \frac{-(x+7)}{3x+7}$$

Can $\frac{-(x+7)}{3x+7}$ be simplified?

No

19
Simplify. $\dfrac{(2x^2-5)(3x^2-7x-7)}{(5x^2-2)(-3x^2+7x+7)}$

$\dfrac{-(2x^2-5)}{(5x^2-2)}$

20
Simplify. $\dfrac{(9-5x)(7x+2)}{(3+x)(7x+2)}$

$\dfrac{(9-5x)}{(3+x)}$

21
Simplify. $\dfrac{(x^2-13x+2)(x-7)}{(x+7)(x^2-13x+2)}$

$\dfrac{x-7}{x+7}$

22
Factoring skills learned in Chapter 6 become important in the following problems. Each polynomial needs to be completely factored.

Simplify $\dfrac{(2x+5)(4x-3)}{20x^2-15x}$ by first factoring the denominator.

$\dfrac{(2x+5)(4x-3)}{5x(4x-3)} = \dfrac{(2x+5)}{5x}$

23
Simplify $\dfrac{(5x-1)(x-1)}{x^2-1}$ by first factoring the denominator.

$\dfrac{(5x-1)(x-1)}{(x+1)(x-1)} = \dfrac{(5x-1)}{(x+1)}$

24
Simplify $\dfrac{3x+6}{x^2+7x+10}$ by first factoring both the numerator and denominator.

$\dfrac{3(x+2)}{(x+2)(x+5)} = \dfrac{3}{(x+5)}$

25
Simplify $\dfrac{4x^2-9x+2}{x^2-4}$ by first factoring both the numerator and denominator.

$\dfrac{(4x-1)(x-2)}{(x+2)(x-2)} = \dfrac{(4x-1)}{(x+2)}$

26
Simplify $\dfrac{x^2-9}{x^2-8x-33}$ by first factoring both the numerator and denominator.

$\dfrac{(x+3)(x-3)}{(x-11)(x+3)} = \dfrac{(x-3)}{(x-11)}$

27
Simplify $\dfrac{x^2-49}{x^2-14x+49}$ by first factoring both the numerator and denominator.

$\dfrac{(x+7)(x-7)}{(x-7)^2} = \dfrac{(x+7)}{(x-7)}$

28

$2x^3 - 8x^2 - x + 4$ is factored by grouping.

$$2x^3 - 8x^2 - x + 4$$
$$2x^3 - 8x^2 \quad - x + 4$$
$$2x^2(x - 4) - 1(x - 4)$$
$$(x - 4)(2x^2 - 1)$$

Factor. $4x^3 - 10x^2 - 2x + 5$

$(2x - 5)(2x^2 - 1)$

29

Simplify. $\dfrac{2x^3 - 8x^2 - x + 4}{4x^3 - 10x^2 - 2x + 5}$

$\dfrac{(x - 4)}{(2x - 5)}$

30

$125x^3 + 27$ is factored as the sum of two cubes.

$$125x^3 + 27$$
$$(5x)^3 + (3)^3$$
$$(5x + 3)(25x^2 - 15x + 9)$$

Factor. $64x^3 - 27$

$(4x - 3)(16x^2 + 12x + 9)$

31

Simplify. $\dfrac{64x^3 - 27}{16x^2 + 12x + 9}$

$(4x - 3)$

32

Simplify. $\dfrac{5x - 7}{49 - 25x^2}$

$\dfrac{5x - 7}{(7 + 5x)(7 - 5x)} = \dfrac{-1}{(7 + 5x)}$

33

Simplify. $\dfrac{4x^2 - 9}{8x^3 - 27}$

$\dfrac{2x + 3}{4x^2 + 6x + 9}$

34

Simplify. $\dfrac{x^2 - 7x + 12}{x^2 + 2x - 15}$

$\dfrac{x - 4}{x + 5}$

260   CHAPTER 7

FEEDBACK UNIT 1

This quiz reviews the preceding unit. Answers are at the back of the book.

1. $\dfrac{10x^2y^3}{5xy^2} = $ _____

2. $\dfrac{-20x^5y^7z^2}{10x^3y^3z^3} = $ _____

3. $\dfrac{4x^2yz^5}{10x^4y^5z^3} = $ _____

4. $\dfrac{-5x^3z^5}{10x^3y^3z^3} = $ _____

5. $\dfrac{-72x^8yz^5}{12x^8y^4z^2} = $ _____

6. $\dfrac{42x^3yz^6}{42x^3y^2z^6} = $ _____

7. $\dfrac{x(3x-1)}{2x^2(1-3x)} = $ _____

8. $\dfrac{(7x-3)(x+5)}{7x(x+5)} = $ _____

9. $\dfrac{x^2 - 9x + 18}{x^2 - 36} = $ _____

10. $\dfrac{6x^2 + 13x - 5}{4x^2 + 10x} = $ _____

11. $\dfrac{3x - 9y + xz - 3yz}{x^3 - 27y^3} = $ _____

12. $\dfrac{8x^3 + 1}{4x^2 + 4x + 1} = $ _____

13. $\dfrac{9 - x^2}{x^2 - 10x + 21} = $ _____

14. $\dfrac{5x^2 + 3x - 14}{25x^2 - 49} = $ _____

UNIT 2: MULTIPLYING/DIVIDING POLYNOMIAL FRACTIONS

The following mathematical terms are crucial to an understanding of this unit.

   Cancellation   Complex fraction

1
To multiply two rational numbers, the numerators are multiplied and the denominators are multiplied.

$$\dfrac{5}{7} \cdot \dfrac{2}{3} = \dfrac{5 \cdot 2}{7 \cdot 3} = \dfrac{10}{21} \quad \text{and} \quad \dfrac{3x}{10} \cdot \dfrac{7}{4y} = $$ _____ $\dfrac{21x}{40y}$

POLYNOMIAL FRACTIONS   261

2
To multiply two polynomial fractions, the numerators are multiplied and the denominators are multiplied.

$$\frac{5x}{x-3} \cdot \frac{x+4}{x-7} = \frac{5x(x+4)}{(x-3)(x-7)}$$

$$\frac{2x-5}{3x} \cdot \frac{4}{x+5} = \underline{}$$

$\dfrac{4(2x-5)}{3x(x+5)}$

3
A useful shortcut in multiplication of fractions is **cancellation**. Canceling simplifies fractions before they are multiplied.

$$\frac{4x(x-7)}{7} \cdot \frac{6x}{5(x-7)} = \frac{4x\cancel{(x-7)}}{7} \cdot \frac{6x}{5\cancel{(x-7)}} = \frac{24x^2}{35}$$

What factor was cancelled out of the multiplication shown above?

$(x - 7)$

4
The canceling process is illustrated in the multiplication shown below.

$$\frac{10x^2}{21y} \cdot \frac{14y^5}{15x^3} = \frac{\overset{2}{\cancel{10x^2}}}{\underset{3}{\cancel{21y}}} \cdot \frac{\overset{2y^4}{\cancel{14y^5}}}{\underset{3x}{\cancel{15x^3}}} = \frac{2}{3} \cdot \frac{2y^4}{3x} = \frac{4y^4}{9x}$$

Use canceling to simplify the multiplication below.

$$\frac{9x^5}{10y^2} \cdot \frac{5y}{6x^6} = \underline{}$$

$\dfrac{3}{2y} \cdot \dfrac{1}{2x} = \dfrac{3}{4xy}$

5

$$\frac{18x^2}{11z^2} \cdot \frac{22z^4}{27xy^4} = \underline{}$$

$\dfrac{2x}{1} \cdot \dfrac{2z^2}{3y^4} = \dfrac{4xz^2}{3y^4}$

6

$$\frac{-8x^4y}{15xz^2} \cdot \frac{-25yz^4}{16x^2} = \underline{}$$

$\dfrac{5xy^2z^2}{6}$

7
Use canceling on the problem shown below. The common factor is a binomial.

$$\frac{3x(x-7)}{x+5} \cdot \frac{5}{x-7} = \underline{}$$

$\dfrac{3x}{x+5} \cdot \dfrac{5}{1} = \dfrac{15x}{x+5}$

262 CHAPTER 7

8

When multiplying polynomial fractions, cancellation can simplify the process.

$$\frac{x-3}{5} \cdot \frac{7}{x-3} = \frac{\cancel{(x-3)}}{5} \cdot \frac{7}{\cancel{(x-3)}} = \frac{1}{5} \cdot \frac{7}{1} = \frac{7}{5}$$

Multiply $\frac{x+2}{4} \cdot \frac{3}{x+2}$ by dividing out the common factor $(x+2)$.

$\frac{3}{4}$

9

$\frac{2(x-5)}{x+7} \cdot \frac{3}{(x-5)} = \frac{6}{x+7}$ because the common factor $(x-5)$ can be divided out.

$\frac{x-6}{xy} \cdot \frac{xy^3}{(x+3)(x-6)} = $ _____

$\frac{1}{1} \cdot \frac{y^2}{(x+3)} = \frac{y^2}{x+3}$

10

$\frac{6x}{5(3x-2)} \cdot \frac{15(3x-2)}{8x(x-1)} = $ _____

$\frac{3}{1} \cdot \frac{3}{4(x-1)} = \frac{9}{4(x-1)}$

11

To multiply $\frac{3x-6}{4} \cdot \frac{5}{x^2-4}$, first factor wherever possible.

$$\frac{3x-6}{4} \cdot \frac{5}{x^2-4} = \frac{3\cancel{(x-2)}}{4} \cdot \frac{5}{\cancel{(x-2)}(x+2)} = \frac{15}{4(x+2)}$$

Multiply $\frac{8x+16}{5} \cdot \frac{1}{x^2-4}$ by first factoring where possible.

$\frac{8}{5(x-2)}$

12

Complete the last step in multiplying $\frac{2x+6}{x^2-9x+14} \cdot \frac{x-7}{x+3}$.

$$\frac{2x+6}{x^2-9x+14} \cdot \frac{x-7}{x+3} = \frac{2\cancel{(x+3)}}{\cancel{(x-7)}(x-2)} \cdot \frac{\cancel{x-7}}{\cancel{x+3}} = $$ _____

$\frac{2}{x-2}$

13
Multiply $\dfrac{x^2 - 9x + 20}{x^2 - 16} \cdot \dfrac{3x + 12}{x^2 + 3x - 40}$ by first factoring each polynomial.

$\dfrac{3}{x + 8}$

14
Multiply $\dfrac{25x^2 - 49}{5x^2 + 7x} \cdot \dfrac{4x^2}{7 - 5x}$ by first factoring each polynomial.

$-4x$

15
$\dfrac{x^3 - 27}{x - 5} \cdot \dfrac{3x}{x^2 + 3x + 9} = \underline{\qquad}$

$\dfrac{3x(x - 3)}{x - 5}$

16
$\dfrac{x^2 - 2x}{5x} \cdot \dfrac{x^2 + 2x + 4}{x^3 - 8} = \underline{\qquad}$

$\dfrac{1}{5}$

17
$\dfrac{49x^2 - 36}{7x^2 + 20x + 12} \cdot \dfrac{2x^2 - 5x - 3}{14x^2 - 5x - 6} = \underline{\qquad}$

$\dfrac{x - 3}{x + 2}$

18
To divide 5 by $\dfrac{3}{4}$, 5 is multiplied by the reciprocal of $\dfrac{3}{4}$.

$5 \div \dfrac{3}{4} = 5 \cdot \dfrac{4}{3} = \dfrac{20}{3}$

Divide 6 by $\dfrac{5}{7}$.

$6 \cdot \dfrac{7}{5} = \dfrac{42}{5}$

19
To divide $\dfrac{6x^2}{11y}$ by $\dfrac{9x}{5y^3}$, invert $\dfrac{9x}{5y^3}$ and multiply.

$\dfrac{6x^2}{11y} \div \dfrac{9x}{5y^3} = \dfrac{6x^2}{11y} \cdot \dfrac{5y^3}{9x} = \dfrac{\cancel{6x^2}^{2x}}{11y} \cdot \dfrac{\cancel{5y^3}^{5y^2}}{\cancel{9x}_{3}} = \dfrac{2x}{11} \cdot \dfrac{5y^2}{3} = \dfrac{10xy^2}{33}$

Divide $\dfrac{8x^2y}{9z^4}$ by $\dfrac{10xy^5}{21z^4}$.

$\dfrac{8x^2y}{9z^4} \cdot \dfrac{21z^4}{10xy^5} = \dfrac{28x}{15y^4}$

20

Division of fractions is accomplished by inverting the divisor and multiplying.

$$\frac{-7x^2}{15y^3} \div \frac{5x^5}{9y} = \underline{\hspace{1in}}$$

$$\frac{-7x^2}{15y^3} \cdot \frac{9y}{5x^5} = \frac{-21}{25x^3y^2}$$

21

$$\frac{4x^3y}{-8xy^2} \div \frac{6xy^4}{-12xy^4} = \underline{\hspace{1in}}$$

$$\frac{x^2}{y}$$

22

$$\frac{-7a^2x}{5x} \div \frac{28a^2}{15x^3b} = \underline{\hspace{1in}}$$

$$\frac{-3x^3b}{4}$$

23

$$\frac{-3a^2bc^3}{20xy^3} \div \frac{-12abc^2}{5x^2y^2} = \underline{\hspace{1in}}$$

$$\frac{acx}{16y}$$

24

The division of polynomial fractions is completed by writing each division as a multiplication. Write the problem below as a multiplication.

$$\frac{6x^2+21x}{x-3} \div \frac{2x+7}{x^2-x-6} = \underline{\hspace{0.7in}} \cdot \underline{\hspace{0.7in}}$$

$$\frac{6x^2+21x}{x-3} \cdot \frac{x^2-x-6}{2x+7}$$

25

Complete the division problem shown below by factoring each polynomial and canceling where possible.

$$\frac{6x^2+21x}{x-3} \div \frac{2x+7}{x^2-x-6}$$

$$\frac{6x^2+21x}{x-3} \cdot \frac{x^2-x-6}{2x+7}$$

$$\underline{\hspace{1.5in}}$$

$$\frac{3x(2x+7)}{x-3} \cdot \frac{(x+2)(x-3)}{2x+7}$$

$$\underline{\hspace{1.5in}}$$

$$3x(x+2)$$

POLYNOMIAL FRACTIONS 265

26

Divide $\dfrac{-3x+18}{x^2-8x+12}$ by $\dfrac{3x+12}{x-2}$ by first writing the problem as a multiplication.

$\dfrac{-3(x-6)}{(x-2)(x-6)} \cdot \dfrac{x-2}{3(x+4)}$

$\dfrac{-1}{x+4}$

27

Divide $\dfrac{x^2-3x+2}{x^2+4x-5}$ by $\dfrac{2-x}{x+5}$ by first writing the problem as a multiplication.

$\dfrac{(x-1)(x-2)}{(x+5)(x-1)} \cdot \dfrac{x+5}{2-x}$

-1

28

$\dfrac{-14x+21}{x^2+3x-4} \div \dfrac{2x^2-x-3}{x^2-1} = $ _____

$\dfrac{-7}{x+4}$

29

$\dfrac{x^2+6x+9}{2x^2-13x+20} \div \dfrac{x^2-x-12}{-4x+10} = $ _____

$\dfrac{-2(x+3)}{(x-4)^2}$

30

$\dfrac{x^3-1}{2x^2-x-1} \div \dfrac{3x^2+3x+3}{4x^2-1} = $ _____

$\dfrac{2x-1}{3}$

31

One interpretation of $\dfrac{5}{3}$ is $5 \div 3$.

Similarly, $\dfrac{\frac{2}{3}}{7}$ means $\dfrac{2}{3} \div 7$ and $\dfrac{\frac{6}{5}}{11}$ means _____.

$\dfrac{6}{5} \div 11$

32

$\dfrac{4}{\frac{7}{9}}$ means $4 \div \dfrac{7}{9}$ and $\dfrac{-2}{\frac{5}{4}}$ means _____.

$-2 \div \dfrac{5}{4}$

33

$\dfrac{\frac{8}{3}}{\frac{7}{4}}$ is a **complex fraction** which means $\dfrac{8}{3} \div \dfrac{7}{4}$.

Write the complex fraction that means $\dfrac{1}{3} \div \dfrac{2}{5}$.

$\dfrac{\frac{1}{3}}{\frac{2}{5}}$

34

The complex fraction $\dfrac{\frac{7x}{3-x}}{\frac{12x^3}{x-3}}$ can be written as the division problem: $\dfrac{7x}{3-x} \div \dfrac{12x^3}{x-3}$. Invert the divisor and multiply.

$\dfrac{7x}{3-x} \cdot \dfrac{x-3}{12x^3} = \dfrac{-7}{12x^2}$

35

$\dfrac{\frac{a-b}{2x^2}}{\frac{b-a}{6x^3}} = \dfrac{a-b}{2x^2} \div \dfrac{b-a}{6x^3} = \dfrac{a-b}{2x^2} \cdot \dfrac{6x^3}{b-a} = $ _____

$-3x$

36

$\dfrac{\frac{-4x+12}{5x}}{\frac{24-8x}{10x}} = \dfrac{-4x+12}{5x} \div \dfrac{24-8x}{10x} = $ _____

$\dfrac{-4(x-3)}{5x} \cdot \dfrac{10x}{8(3-x)} = 1$

37

$\dfrac{\frac{x-3}{4x+7}}{x^2-4x+3} = \dfrac{x-3}{4x+7} \div \dfrac{x^2-4x+3}{1} = $ _____

$\dfrac{1}{(4x+7)(x-1)}$

38

$\dfrac{\frac{4x^2-6}{x^2-11x+2}}{\frac{2}{x^2-11x+2}} = \dfrac{4x^2-6}{x^2-11x+2} \div \dfrac{2}{x^2-11x+2} = $ _____

$2x^2 - 3$

POLYNOMIAL FRACTIONS   267

39. $\dfrac{\dfrac{x-6}{3x^3}}{\dfrac{x^2-8x+12}{12x}} = \underline{\qquad}$ $\dfrac{4}{x^2(x-2)}$

40. $\dfrac{\dfrac{9x^2-16}{10x+5}}{\dfrac{3x^2+x-4}{2x^2-x-1}} = \underline{\qquad}$ $\dfrac{3x-4}{5}$

41. $\dfrac{\dfrac{16x^2-8x+1}{x^3+8}}{\dfrac{8x^2-14x+3}{7x+14}} = \underline{\qquad}$ $\dfrac{7(4x-1)}{(2x-3)(x^2-2x+4)}$

FEEDBACK UNIT 2

This quiz reviews the preceding unit. Answers are at the back of the book.

1. $\dfrac{5x^2-35}{2x+3} \cdot \dfrac{x+9}{x^3-7x} = \underline{\qquad}$

2. $\dfrac{x^2+5x-24}{x^2+9x+8} \cdot \dfrac{x^2-1}{5x-15} = \underline{\qquad}$

3. $\dfrac{x^2-3x+9}{4x^2} \cdot \dfrac{x^3+3x^2}{x^3+27} = \underline{\qquad}$

4. $\dfrac{12x^2+23x-9}{2x-3} \cdot \dfrac{4x^2-12x+9}{6x^2-11x+3} = \underline{\qquad}$

5. $\dfrac{x^3-5x^2-x+5}{x^3-1} \cdot \dfrac{x-2}{x^2-4x-5} = \underline{\qquad}$

6. $\dfrac{100x^2-y^2}{5x-10y} \div \dfrac{10x^2+9xy-y^2}{x^2-xy-2y^2} = \underline{\qquad}$

7. $\dfrac{4x^2-9}{x^2+13x+30} \div \dfrac{4x^2+12x+9}{2x^2+9x+9} = \underline{\qquad}$

8. $\dfrac{x^2-12x+32}{16-x^2} \div \dfrac{x^2-x-56}{x^2+11x+28} = \underline{\qquad}$

9. $\dfrac{\dfrac{x-2}{x^2-9x+14}}{\dfrac{3}{x^2-49}} = \underline{\qquad}$

10. $\dfrac{\dfrac{x^2-3x}{x^2-12x+27}}{x+2} = \underline{\qquad}$

Unit 3: Writing Polynomial Fractions with a Common Denominator

The following mathematical term is crucial to an understanding of this unit.

Common denominator

1

The addition of $\frac{5}{x} + \frac{13}{x}$ is relatively easy because the fractions have a **common denominator**, x.

$$\frac{5}{x} + \frac{13}{x} = \frac{5+13}{x} = \frac{18}{x}$$

Similarly, $\frac{7}{x} + \frac{5}{x} = $ _____

$\frac{12}{x}$

2

To add $\frac{6x-1}{x^2-17}$ and $\frac{5x+7}{x^2-17}$ only the numerators need to be added because the denominators are the same.

$$\frac{6x-1}{x^2-17} + \frac{5x+7}{x^2-17} = \frac{(6x-1)+(5x+7)}{x^2-17} = \frac{11x+6}{x^2-17}$$

Add. $\frac{5x-9}{x+7} + \frac{6-3x}{x+7} = $ _____

$\frac{2x-3}{x+7}$

3

Subtraction of fractions is relatively easy when the denominators are the same.

$$\frac{7}{3x} - \frac{2x-5}{3x} = \frac{7-(2x-5)}{3x} = \frac{7-2x+5}{3x} = \frac{12-2x}{3x}$$

Complete the following subtraction problem. Pay close attention to the signs in the numerator.

$$\frac{6x-1}{x-1} - \frac{4x-3}{x-1} = \frac{(6x-1)-(4x-3)}{x-1} = $$ _____

$\frac{6x-1-4x+3}{x-1} = \frac{2x+2}{x-1}$

POLYNOMIAL FRACTIONS

4

$$\frac{x-7}{x^2+5} - \frac{2-7x}{x^2+5} = \underline{\hspace{2cm}}$$

$\dfrac{8x-9}{x^2+5}$

5

$$\frac{x^2-9}{x^2-4x+10} + \frac{x-5}{x^2-4x+10} = \underline{\hspace{2cm}}$$

$\dfrac{x^2+x-14}{x^2-4x+10}$

6

$$\frac{x^2-17x-3}{x^2+3x-7} - \frac{x^2+4x-3}{x^2+3x-7} = \underline{\hspace{2cm}}$$

$\dfrac{-21x}{x^2+3x-7}$

7

The addition of $\dfrac{7}{6x}$ and $\dfrac{5}{4x^2}$ requires more than just adding the numerators because the denominators are

_____ (the same, different).

different

8

The addition of $\dfrac{7}{6x} + \dfrac{5}{4x^2}$ requires that both fractions need to be rewritten so that they have the same denominator. Begin this process by finding a common factor for the denominators.

$2x$

9

The denominators of $\dfrac{7}{6x} + \dfrac{5}{4x^2}$ have 2x as their common factor. To find a common denominator for the fractions, multiply the 2 denominators and divide by their common factor.

$$\frac{6x \cdot 4x^2}{2x} = \underline{\hspace{2cm}}$$

$12x^2$

10

The lowest common denominator for $\dfrac{7}{6x} + \dfrac{5}{4x^2}$ is $12x^2$.

Write $\dfrac{7}{6x}$ with $12x^2$ as its denominator.

$\dfrac{7}{6x} \cdot \dfrac{2x}{2x} = \dfrac{14x}{12x^2}$

11

The lowest common denominator for $\frac{7}{6x} + \frac{5}{4x^2}$ is $12x^2$.

Write $\frac{5}{4x^2}$ with $12x^2$ as its denominator.

$\frac{5}{4x^2} \cdot \frac{3}{3} = \frac{15}{12x^2}$

12

$$\frac{7}{6x} + \frac{5}{4x^2} = \frac{14x}{12x^2} + \frac{15}{12x^2} = \frac{14x + 15}{12x^2}$$

The problem above illustrates the most important steps for adding fractions with different denominators.

1. Find a common denominator. $\frac{6x \cdot 4x^2}{2x} = 12x^2$
2. Write each fraction with the common denominator.
3. Add the numerators while maintaining the common denominator.

Can $\frac{14x + 15}{12x^2}$ be simplified?   No

13

Begin the process for adding $\frac{5}{3x^2} + \frac{7}{6x}$ by finding the common factor of the denominators. $3x$

14

Find the lowest common denominator for $\frac{5}{3x^2} + \frac{7}{6x}$ by dividing the product of the 2 denominators by their common factor.

$\frac{3x^2 \cdot 6x}{3x} = 6x^2$

15

Complete the addition shown below by writing each fraction with $6x^2$ as its denominator and adding the new numerators.

$\frac{5}{3x^2} + \frac{7}{6x} = $ _____

$\frac{10}{6x^2} + \frac{7x}{6x^2} = \frac{10 + 7x}{6x^2}$

16

Find the common denominator for $\frac{4}{5x^2} + \frac{3}{2x^2}$ by first finding a common factor of the original denominators.

$\frac{5x^2 \cdot 2x^2}{x^2} = 10x^2$

17
Find the common denominator for $\frac{x-3}{6x} + \frac{3}{8x^2}$ by first finding a common factor of the original denominators.

$$\frac{6x \cdot 8x^2}{2x} = 24x^2$$

18
Find the common denominator for $\frac{x-6}{3x} + \frac{x+1}{4x^2}$ by first finding a common factor of the original denominators.

$$\frac{3x \cdot 4x^2}{x} = 12x^2$$

19
Finding common denominators follows the same process whether the denominators are monomials or polynomials. Find the lowest common denominator for the problem below by dividing the product of the denominators by their common factor.

$$\frac{5}{(x+9)(x+1)} + \frac{3}{(x-1)(x+1)}$$

$$\frac{(x+9)(x+1) \cdot (x-1)(x+1)}{(x+1)}$$

$$(x+9)(x+1)(x-1)$$

20
To add fractions with polynomial denominators, begin by factoring each denominator.

$$\frac{x+3}{x^2-7x+12} + \frac{x-2}{x^2-3x-4}$$

$$\frac{x+3}{(x-3)(x-4)} + \frac{x-2}{(x-4)(x+1)}$$

Find the common denominator for the problem above. It must contain all the factors found in the denominators.

$$(x-3)(x-4)(x+1)$$

21
To add fractions with polynomial denominators, begin by factoring each denominator.

$$\frac{4}{x^2+8x-9} + \frac{2x-1}{x^2-81}$$

$$\frac{4}{(x-1)(x+9)} + \frac{2x-1}{(x-9)(x+9)}$$

Find the common denominator for the problem above. It must contain all the factors found in the denominators.

$$(x-1)(x+9)(x-9)$$

22

To add fractions with polynomial denominators, begin by factoring each denominator.

$$\frac{x-7}{x^2-10x+24} + \frac{5x}{x^2-2x-8}$$

$$\frac{x-7}{(x-4)(x-6)} + \frac{5x}{(x-4)(x+2)}$$

Find the common denominator for the problem above. It must contain all the factors found in the denominators.

$(x-6)(x-4)(x+2)$

23

$$\frac{5}{x^2+10x+9} + \frac{3}{x^2-1} = \frac{5}{(x+9)(x+1)} + \frac{3}{(x-1)(x+1)}$$

$$\frac{(x+9)(x+1) \cdot (x-1)(x+1)}{(x+1)}$$

Find the common denominator for the problem above by dividing the product of the denominators by their common factor.

$(x+9)(x+1)(x-1)$

24

$$\frac{6}{x^3+8} + \frac{7}{x^2-4}$$

$$\frac{6}{(x+2)(x^2-2x+4)} + \frac{7}{(x+2)(x-2)}$$

Find the common denominator for the problem above by dividing the product of the denominators by their common factor.

$(x+2)(x^2-2x+4)(x-2)$

25

$$\frac{6}{x^2-9x+20} + \frac{7}{x^2-16}$$

Find the common denominator for the problem above by dividing the product of the denominators by their common factor.

$(x-4)(x-5)(x+4)$

26

$$\frac{2}{3x^2+7x-6} + \frac{5}{3x^2+x-2}$$

Find the common denominator for the problem above by dividing the product of the denominators by their common factor.

$(x+3)(3x-2)(x+1)$

27
Find a common denominator for the problem below.

$$\frac{11}{12x-18} + \frac{1}{4x^2-16x+15}$$

$6(2x-3)(2x-5)$

28
Find a common denominator for the problem below.

$$\frac{5x-2}{4x^2-25} - \frac{x-3}{4x^2-20x+25}$$

$(2x+5)(2x-5)^2$

29
Find a common denominator for the problem below.

$$\frac{x^2-x+5}{x^3-27} - \frac{x-4}{x^2-6x+9}$$

$(x-3)^2(x+3x+9)$

FEEDBACK UNIT 3

This quiz reviews the preceding unit. Answers are at the back of the book.

1. $\dfrac{13}{x} + \dfrac{6}{x} = $ _____

2. $\dfrac{7x-2}{2x+3} + \dfrac{2x-9}{2x+3} = $ _____

3. $\dfrac{3x-1}{x^2-17} - \dfrac{x^2-5x}{x^2-17} = $ _____

4. $\dfrac{x^2+4x+15}{(3x+4)(x-7)} + \dfrac{x^2-3x+11}{(3x+4)(x-7)} = $ _____

5. $\dfrac{x^2-7x+12}{(x-5)(x-3)} - \dfrac{x^2-7x+10}{(x-5)(x-3)} = $ _____

For problems 6-10 find the common denominator for each problem.

6. $\dfrac{x-8}{12x^2} + \dfrac{2x-13}{8x^3}$

7. $\dfrac{x+1}{x^2+16x+63} + \dfrac{x^2+2x}{x^2+5x-36}$

8. $\dfrac{x+9}{6x^2+7x+2} - \dfrac{x^2-3}{4x^2-1}$

9. $\dfrac{x^2+7x-4}{25x^2+10x+1} - \dfrac{x-5}{25x^2-1}$

10. $\dfrac{5x}{x^3-64} + \dfrac{2x-1}{x^2+4x+16}$

Unit 4: Adding/Subtracting Polynomial Fractions

1
Begin the process for adding $\dfrac{x-8}{12x^2} + \dfrac{2x-13}{8x^3}$ by finding the common denominator.

$24x^3$

2
$$\dfrac{x-8}{12x^2} + \dfrac{2x-13}{8x^3}$$

$$\dfrac{x-8}{3 \cdot 4x^2} \cdot \dfrac{2x}{2x} + \dfrac{2x-13}{2x \cdot 4x^2} \cdot \dfrac{?}{?}$$

To give the first fraction a denominator of $24x^3$, it is multiplied by $\dfrac{2x}{2x}$. What needs to be multiplied by the second fraction?

$\dfrac{3}{3}$

3
$$\dfrac{x-8}{12x^2} + \dfrac{2x-13}{8x^3}$$

$$\dfrac{x-8}{3 \cdot 4x^2} \cdot \dfrac{2x}{2x} + \dfrac{2x-13}{2x \cdot 4x^2} \cdot \dfrac{3}{3}$$

$$\dfrac{2x^2 - 16x}{24x^3} + \dfrac{6x - 39}{24x^3}$$

_____

Complete the addition.

$\dfrac{2x^2 - 10x - 39}{24x^3}$

4
Begin the process for adding $\dfrac{x+1}{x^2+16x+63} + \dfrac{x^2+2x}{x^2+5x-36}$ by finding the common denominator.

$(x+7)(x+9)(x-4)$

Polynomial Fractions

5

$$\frac{x+1}{(x+7)(x+9)} + \frac{x^2+2x}{(x+9)(x-4)}$$

$$\frac{x+1}{(x+7)(x+9)} \cdot \frac{(x-4)}{(x-4)} + \frac{x^2+2x}{(x+9)(x-4)} \cdot \frac{?}{?}$$

To give the first fraction a denominator of $(x+7)(x+9)(x-4)$, it is multiplied by $\frac{(x-4)}{(x-4)}$. What needs to be multiplied by the second fraction?

$$\frac{(x+7)}{(x+7)}$$

6

$$\frac{x+1}{(x+7)(x+9)} + \frac{x^2+2x}{(x+9)(x-4)}$$

$$\frac{x+1}{(x+7)(x+9)} \cdot \frac{(x-4)}{(x-4)} + \frac{x^2+2x}{(x+9)(x-4)} \cdot \frac{(x+7)}{(x+7)}$$

$$\frac{x^2-3x-4}{(x+7)(x+9)(x-4)} + \frac{x^3+9x^2+14x}{(x+9)(x-4)(x+7)}$$

_____

Complete the addition.

$$\frac{x^3+10x^2+11x-4}{(x+9)(x-4)(x+7)}$$

7

Begin the process for subtracting $\dfrac{x+9}{6x^2+7x+2} - \dfrac{x^2-3}{4x^2-1}$ by writing the fractions with the same denominator.

$$\frac{(x+9)(2x-1)}{(3x+2)(2x+1)(2x-1)} -$$

$$\frac{(x^2-3)(3x+2)}{(3x+2)(2x+1)(2x-1)}$$

8

$$\frac{x+9}{6x^2+7x+2} - \frac{x^2-3}{4x^2-1}$$

$$\frac{(x+9)(2x-1)}{(3x+2)(2x+1)(2x-1)} - \frac{(x^2-3)(3x+2)}{(3x+2)(2x+1)(2x-1)}$$

$$\frac{2x^2+17x-9-(3x^3+2x^2-9x-6)}{(3x+2)(2x+1)(2x-1)}$$

_____

Complete the subtraction.

$$\frac{-3x^3+26x-3}{(3x+2)(2x+1)(2x-1)}$$

276   CHAPTER 7

9

Begin the process for subtracting $\dfrac{x^2+7x-4}{25x^2+10x+1} - \dfrac{x-5}{25x^2-1}$ by writing the fractions with the same denominator.

$$\dfrac{(x^2+7x-4)(5x-1)}{(5x+1)^2(5x-1)} -$$

$$\dfrac{(x-5)(5x+1)}{(5x+1)^2(5x-1)}$$

10

$$\dfrac{x^2+7x-4}{25x^2+10x+1} - \dfrac{x-5}{25x^2-1}$$

$$\dfrac{(x^2+7x-4)(5x-1)}{(5x+1)^2(5x-1)} - \dfrac{(x-5)(5x+1)}{(5x+1)^2(5x-1)}$$

Complete the subtraction.

$$\dfrac{5x^3+34x^2-27x+4-(5x^2-24x-5)}{(5x+1)^2(5x-1)}$$

$$\dfrac{5x^3+29x^2-3x+9}{(5x+1)^2(5x-1)}$$

11

Write the fractions of $\dfrac{5x}{x^3-64} + \dfrac{2x-1}{x^2+4x+16}$ with the same denominator.

$$\dfrac{5x}{(x^2+4x+16)(x-4)} +$$

$$\dfrac{(2x-1)(x-4)}{(x^2+4x+16)(x-4)}$$

12

Complete the following addition.

$$\dfrac{5x}{x^3-64} + \dfrac{2x-1}{x^2+4x+16}$$

$$\dfrac{5x}{(x^2+4x+16)(x-4)} + \dfrac{(2x-1)(x-4)}{(x^2+4x+16)(x-4)}$$

$$\dfrac{2x^2-4x+4}{(x^2+4x+16)(x-4)}$$

13
Write the fractions of $\dfrac{7x-2}{3x^2-19x+20} + \dfrac{2x+3}{3x^2+14x-24}$ with the same denominator.

$\dfrac{(7x-2)(x+6)}{(x-5)(3x-4)(x+6)} +$

$\dfrac{(2x+3)(x-5)}{(x-5)(3x-4)(x+6)}$

14
Complete the following addition.

$$\dfrac{7x-2}{3x^2-19x+20} + \dfrac{2x+3}{3x^2+14x-24}$$

$$\dfrac{(7x-2)(x+6)}{(x-5)(3x-4)(x+6)} + \dfrac{(2x+3)(x-5)}{(x-5)(3x-4)(x+6)}$$

_____

$\dfrac{9x^2+33x-27}{(x-5)(3x-4)(x+6)}$

15
Write the fractions of $\dfrac{3x-1}{2x^2+11x-6} - \dfrac{x+3}{x^2+4x-12}$ with the same denominator.

$\dfrac{(3x-1)(x-2)}{(2x-1)(x+6)(x-2)} -$

$\dfrac{(x+3)(2x-1)}{(2x-1)(x+6)(x-2)}$

16
Complete the following subtraction.

$$\dfrac{3x-1}{2x^2+11x-6} - \dfrac{x+3}{x^2+4x-12}$$

$$\dfrac{(3x-1)(x-2)}{(2x-1)(x+6)(x-2)} - \dfrac{(x+3)(2x-1)}{(2x-1)(x+6)(x-2)}$$

_____

$\dfrac{x^2-12x+5}{(2x-1)(x+6)(x-2)}$

17
Complete all steps for the following problem. Leave the denominator of the final result in factored form.

$$\dfrac{6x-5}{x^3+8} + \dfrac{7}{x^2-4} = \underline{\hspace{2cm}}$$

$\dfrac{13x^2-31x+38}{(x^2-2x+4)(x+2)(x-2)}$

18
Complete all steps for the following problem. Leave the denominator of the final result in factored form.

$$\frac{x+5}{3x^2+17x-6} - \frac{x-3}{x^2+15x+54} = \underline{\hspace{2cm}}$$

$\dfrac{-2x^2+24x+42}{(3x-1)(x+6)(x+9)}$

19
Any complex fraction may be written as a division problem. Notice the need for inserting parentheses.

$$\frac{\dfrac{5}{x+3} - \dfrac{4}{x+2}}{\dfrac{x^2-6x+8}{x^2-9}} = \left(\frac{5}{x+3} - \frac{4}{x+2}\right) \div \frac{x^2-6x+8}{x^2-9}$$

The next step for this problem is _____ (subtraction, division).

subtraction

20
Perform the next step for the problem below.

$$\left(\frac{5}{x+3} - \frac{4}{x+2}\right) \div \frac{x^2-6x+8}{x^2-9} = \underline{\hspace{2cm}}$$

$\dfrac{x-2}{(x+3)(x+2)} \div \dfrac{x^2-6x+8}{x^2-9}$

21
Finish the problem shown below.

$$\frac{x-2}{(x+3)(x+2)} \div \frac{x^2-6x+8}{x^2-9} = \underline{\hspace{2cm}}$$

$\dfrac{x-2}{(x+3)(x+2)} \cdot \dfrac{(x+3)(x-3)}{(x-2)(x-4)}$

$\dfrac{(x-3)}{(x+2)(x-4)}$

22
Write $\dfrac{x^2+2xy+y^2}{\dfrac{x}{y} - \dfrac{y}{x}}$ as a division problem.

$x^2+2xy+y^2 \div \left(\dfrac{x}{y} - \dfrac{y}{x}\right)$

23

$$\frac{x^2+2xy+y^2}{\dfrac{x}{y} - \dfrac{y}{x}} = x^2+2xy+y^2 \div \left(\frac{x}{y} - \frac{y}{x}\right)$$

Do the subtraction indicated for the problem above.

$x^2+2xy+y^2 \div \dfrac{x^2-y^2}{xy}$

24
Finish the problem shown below.

$$x^2 + 2xy + y^2 \div \frac{x^2 - y^2}{xy} = \underline{}$$

$$\frac{(x+y)^2}{1} \cdot \frac{xy}{(x+y)(x-y)}$$

$$\frac{xy(x+y)}{(x-y)}$$

25
Complete the following problem.

$$\frac{\frac{4}{3x} - \frac{1}{x-2}}{\frac{x^2 - 6x - 16}{x^2 - 4}} = \left(\frac{4}{3x} - \frac{1}{x-2}\right) \div \frac{x^2 - 6x - 16}{x^2 - 4} = \underline{}$$

$$\frac{x-8}{3x(x-2)} \cdot \frac{(x+2)(x-2)}{(x+2)(x-8)}$$

$$\frac{1}{3x}$$

26

$$\frac{\frac{3}{5x^2} - \frac{1}{10x}}{\frac{x^2 - 2x - 24}{2x^2 + 8x}} = \underline{}$$

$$\frac{6-x}{10x^2} \cdot \frac{2x(x+4)}{(x+4)(x-6)} = \frac{-1}{5x}$$

27

$$\frac{\frac{x}{4y} - \frac{y}{x}}{\frac{1}{y} + \frac{2}{x}} = \underline{}$$

$$\frac{x^2 - 4y^2}{4xy} \cdot \frac{xy}{x + 2y} = \frac{x - 2y}{4}$$

Feedback Unit 4

This quiz reviews the preceding unit. Answers are at the back of the book.

1. $\dfrac{5x}{16y^2} + \dfrac{3x}{10y^3} = $ _____

2. $\dfrac{9}{8xy} - \dfrac{7}{2y} = $ _____

3. $\dfrac{x+2}{x^2-8x+7} + \dfrac{x-2}{x^2-4x-21} = $ _____

4. $\dfrac{4x}{x^2+11x+24} - \dfrac{5}{x^2-64} = $ _____

5. $\dfrac{9}{x^3-125} + \dfrac{4x}{x^2+5x+25} = $ _____

6. $\dfrac{3x-2}{2x^2+3x-9} - \dfrac{x+4}{x^2-9} = $ _____

7. $\dfrac{\dfrac{x}{2}+\dfrac{4}{3}}{\dfrac{3x^2}{4}-\dfrac{16}{3}} = $ _____

8. $\dfrac{\dfrac{x^2+5x+6}{x+3}}{\dfrac{x}{2}-\dfrac{2}{x}} = $ _____

Unit 5: Applications

In this Applications Section, the format of the text has been altered. Answers for the problems appear beneath them rather than in the right-hand column. Your studying emphasis should be on learning the best procedures to follow with word problems.

1
The word problems of this unit all depend upon the following idea:

If a person can do a job in 8 days then:

The person can do $\frac{1}{8}$ of the job in 1 day.

The person can do $\frac{2}{8}$ of the job in 2 days.

The person can do $\frac{3}{8}$ of the job in 3 days.

The person can do $\frac{4}{8}$ of the job in 4 days.

The person can do _____ of the job in 5 days.

Answer: $\frac{5}{8}$

2
If a painter can paint a house in 5 days, what fraction of the job can be completed in 3 days?

Answer: $\frac{3}{5}$

3

If a painter can paint a house in 5 days, what fraction of the job can be completed in n days?

Answer: $\dfrac{n}{5}$

4

One painter can paint a house in 5 days and another can do the same job in 4 days. This means that in one day the 1st painter would finish $\dfrac{1}{5}$ of the job and the 2nd would finish $\dfrac{1}{4}$ of the job. Working together the painters would finish $\dfrac{1}{5} + \dfrac{1}{4}$ ($\dfrac{9}{20}$) of the job. If the painters worked together for 2 days, how much of the job would they finish?

Answer:

$\dfrac{2}{5} + \dfrac{2}{4}$ ($\dfrac{9}{10}$) of the house painting will be finished in 2 days when the painters work together.

5

The 2 painters of frame 4 have almost finished the house painting in 2 days; they have done $\dfrac{9}{10}$ of it. If n represents the number of days needed to completely finish the painting, then $\dfrac{n}{5}$ represents the fraction of the job done by the 1st painter and _____ represents the fraction of the job done by the 2nd painter.

Answer: $\dfrac{n}{4}$

6

The 2 painters of frame 4 when working together do $\frac{1}{5} + \frac{1}{4}$ ($\frac{9}{20}$) of the job in 1 day and $\frac{2}{5} + \frac{2}{4}$ ($\frac{9}{10}$) of the job in 2 days. What fraction of the job will they finish in n days?

Answer: $\frac{n}{5} + \frac{n}{4}$

7

The 2 painters of frame 4 when working together do $\frac{n}{5} + \frac{n}{4}$ of the job in n days. If n represents the number of days to complete the job, then this gives the equation $\frac{n}{5} + \frac{n}{4} = 1$ because they have finished 1 full job. Solve the equation by first multiplying each term by 20, the common denominator.

Answer:

$\frac{n}{5} + \frac{n}{4} = 1$ is equivalent to $20 \cdot \frac{n}{5} + 20 \cdot \frac{n}{4} = 20 \cdot 1$ or $4n + 5n = 20$.

$9n = 20$ has $\frac{20}{9}$ as its solution, so $2\frac{2}{9}$ days is the time needed for the painters, working together, to paint the 1 house.

8

Two pipes are attached to a water tank. One pipe can fill the tank in 4 hours. The other pipe can empty the tank in 6 hours. Give 2 fractions: one for the amount of the tank that can be filled in one hour and the other for the amount of the tank that is emptied in one hour.

Answer:

The Fill (F) pipe will put $\frac{1}{4}$ of a tank into it each hour.

The Empty (E) pipe will remove $\frac{1}{6}$ of a tank from it each hour.

9

The pipes of frame 8 are working against each other. The fill pipe is putting $\frac{1}{4}$ of a tank into it each hour, while the empty pipe is removing $\frac{1}{6}$ of a tank from it each hour. Write a subtraction showing the amount of the tank that is full after one hour.

Answer:
$\frac{1}{4} - \frac{1}{6} = \frac{3}{12} - \frac{2}{12} = \frac{1}{12}$

Working against each other the pipes have $\frac{1}{12}$ of the tank filled after 1 hour.

10

Two pipes are attached to a water tank. One pipe can fill the tank in 4 hours. The other pipe can empty the tank in 6 hours. If the Fill pipe is open for n hours, it will fill $\frac{n}{4}$ of the tank. If the Empty pipe is open for n hours, it will empty _____ of the tank.

Answer: $\frac{n}{6}$

11

Two pipes are attached to a water tank. One pipe can fill the tank in 4 hours. The other pipe can empty the tank in 6 hours. If n represents the number of hours needed to fill the tank when both pipes are open, what equation may be written?

Answer: $\frac{n}{4} - \frac{n}{6} = 1$

The expression $\frac{n}{4} - \frac{n}{6}$ is equal to 1 because there is 1 full tank after n hours.

12

Two pipes are attached to a water tank. One pipe can fill the tank in 4 hours. The other pipe can empty the tank in 6 hours. Solve the equation $\frac{n}{4} - \frac{n}{6} = 1$ to find the number of hours needed to fill the tank when both pipes are open.

Answer:

$\frac{n}{4} - \frac{n}{6} = 1$ is equivalent to $12 \cdot \frac{n}{4} - 12 \cdot \frac{n}{6} = 12 \cdot 1$ or $3n - 2n = 12$.

The solution is 12; it takes 12 hours for the tank to be full.

FEEDBACK UNIT 5 FOR APPLICATIONS

This quiz reviews the preceding unit. Answers are at the back of the book.

Show all steps in the 3-step process for solving each of the following.

1. If a cleaner can iron a shirt in 7 minutes, what fraction of the shirt can be ironed in 1 minute?

2. If a cleaner can iron a shirt in 7 minutes, what fraction of the shirt can be ironed in n minutes?

3. Bricklayer #1 can do a wall in 10 days. Bricklayer #2 could do the same wall in 6 days. Write 2 fractions: One for the amount of the wall that Bricklayer #1 can do in 1 day. The other for the amount of the wall that Bricklayer #2 can do in 1 day.

4. Bricklayer #1 can do a wall in 10 days. Bricklayer #2 could do the same wall in 6 days. Write an addition of 2 fractions that indicates the amount of the wall that can be completed in 1 day if the bricklayers work together.

5. Bricklayer #1 can do a wall in 10 days. Bricklayer #2 could do the same wall in 6 days. Write 2 fractions: One for the amount of the wall that Bricklayer #1 can do in n days. The other for the amount of the wall that Bricklayer #2 can do in n days.

6. Bricklayer #1 can do a wall in 10 days. Bricklayer #2 could do the same wall in 6 days. Write an addition of 2 fractions that indicates the amount of the wall that can be completed in n days if the bricklayers work together.

7. Bricklayer #1 can do a wall in 10 days. Bricklayer #2 could do the same wall in 6 days. How many days will it take to complete the wall by working together?

Summary for Chapter 7

The following mathematical terms are crucial to an understanding of this chapter.

> Reduce (fraction)
> Opposite (of a polynomial)
> Cancellation
> Common denominator
> Simplify (polynomial fraction)
> Complex fraction

Problems with polynomial fractions are completed by using the same concepts as fractional problems with rational numbers.

Polynomial fractions are simplified by dividing out common factors:

$$\frac{x^2 - 9x - 10}{x^2 - 1} = \frac{\cancel{(x+1)}(x-10)}{\cancel{(x+1)}(x-1)} = \frac{(x-10)}{(x-1)}$$

Polynomial fractions are multiplied by cancelling where possible and then separately multiplying numerators and denominators:

$$\frac{x^3 - 5x^2}{x^2 - 7x + 12} \cdot \frac{x-4}{x-5} = \frac{x^2(x-5)}{(x-3)(x-4)} \cdot \frac{x-4}{x-5} = \frac{x^2}{x-3}$$

Polynomial fractions are added/subtracted with a common denominator:

$$\frac{x+3}{x^2-x} + \frac{x-8}{x^2-5x+4} = \frac{x+3}{x(x-1)} + \frac{x-8}{(x-1)(x-4)} = \frac{(x+3)(x-4) + x(x-8)}{x(x-1)(x-4)} = \frac{2x^2 - 9x - 12}{x(x-1)(x-4)}$$

Complex fractions are division problems:

$$\frac{\frac{2}{3} - \frac{5}{4x}}{\frac{x-2}{8x^2}} = \left(\frac{2}{3} - \frac{5}{4x}\right) \div \frac{x-2}{8x^2} = \frac{8x - 15}{12x} \cdot \frac{8x^2}{x-2} = \frac{2x(8x-15)}{3(x-2)}$$

Chapter 7 Mastery Test

The following questions test the objectives of Chapter 7. Answers are at the back of the book. The number in parentheses which follows each problem indicates the unit in which it can be learned.

1. $\dfrac{12xy^3z^5}{18x^3yz^8} = $ _____ (1)

2. $\dfrac{-6x^4yz^3}{-4xy^2z^2} = $ _____ (1)

3. $\dfrac{6 - 18x}{3x - 1} = $ _____ (1)

4. $\dfrac{5x - x^2}{x^2 - 10x + 25} = $ _____ (1)

5. $\dfrac{x^3 - 64}{x^2 - 13x + 36} = $ _____ (1)

6. $\dfrac{25 - x^2}{2x^2 - 9x - 5} = $ _____ (1)

7. $\dfrac{7x - 5}{x^2 - 5x - 6} \cdot \dfrac{x^2 - 1}{7x - 5} = $ _____ (2)

8. $\dfrac{4y^2 - x^2}{x^2 + xy - 2y^2} \cdot \dfrac{2x^2 - 5xy + 3y^2}{2x^2 - 7xy + 6y^2} = $ _____ (2)

9. $\dfrac{9x - 2}{4x^2 + 4x + 1} \div \dfrac{81x^2 - 4}{6x^2 - 7x - 5} = $ _____ (2)

10. $\dfrac{\frac{5x - 3}{xy}}{\frac{7}{x}} = $ _____ (2)

11. $\dfrac{\frac{x - 5}{x^2 - 8x + 7}}{\frac{x + 2}{6x - 42}} = $ _____ (2)

12. $\dfrac{x^2 + 3x + 13}{x^3 - 7} + \dfrac{2x^2 - 11x - 1}{x^3 - 7} = $ _____ (3)

13. $\dfrac{x - 4}{x^2 - 10x + 21} + \dfrac{2x}{x^2 - 49} = $ _____ (4)

14. $\dfrac{x - 8}{x^3 + 27} + \dfrac{5}{x^2 - 9} = $ _____ (4)

15. $\dfrac{6x}{3x^2 + 5x + 2} - \dfrac{x + 7}{9x^2 - 4} = $ _____ (4)

16. $\dfrac{x + 2}{4x^2 + 12x + 9} - \dfrac{x - 5}{2x^2 - x - 6} = $ _____ (4)

17. $\dfrac{\frac{x + 7}{2} - \frac{4}{x}}{\frac{x^2 - 64}{6x^2}} = $ _____ (4)

18. $\dfrac{\frac{4x^2 + 5xy - 6y^2}{x + 2y}}{\frac{4}{y} - \frac{3}{x}} = $ _____ (4)

19. Two pipes can be used to fill a large tank. Pipe #1 would fill the tank in 10 hours. Pipe #2 would fill the tank in 12 hours. Write 2 fractions: One for the amount of the tank filled by Pipe #1 in 1 hour, the other for the amount of the tank filled by Pipe #1 in n hours. (5)

20. Two pipes can be used to fill a large tank. Pipe #1 would fill the tank in 10 hours. Pipe #2 would fill the tank in 12 hours. Write 2 addition expressions: One for the amount of the tank filled by both pipes in 1 hour, the other for the amount of the tank filled by both pipes in n hours. (5)

Chapter 8 Objectives

The following problems illustrate the objectives of this chapter. At this time you are not expected to know how to do these problems. However, if all these problems are thoroughly understood, proceed directly to the Chapter Mastery Test. The number in parentheses which follows each problem indicates the unit in which it can be learned.

Solve each of the following.

1. $4x - 5 = 27$ (1)
2. $8x - 3\sqrt{3} = 5\sqrt{3}$ (1)
3. $\frac{2}{7}x = \frac{3}{4}$ (1)
4. $16 = -5x$ (1)
5. $\sqrt{3} \cdot x = 8$ (1)
6. $\sqrt{7} \cdot x + \sqrt{5} = 11$ (1)
7. $(3 - i)x + (2 + 2i) = (4 + i)$ (1)
8. $7 - (2x + 1) = 3x - (4x - 5)$ (2)
9. $\frac{3}{4}x - \frac{1}{6} = \frac{3}{10}$ (2)
10. $\frac{4}{x} + \frac{2}{3} = \frac{-1}{2}$ (2)
11. $\sqrt{x + 3} = 7$ (2)
12. $x^2 + 9x - 22 = 0$ (3)
13. $x^2 - 81 = 0$ (3)
14. $x^2 - 21x - 72 = 0$ (3)
15. $x^2 - 16x + 28 = 0$ (3)
16. $9x^2 - 64 = 0$ (3)
17. $x^2 - 11x + \frac{121}{4} = 0$ (3)
18. $\frac{5}{9} - \frac{1}{x} = \frac{x-1}{9}$ (3)
19. $\sqrt{x + 4} - 1 = \sqrt{x - 1}$ (3)
20. $x^2 + 9x + 2 = 0$ (5)
21. $x^2 - 10x - 4 = 0$ (5)
22. $x^2 + 4x - 14 = 0$ (5)
23. $x^2 + 4x + 10 = 0$ (5)
24. $x^3 - 64 = 0$ (5)
25. (-2, ____) is a solution for $3x - y = 7$. (6)
26. (3, ____) is a solution for $2x + 3y = -4$. (6)
27. (7, ____) is a solution for $y = \frac{3x}{7} + 5$. (6)
28. (-3, ____) is a solution for $y = x^2 + 5x - 4$. (6)
29. (2, ____) is a solution for $x^2 + 9y^2 = 9$. (6)
30. (-6, ____) is a solution for $x^2 + y^2 = 100$. (6)

CHAPTER 8
EQUATION SOLVING

UNIT 1: EQUIVALENT LINEAR EQUATIONS

The following mathematical terms are crucial to an understanding of this unit.

- Statement
- Complex numbers
- Solve an equation
- Quadratic equation
- Root
- Additive Property for Equivalent Equations
- Multiplicative Property for Equivalent Equations
- Linear equation
- Replacement set
- Truth set
- Solution
- Equivalent equations
- Reciprocal
- Opposite

1
 $7 + 3 = 21$ is a **statement** which is false because $7 + 3$ does not equal 21. Is $4 + 5 = 9$ a false statement?

No, it is a true statement.

2

$x + 13 = 8$ is a **linear equation**. It is neither true nor false. It becomes a statement when x is replaced by a number. Is $5x = 10$ true?

No, it is not a statement.

3

Complex numbers are of the form $c + di$ where c and d are real numbers and i is the imaginary number such that $i^2 = -1$. $7 = 7 + 0i$ so 7 is a complex number. Is -13 a complex number?

Yes, $-13 = -13 + 0i$

4

Is every real number a complex number?

Yes

5

Every equation has a set of elements, implied or explicitly stated, that is the **replacement set** for its variable(s). Throughout the remainder of this unit the replacement set will be the set of complex numbers. Is $\sqrt{2} - \frac{3}{4}i$ an acceptable replacement for a variable in this chapter?

Yes, $\sqrt{2} - \frac{3}{4}i$ is a complex number

6

To "**solve an equation**" means to find all the complex numbers that can be used as replacements for x to make the equation into a true statement. List every complex number that solves the linear equation $x + 3 = 10$.

7

7

The one and only complex number that will make $x + 3 = 10$ a true statement is 7. All other complex numbers would make $x + 3 = 10$ a _____ statement.

false

8

Solving $x + 3 = 10$ separates the set of complex numbers into two subsets. One is the **truth set** {7}. The other is the set of all complex numbers except 7; every element of this second set would make $x + 3 = 10$ a _____ statement.

false

9
The equation $x^2 = 9$ is a **quadratic equation** and has two elements in its truth set, {3, -3}. Each element of the truth set is a **solution** or **root** of its equation.
Is -4 a solution of $x^2 = 16$?

Yes

10
The solution of $5x = 10$ is relatively easy to find because it states: 5 times some number is 10.
What is the solution of $5x = 10$?

2

11
Equations like $x = 9$, $x = \frac{3}{4}$, and $x = \sqrt{3}$ have the obvious solutions 9, $\frac{3}{4}$, and $\sqrt{3}$. What is the obvious solution of $x = -8$?

-8

12
Two linear equations are **equivalent** when they have identical truth sets. Are $5x = 10$ and $x = 2$ equivalent?

Yes, both have {2} as their truth set.

13
$4x = 12$ and $x = 3$ are equivalent equations; both have {3} as a truth set. Are $5x = 30$ and $x + 7 = 13$ equivalent equations?

Yes

14
When two equations are equivalent, any solution of one equation _____ (is, isn't) a solution of the other.

is

15
The primary method for solving linear equations is to generate an equivalent equation which has an obvious truth set. $7x - 8 = 4x + 25$ and $x = 11$ are equivalent equations. What is the truth set of $7x - 8 = 4x + 25$?

{11}

16

There are two basic methods for generating equivalent linear equations.

    a.   Any number can be added to/subtracted from both sides of an equation.

    b.   Any number except 0 can be multiplied by/divided into both sides of an equation.

If both sides of an equation are squared, will the newly generated equation always be equivalent?

    No, $x = 3$ is not equivalent to $x^2 = 9$

17

To generate an equivalent linear equation, there are two methods that can be used without question.

Additive Property for Equivalent Equations:
    Any number may be added to/subtracted from both sides of an equation.

Multiplicative Property for Equivalent Equations:
    Any number except 0 may be multiplied by/divided into both sides of an equation.

Can $\sqrt{5} + 7$ be added to each side of an equation with assurance that an equivalent equation will be generated?

    Yes

18

Add -7 to both sides of $12x + 7 = 34$. What equivalent equation is generated?

    $12x = 27$

19

Multiply both sides of $12x = 27$ by $\frac{1}{12}$. What equivalent equation is generated?

    $x = \frac{9}{4}$

20

    $12x + 7 = 34$, $12x = 27$, and $x = \frac{9}{4}$ are equivalent equations. What is the solution of $12x + 7 = 34$?

    $\frac{9}{4}$

EQUATION SOLVING   293

21
The first step in solving 4 − 3x = 14 is to add -4, the **opposite** of 4, to both sides of the equation.

$$\begin{aligned} 4 - 3x &= 14 \\ -4 & -4 \\ \hline -3x &= 10 \end{aligned}$$

Are 4 − 3x = 14 and -3x = 10 equivalent?   Yes

22
To solve -3x = 10, both sides of the equation are multiplied by $\frac{-1}{3}$ which is the **reciprocal** of -3.

$$-3x = 10$$
$$\frac{-1}{3} \cdot -3x = \frac{-1}{3} \cdot 10$$
$$x = \frac{-10}{3}$$

Are 4 − 3x = 14, -3x = 10, and x = $\frac{-10}{3}$ equivalent?   Yes

23
Find the solution of 7x + 4 = 9 by using the opposite of 4 and the reciprocal of 7. $\frac{5}{7}$

24
Find the solution of 4x + 7 = -7 by using the opposite of 7 and the reciprocal of 4. $\frac{-7}{2}$

25
Find the solution of $\frac{2}{3}$x = 5 by using the reciprocal of $\frac{2}{3}$. $\frac{15}{2}$

26
Linear equations of the form ax + b = c where a, b, and c are rational numbers, a ≠ 0, always have exactly one solution and it will be a rational number. Find all the solutions of $\frac{2}{3}$x + 1 = 8. $\frac{21}{2}$

27

When the coefficients and/or constants of a linear equation are not rational numbers, then the solution often is not rational. To solve $3x + \sqrt{5} = 7$, the following steps are used.

$$
\begin{aligned}
3x + \sqrt{5} &= 7 \\
-\sqrt{5} &\quad -\sqrt{5} \\
3x &= 7 - \sqrt{5} \\
\tfrac{1}{3} \cdot 3x &= \tfrac{1}{3} \cdot (7 - \sqrt{5}) \\
x &= \tfrac{7 - \sqrt{5}}{3}
\end{aligned}
$$

Are $3x + \sqrt{5} = 7$ and $x = \tfrac{7-\sqrt{5}}{3}$ equivalent equations?   Yes

28

Find the solution of $4x - \sqrt{7} = 6$ by adding opposites and multiplying by reciprocals. $\tfrac{6+\sqrt{7}}{4}$

29

The reciprocal of $\sqrt{5}$ is $\tfrac{1}{\sqrt{5}}$. In solving $\sqrt{5} \cdot x - \sqrt{3} = 4$, notice the way the reciprocal of $\sqrt{5}$ is used.

$$
\begin{aligned}
\sqrt{5} \cdot x - \sqrt{3} &= 4 \\
+\sqrt{3} &\quad +\sqrt{3} \\
\sqrt{5} \cdot x &= 4 + \sqrt{3} \\
\tfrac{1}{\sqrt{5}} \cdot \sqrt{5} \cdot x &= \tfrac{1}{\sqrt{5}} \cdot (4 + \sqrt{3}) \\
x &= \tfrac{4+\sqrt{3}}{\sqrt{5}}
\end{aligned}
$$

Simplify the solution by rationalizing the denominator of $\tfrac{4+\sqrt{3}}{\sqrt{5}}$. $\tfrac{4+\sqrt{3}}{\sqrt{5}} \cdot \tfrac{\sqrt{5}}{\sqrt{5}} = \tfrac{4\sqrt{5}+\sqrt{15}}{5}$

30

Find the solution of $\sqrt{7} \cdot x + 12 = \sqrt{2}$ by adding opposites and multiplying by reciprocals. $\tfrac{\sqrt{14} - 12\sqrt{7}}{7}$

31
When the constants and/or coefficients of a linear equation are complex numbers, the solution process still depends upon adding opposites and multiplying by reciprocals.

$$(3 - 2i)x + (4 - 3i) = (8 - 5i)$$
$$\underline{(-4 + 3i) (-4 + 3i)}$$
$$(3 - 2i)x = (4 - 2i)$$
$$(3 - 2i)x \cdot \frac{1}{3 - 2i} = (4 - 2i) \cdot \frac{1}{3 - 2i}$$
$$x = \frac{4 - 2i}{3 - 2i}$$

Simplify the solution by multiplying the numerator and denominator by the conjugate of $3 - 2i$.
[Note: Recall that $i^2 = -1$.]

$\dfrac{4 - 2i}{3 - 2i} \cdot \dfrac{3 + 2i}{3 + 2i} = \dfrac{16 + 2i}{13}$

32
Find the solution of $(4 - i)x + (-2 - 3i) = (5 + 2i)$ by adding opposites and multiplying by reciprocals.

$\dfrac{7 + 5i}{4 - i} \cdot \dfrac{4 + i}{4 + i} = \dfrac{23 + 27i}{17}$

33
Linear equations of the form $ax + b = c$ are solved by adding opposites and multiplying by reciprocals. These processes _____ (are, are not) guaranteed to generate equivalent equations.

are

FEEDBACK UNIT 1

This quiz reviews the preceding unit. Answers are at the back of the book.

Find a solution for each of the following linear equations.

1. $6x - 5 = 12$
2. $8 - 3x = 15$
3. $10 = 4 + 5x$
4. $11 = 15 - 2x$

5. $\frac{4}{5}x = \frac{2}{3}$
6. $\frac{-5}{7}x = 3$
7. $8 = -3x$
8. $\frac{x}{4} = \frac{-3}{5}$

9. $\sqrt{5} \cdot x = 4$
10. $\sqrt{2} \cdot x + \sqrt{7} = 8$
11. $\sqrt{3} \cdot x - 11 = 2\sqrt{6}$
12. $(2 + 5i)x + (4 - 3i) = (-1 + 3i)$

Unit 2: Generating Linear Equations

The following mathematical terms are crucial to an understanding of this unit.

Solve Identity
Least common multiple (LCM)

1
To **solve** (find all solutions of) $4x - 3 = 9x - 1$, the following steps are used.

$$
\begin{array}{rcl}
4x - 3 &=& 9x - 1 \\
-9x && -9x \\
\hline
-5x - 3 &=& -1 \\
+3 && +3 \\
\hline
-5x &=& 2 \\
x &=& \dfrac{-2}{5}
\end{array}
$$

Since $4x - 3 = 9x - 1$ and $x = \dfrac{-2}{5}$ are equivalent equations, what is the solution of $4x - 3 = 9x - 1$?

$\dfrac{-2}{5}$

2
Solve $5x + 6 = 2 - 3x$ by adding opposites and multiplying by reciprocals.

$8x = -4$

$x = \dfrac{-1}{2}$

3
Solve $4x - 7 = 2x + 3$ by adding opposites and multiplying by reciprocals.

$2x = 10$

5

4
Solve. $5x - 2 = 3x + 7$

$\dfrac{9}{2}$

5

Solve. $5x - 4 = 12x - 1$

$\frac{-3}{7}$

6

The first step in solving $4 - (3x - 2) = 2x + 3(x - 5)$ is to simplify, separately, the two sides of the equation.

$$4 - (3x - 2) = 2x + 3(x - 5)$$
$$4 - 3x + 2 = 2x + 3x - 15$$
$$-3x + 6 = 5x - 15$$

Complete the solution of $4 - (3x - 2) = 2x + 3(x - 5)$.

$-8x = -21$

$x = \frac{21}{8}$

7

Solve $2(5x - 2) - 7x = x - (6x - 5)$, by first simplifying each side of the equation.

$3x - 4 = -5x + 5$

$x = \frac{9}{8}$

8

Solve. $2(3x - 1) + 5x = 4 - (5x + 1)$

$11x - 2 = 3 - 5x$

$x = \frac{5}{16}$

9

Solve. $6(x + 2) - 7 = 3x + 5$

$3x = 0$

$x = 0$

10

In Unit 1 it was stated that every linear equation of the form $ax + b = c$ where $a \neq 0$ has exactly one solution. The restriction that the coefficient of x must not be zero becomes very important in the example shown below.

$$3x - 5 + x = 4(1 + x) - 9$$
$$4x - 5 = 4 + 4x - 9$$
$$4x - 5 = 4x - 5$$

Is there only one solution for $4x - 5 = 4x - 5$?

No

11

Equations such as 4x – 5 = 4x – 5 become true for all numbers in their replacement set because the equation is an **identity**. If the set of complex numbers is the replacement set, what is/are the solutions of
2x + 7 = 4x – 2(x – 3) + 1?

Every complex number

12

Solve. 4(x + 2) = 6(1 + x) – 2(x – 1)

Every complex number

13

It is true that every linear equation of the form ax + b = c where a ≠ 0 has exactly one solution. Again, the restriction that the coefficient of x must not be zero becomes important in the example shown below.

$$6x - 3 - 5x = 7(x + 5) - 6x$$
$$x - 3 = 7x + 35 - 6x$$
$$x - 3 = x + 35$$

Is there any solution for x – 3 = x + 35?

No

14

Equations such as x – 3 = x + 35 have no solutions. There is no complex number that can replace x and make a true statement. If the set of complex numbers is the replacement set, what is/are the solutions of 4x + 2 = 4x + 7?

There is no solution.

15

Solve. 4x – 8 = 2(2x + 4)

There is no solution.

16

Solve. 5(x – 1) – 3 = 3x – (x + 5)

1

17

Solve. 2(x – 1) + x = x – 2 + 2x

Every complex number

18

Solve. 3x – (x – 2) = -7(2 – 2x)

$\dfrac{4}{3}$

19

To solve $\frac{2}{9}x - \frac{5}{6} = \frac{1}{2}$ the process can begin with adding $\frac{5}{6}$, but it is easier to first multiply each term by 18, which is the **least common multiple (LCM)** of the denominators, 9, 6, and 2.

$$\frac{2}{9}x - \frac{5}{6} = \frac{1}{2}$$

$$18 \cdot \frac{2}{9}x - 18 \cdot \frac{5}{6} = 18 \cdot \frac{1}{2}$$

$$4x - 15 = 9$$

Are $\frac{2}{9}x - \frac{5}{6} = \frac{1}{2}$ and $4x - 15 = 9$ equivalent equations?

Yes

20

$\frac{2}{9}x - \frac{5}{6} = \frac{1}{2}$ and $4x - 15 = 9$ are equivalent equations because both sides of the first equation were multiplied by a non-zero number, 18. What is the solution for $\frac{2}{9}x - \frac{5}{6} = \frac{1}{2}$?

$4x = 24, x = 6$

21

If both sides of $\frac{5}{6}x - \frac{3}{4} = \frac{7}{8}$ are multiplied by the same non-zero number, will an equivalent equation be generated?

Yes

22

What non-zero number can be multiplied by both sides of $\frac{5}{6}x - \frac{3}{4} = \frac{7}{8}$ to generate an equation with integers as its coefficient and constants?

24, the LCM of 6, 4, 8

23

Solve. $\frac{5}{6}x - \frac{3}{4} = \frac{7}{8}$

$20x - 18 = 21$

$x = \frac{39}{20}$

24

What non-zero number can be multiplied by both sides of $\frac{5}{8}x - \frac{3}{10} = \frac{2}{5}$ to generate an equation with integers as its coefficient and constants?

40

25
Solve. $\frac{2}{5}x - \frac{1}{3} = \frac{3}{10}$

$12x - 10 = 9$
$x = \frac{19}{12}$

26
Solve. $\frac{3}{4}x - \frac{5}{2} = \frac{7}{3}$

$9x - 30 = 28$
$x = \frac{58}{9}$

27
Solve. $\frac{3}{2}x - \frac{5}{4} = \frac{1}{5}$

$30x - 25 = 4$
$x = \frac{29}{30}$

28
There is no problem with generating an equivalent equation for $\frac{3}{2}x - \frac{5}{4} = \frac{1}{5}$ because each term can be multiplied by 20 and 20 is not equal to 0. This is not quite the case for an equation which has a variable in a denominator. What should be multiplied by both sides of $\frac{7}{5x} - \frac{2}{3} = \frac{1}{6}$?

$30x$

29
If each side of $\frac{7}{5x} - \frac{2}{3} = \frac{1}{6}$ is multiplied by 30x, the new equation generated can be guaranteed to be equivalent only if 30x is not equal to 0. This needs to be noted when multiplying each term by 30x.

$$\frac{7}{5x} - \frac{2}{3} = \frac{1}{6}$$

$$30x \cdot \frac{7}{5x} - 30x \cdot \frac{2}{3} = 30x \cdot \frac{1}{6} \qquad 30x \neq 0$$

$$42 - 20x = 5x$$

$$42 = 25x$$

$$\frac{42}{25} = x$$

When x is replaced by $\frac{42}{25}$, $30x \neq 0$ becomes a _____ (true, false) statement.

true

30

The equations $\frac{7}{5x} - \frac{2}{3} = \frac{1}{6}$ and $\frac{42}{25} = x$ are equivalent because the multiplier, 30x, is not equal to 0 when x is $\frac{42}{25}$. The process for generating equivalent equations by multiplication requires that the multiplier not be equal to _____.

0

31

What can be multiplied by both sides of $\frac{5}{x} - \frac{3}{4} = \frac{5}{12}$ to generate an equation with integers as its coefficient and constants?

12x

32

The process for solving $\frac{5}{x} - \frac{3}{4} = \frac{5}{12}$ is shown below.

$$\frac{5}{x} - \frac{3}{4} = \frac{5}{12}$$

$$12x \cdot \frac{5}{x} - 12x \cdot \frac{3}{4} = 12x \cdot \frac{5}{12} \qquad 12x \neq 0$$

$$60 - 9x = 5x$$

$$60 = 14x$$

$$\frac{30}{7} = x$$

To be certain that $\frac{5}{x} - \frac{3}{4} = \frac{5}{12}$ and $\frac{30}{7} = x$ are equivalent, answer the following question. When $\frac{30}{7}$ replaces x is $12x \neq 0$ a true statement?

Yes. The equations are equivalent.

33

The example below indicates the importance of the last 5 frames.

$$\frac{3}{4x} + \frac{7}{8} = \frac{9}{12x}$$

$$24x \cdot \frac{3}{4x} + 24x \cdot \frac{7}{8} = 24x \cdot \frac{9}{12x} \qquad 24x \neq 0$$

$$18 + 21x = 18$$

$$21x = 0$$

$$x = 0$$

Are the equations $\frac{3}{4x} + \frac{7}{8} = \frac{9}{12x}$ and $x = 0$ equivalent?

No. When x = 0, then the multiplier 24x is also 0.

34

The equation $\frac{3}{4x} + \frac{7}{8} = \frac{9}{12x}$ has no solution. It is an example of a situation in which the generated equation, x = 0, _____ (is, is not) equivalent to the original.

is not

35

Solve $\frac{4}{3x} - \frac{3}{2} = \frac{5}{x}$ by multiplying both sides of the equation by 6x. Make certain the generated equation is equivalent to the original.

$\frac{-22}{9}$

When $x = \frac{-22}{9}$, 6x does not equal 0.

36

Solve $\frac{8}{4x} + \frac{5}{6} = \frac{2}{x}$ by multiplying both sides of the equation by 12x. Make certain the generated equation is equivalent to the original.

No solution. When x = 0, 12x also equals 0.

37

Solve. $\frac{4}{5x} - \frac{3}{2x} = \frac{-7}{10}$

1

38

Solve. $\frac{3}{2x} + \frac{1}{2} = \frac{5}{3}$

$\frac{9}{7}$

39

Another type of situation that may generate a non-equivalent equation is shown below. This equation is solved by squaring each of its sides.

$$\sqrt{x-4} = 11$$
$$(\sqrt{x-4})^2 = (11)^2$$
$$x - 4 = 121$$
$$x = 125$$

If x is replaced by 125 in $\sqrt{x-4} = 11$, will the statement be true?

Yes, $\sqrt{121} = 11$

EQUATION SOLVING 303

40
The equation below is solved by squaring each side.

$$\sqrt{x+5} = -7$$
$$(\sqrt{x+5})^2 = (-7)^2$$
$$x + 5 = 49$$
$$x = 44$$

If x is replaced by 44 in $\sqrt{x+5} = -7$, will the statement be true?

No, $\sqrt{49} = 7$

41
Squaring both sides of an equation does not always generate an equivalent equation. It did for the equation $\sqrt{x-4} = 11$. It _____ (did, did not) for the equation $\sqrt{x+5} = -7$.

did not

42
Try to solve $\sqrt{4-x} = 5$ by squaring both sides. Check the result to see if the process generated an equivalent equation.

$4 - x = 25$
$x = -21$
$\sqrt{4 - (-21)} = 5$ is true.

43
Solve $\sqrt{2x+1} = 7$ by squaring both sides.

$x = 24$
$\sqrt{48 + 1} = 7$ is true.

44
To generate an equivalent linear equation two methods always work. First, the same number can be added to both sides of the equation. Second, the same number, except ____, can be multiplied by both sides.

0

45
Is squaring both sides of an equation always going to generate an equivalent equation?

No

Feedback Unit 2

This quiz reviews the preceding unit. Answers are at the back of the book.

Solve each of the following.

1. $6 - 3(x - 2) = 5x - 1$
2. $4(2x + 5) - 5x = 10 - (x - 1)$
3. $4 - (x - 3) = 5x - (4x - 1)$
4. $2(x + 3) - 5(x - 6) = x + (3x - 1)$
5. $\frac{3}{4}x - \frac{1}{6} = \frac{5}{8}$
6. $\frac{2}{3}x - \frac{1}{5} = \frac{3}{4}$
7. $\frac{-7}{8}x - \frac{3}{7} = \frac{2}{7}$
8. $\frac{5}{x} = \frac{-3}{2x}$
9. $\frac{3}{x} + \frac{5}{6} = \frac{-1}{3}$
10. $\frac{2}{3x} + \frac{1}{6} = \frac{3}{8} - \frac{1}{4x}$
11. $\sqrt{x - 7} = 8$
12. $\sqrt{4 - 3x} = 7$

Unit 3: Polynomial Equations with Rational Solutions

The following mathematical terms are crucial to an understanding of this unit.

- First degree equation
- Quadratic equation
- Cubic equation
- Polynomial equation
- Second degree equation
- Third degree equation
- Fourth degree equation
- Multiplicity

1
A linear equation like $3x - 4 = 13$ is a **first degree equation**. The exponent on its variable is 1. Is $x^2 - 9 = 7$ a first degree equation?

No, its variable has 2 as its exponent.

EQUATION SOLVING 305

2
The equation $x^2 - 9 = 7$ is a **second degree equation.**
The largest exponent on its variable is 2.
Is $x^2 - 3x + 2 = 0$ a second degree equation?

Yes

3
Second degree equations are often called **quadratic equations**. Is $6x^2 - 7x + 2 = 0$ a quadratic equation?

Yes

4
Third degree equations have 3 as the largest exponent on the variable. $x^3 + 2x^2 - 3x + 7 = 0$ is a third degree equation. Is $x^3 - 1 = 0$ a third degree equation?

Yes

5
Third degree equations are often called **cubic equations**.
Is $x^4 - 81 = 0$ a cubic equation?

No, it is a **fourth degree equation**.

6
The equation $5x(x + 7)(3x - 5) = 0$ is a **polynomial equation** in factored form. It has 0 as one side of the equation and the polynomial is factored completely. Is $-3x(5x + 3)(2x - 7)(6x - 7) = 0$ a polynomial equation in factored form?

Yes

7
The degree of a polynomial equation in factored form is the sum of the degrees of its factors. What is the degree of $5x(x + 7)(3x - 5) = 0$?

3

8
The equation $-3x(5x + 3)(2x - 7)(6x - 7) = 0$ is a fourth degree equation. It has 4 factors and each factor has 1 as the exponent on its variable. $(x + 7)(x - 3)^2 = 0$ is a _____ degree equation.

3rd
It has 3 factors.

9
To solve -2x(5x – 1)(2x + 7)(x – 3) = 0, first note that the product of the four factors is 0. If a multiplication has 0 as its product, at least one of the factors must itself be ____.

0

10
Whenever the product of two or more numbers is 0, then at least one of the factors itself must be 0. If xy = 0 then x = 0 or _____.

y = 0

11
To solve -2x(5x – 1)(2x + 7)(x – 3) = 0,

1. Set each factor equal to zero.
   -2x = 0   5x – 1 = 0   2x + 7 = 0   x – 3 = 0

2. Solve each equation separately.
   $x = 0$ $x = \frac{1}{5}$ $x = \frac{-7}{2}$ $x = 3$

The equation -2x(5x – 1)(2x + 7)(x – 3) = 0 has four solutions (roots). They are: ____, ____, ____, ____.

$0, \frac{1}{5}, \frac{-7}{2}, 3$

12
If ab = 0, then a = 0 or b = 0.
If 2x(x + 4) = 0, then 2x = 0 or _____.

x + 4 = 0

13
If x(3x – 1) = 0, then x = 0 or _____.

3x – 1 = 0

14
To solve 5x(x + 9)(3x – 2) = 0, write three equations. The left side of each equation will be a factor of the polynomial. The right side of equation will be _____.

0

15
Complete the solution of 5x(x + 9)(3x – 2) = 0 by finding its three roots.

$$5x(x + 9)(3x – 2) = 0$$
$$5x = 0 \quad x + 9 = 0 \quad 3x – 2 = 0$$

$0, -9, \frac{2}{3}$

EQUATION SOLVING   307

16
    $13(6x - 5)(x - 5) = 0$ is a second degree equation.
Its two solutions are found using $6x - 5 = 0$ and $x - 5 = 0$.
The factor 13 may be disregarded because $13 \neq 0$.
Find the two roots of $2(6x - 5)(x - 5) = 0$.

$\frac{5}{6}, 5$

17
    $-36(4x + 7)(3x + 1)(x - 3) = 0$ is a third degree
equation. Find its three roots.

$\frac{-7}{4}, \frac{-1}{3}, 3$

18
Solve $6x(x + 3)(x - 1)(2x - 7) = 0$.

$0, -3, 1, \frac{7}{2}$

19
A second degree equation with one variable has 2 roots.
A third degree equation with one variable has 3 roots.
An eighth degree equation with one variable has ___ roots.

8

20
The sixth degree equation $(x + 7)(x - 2)^5 = 0$ has
6 roots which could be listed as -7, 2, 2, 2, 2, 2.
Since 2 is a root 5 times, this is often stated as

    "2 has a **multiplicity** of 5" as a root

Find the 14 roots of $(x - 5)^3(x + 9)^2(x + 8)^4(x - 6)^5 = 0$
and show the multiplicity of each solution.

5 multiplicity 3
-9 multiplicity 2
-8 multiplicity 4
6 multiplicity 5

21
To solve $6x^2 - 3x - 45 = 0$, the polynomial must first
be factored.

$$6x^2 - 3x - 45 = 0$$
$$3(2x^2 - x - 15) = 0$$
$$3(x - 3)(2x + 5) = 0$$

Complete the solution of $6x^2 - 3x - 45 = 0$ and find
its two roots.

$x - 3 = 0 \quad 2x + 5 = 0$
$3, \frac{-5}{2}$

22
Complete the solution of the quadratic equation shown below.

$$x^2 - 7x + 12 = 0$$
$$(x - 3)(x - 4) = 0$$

$x - 3 = 0 \quad x - 4 = 0$
3, 4

23
Find all the solutions of $x^2 - 8x + 15 = 0$ by first factoring the trinomial.

3, 5

24
Solve. $x^2 - x - 12 = 0$

-3, 4

25
Complete the solution of $x^2 - 9x = 0$ shown below:

$$x^2 - 9x = 0$$
$$x(x - 9) = 0$$

$x = 0 \quad x - 9 = 0$
0, 9

26
Solve. $x^2 + 8x = 0$

0, -8

27
To solve $x^2 - 9 = 16$, the right side must first be changed to 0. This is accomplished by adding the opposite of 16 to both sides of the equation.

$$x^2 - 9 = 16$$
$$x^2 - 9 - 16 = 16 - 16$$
$$x^2 - 25 = 0$$
$$(x + 5)(x - 5) = 0$$
$$x + 5 = 0 \quad x - 5 = 0$$

What are the two solutions of $x^2 - 9 = 16$?

-5, 5

28
Solve $x^2 = 7x$ by first adding the opposite of $7x$ to both sides of the equation.

$x^2 - 7x = 0$
0, 7

EQUATION SOLVING 309

29
Solve. $x^2 - 2x = 15$

-3, 5

30
The trinomial of the equation $x^2 - 14x + 49 = 0$ is a perfect square. Solve. $x^2 - 14x + 49 = 0$

$(x - 7)^2 = 0$
7

31
Quadratic equations of the form $x^2 + 2bx + b^2 = 0$ are solved by factoring a trinomial perfect square.

$x^2 + 2bx + b^2 = 0$ $x^2 + 18x + 81 = 0$
$x^2 + 2 \cdot bx + b^2 = 0$ $x^2 + 2 \cdot 9x + (9)^2 = 0$
$(x + b)^2 = 0$ $(x + 9)^2 = 0$

What is/are the solution(s) of $x^2 + 18x + 81 = 0$?

-9

32
Quadratic equations of the form $x^2 - 2bx + (-b)^2 = 0$ are solved by factoring a trinomial perfect square.

$x^2 - 2bx + (-b)^2$ $x^2 - 22x + 121 = 0$
$x^2 + 2(-b)x + (-b)^2$ $x^2 + 2(-11)x + (-11)^2 = 0$
$(x - b)^2$ $(x - 11)^2 = 0$

What is/are the solution(s) of $x^2 - 22x + 121 = 0$?

11

33
The form $x^2 + 2bx + b^2 = 0$ is of increased importance when b is not an integer.

$x^2 + 2bx + b^2 = 0$ $x^2 + 5x + \frac{25}{4} = 0$

$x^2 + 2 \cdot bx + b^2 = 0$ $x^2 + 2 \cdot \frac{5}{2} x + \left(\frac{5}{2}\right)^2 = 0$

$(x + b)^2 = 0$ $\left(x + \frac{5}{2}\right)^2 = 0$

What is/are the solution(s) of $x^2 + 5x + \frac{25}{4} = 0$?

$\frac{-5}{2}$

34

$x^2 - 2bx + (-b)^2$ $x^2 - x + \frac{1}{4} = 0$

$x^2 + 2(-b)x + (-b)^2$ $x^2 + 2\left(\frac{-1}{2}\right)x + \left(\frac{-1}{2}\right)^2 = 0$

$(x - b)^2$ $\left(x - \frac{1}{2}\right)^2 = 0$

What is/are the solution(s) of $x^2 - x + \frac{1}{4} = 0$? $\frac{1}{2}$

35
Solve $x^2 + 9x + \frac{81}{4} = 0$ by factoring the trinomial as a perfect square.

$\left(x + \frac{9}{2}\right)^2 = 0$

$\frac{-9}{2}$

36
Solve $x^2 - 3x + \frac{9}{4} = 0$ by factoring the trinomial as a perfect square.

$\left(x - \frac{3}{2}\right)^2 = 0$

$\frac{3}{2}$

37
Solve. $x^2 - 16 = 0$ 4, -4

38
Solve. $2x^2 - x - 3 = 0$ $-1, \frac{3}{2}$

39
Solve. $5x^2 - 14x - 3 = 0$ $\frac{-1}{5}, 3$

40
Solve. $x^2 - 7x + \frac{49}{4} = 0$ $\frac{7}{2}$

EQUATION SOLVING   311

41
Some equations are solved as quadratic equations. This is the case with the example shown below. The first step in solving is to multiply each term by the LCM of the denominators.

$$\frac{5}{2x} - \frac{x}{6} = \frac{1}{x}$$

$$6x \cdot \frac{5}{2x} - 6x \cdot \frac{x}{6} = 6x \cdot \frac{1}{x} \qquad 6x \neq 0$$

$$15 - x^2 = 6$$

Finish solving the equation. $9 = x^2,\ x = 3,\ x = -3$

42
The equation $\frac{5}{2x} - \frac{x}{6} = \frac{1}{x}$ has 2 solutions, 3 and -3. Neither contradicts the requirement that $6x \neq 0$ and both will check in the original equation. The check of 3 is shown below.

$$\frac{5}{2x} - \frac{x}{6} = \frac{1}{x}$$

if x = 3
$$\frac{5}{6} - \frac{3}{6} = \frac{1}{3}$$

$$\frac{1}{3} = \frac{1}{3} \qquad \text{is true}$$

Show the check of -3.

$$\frac{-5}{6} - \frac{-3}{6} = \frac{-1}{3}$$

$$\frac{-5}{6} + \frac{1}{2} = \frac{-1}{3} \quad \text{is true}$$

43

To solve $\frac{3}{x} - \frac{x}{10} = \frac{3}{2x} - \frac{1}{5}$ each term is multiplied by 10x and, after simplifying, the solutions of a quadratic equation are found.

$$\frac{3}{x} - \frac{x}{10} = \frac{3}{2x} - \frac{1}{5}$$

$$10x \cdot \frac{3}{x} - 10x \cdot \frac{x}{10} = 10x \cdot \frac{3}{2x} - 10x \cdot \frac{1}{5} \qquad 10x \neq 0$$

$$30 - x^2 = 15 - 2x$$

Finish solving the equation.

$x^2 - 2x - 15 = 0$
$(x - 5)(x + 3) = 0$
$x = 5, \quad x = -3$

44

Are 5 and -3 acceptable solutions for $\frac{3}{x} - \frac{x}{10} = \frac{3}{2x} - \frac{1}{5}$ with the restriction that $10x \neq 0$?

Yes

45

Complete the solution of the equation below.

$$\frac{x+1}{6} + \frac{1}{x} = \frac{5}{12} + \frac{8}{3x}$$

$$12x \cdot \frac{x+1}{6} + 12x \cdot \frac{1}{x} = 12x \cdot \frac{5}{12} + 12x \cdot \frac{8}{3x} \qquad 12x \neq 0$$

$$2x(x + 1) + 12 = 5x + 32$$

_____

_____

$2x^2 - 3x - 20 = 0$

$(x - 4)(2x + 5) = 0$

$x = 4, \quad x = \frac{-5}{2}$

46

Solve. $\frac{2}{x} + \frac{x-6}{4} = \frac{2x+1}{4x}$

7, 1

47

Another type of equation that is solved as a quadratic is shown below. Notice that each side of the equation is squared to eliminate the radical term on the right side of the equation.

$$\sqrt{3x} + 1 = 2\sqrt{x+1}$$
$$(\sqrt{3x} + 1)^2 = (2\sqrt{x+1})^2$$
$$(\sqrt{3x} + 1)(\sqrt{3x} + 1) = (2\sqrt{x+1})(2\sqrt{x+1})$$
$$3x + 2\sqrt{3x} + 1 = 4(x+1)$$
$$2\sqrt{3x} = x + 3$$

The equation is simplified so that the radical term is isolated on one side of the equation. This makes it possible to square both sides and eliminate the radical.

$$(2\sqrt{3x})^2 = (x+3)^2$$
$$4 \cdot 3x = x^2 + 6x + 9$$
$$0 = x^2 - 6x + 9$$

Finish solving the equation.

$(x - 3)^2 = 0, x = 3$

48

When an equation solving radicals is solved by squaring each side, any solution needs to be checked in the original equation. Check 3 as a solution for $\sqrt{3x} + 1 = 2\sqrt{x+1}$.

$\sqrt{9} + 1 = 2\sqrt{4}$
simplifies to the true statement $3 + 1 = 2 \cdot 2$

49

To solve an equation with radical terms, isolate a radical on one side of the equation, square both sides, and repeat the process if another radical term is present.

$$1 + \sqrt{x} = \sqrt{2x-7}$$
$$(1 + \sqrt{x})^2 = (\sqrt{2x-7})^2$$
$$1 + 2\sqrt{x} + x = 2x - 7$$
$$2\sqrt{x} = x - 8$$
$$(2\sqrt{x})^2 = (x-8)^2$$
$$4 \cdot x = x^2 - 16x + 64$$
$$0 = x^2 - 20x + 64$$

What are the solutions of the quadratic equation?

$(x - 4)(x - 16) = 0$
$x = 4, x = 16$

50

The solutions of $0 = x^2 - 20x + 64$ are 4 and 16, but these roots need to be checked in the original equation. Check 4 and 16 in $1 + \sqrt{x} = \sqrt{2x-7}$ to see if they are solutions.

16 checks. $1 + 4 = 5$
4 fails to check. $1 + 2 \neq 1$

51

In solving $1 + \sqrt{x} = \sqrt{2x-7}$ the equation $0 = x^2 - 20x + 64$ was generated. Are the equations equivalent?

No, 16 is the only root of $1 + \sqrt{x} = \sqrt{2x-7}$

52

To solve $\sqrt{x-2} + \sqrt{x+3} = 5$ first isolate a radical term on one side of the equation. After squaring and simplifying, that process will need to be repeated.

Solve. $\sqrt{x-2} + \sqrt{x+3} = 5$

6

53

Solve. $\sqrt{2x} = 3 + \sqrt{9-x}$

8; 0 will not check

Feedback Unit 3

This quiz reviews the preceding unit. Answers are at the back of the book.

Find solutions for each of the following equations.

1. $(x-6)(x+1) = 0$

2. $(x+4)(x-2)^2 = 0$

3. $4x(x-3)(3x+7) = 0$

4. $3(x+5)(2x-11) = 0$

5. $x^2 + 9x - 22 = 0$

6. $x^2 - 81 = 0$

7. $4x^2 - 11x + 6 = 0$

8. $6x^3 - x^2 - 12x = 0$

9. $x^2 - 5x + \frac{25}{4} = 0$

10. $x^2 + x + \frac{1}{4} = 0$

11. $\frac{3}{x} - \frac{x-4}{3} = \frac{x-1}{3x}$

12. $\sqrt{x+6} = \sqrt{7-x} - 1$

Unit 4: Completing the Square

The following mathematical term is crucial to an understanding of this unit.

Completing the square

1
In the last unit, polynomial equations were solved using the fact that when the product of two numbers is 0 then at least one of the numbers must be _____.

0

2
In this unit, polynomial equations are solved using the fact that if $x^2 = y^2$ then $x = y$ or $x = -y$.
For example, if $x^2 = 25$ then $x = 5$ or $x = -5$.
Similarly, if $x^2 = 49$ then $x = 7$ or _____.

$x = -7$

3
There are two numbers that can make $x^2 = 100$ true.
What are the two solutions of $x^2 = 100$?

10, -10

4
The two solutions of $x^2 = 5$ are $\sqrt{5}$ and $-\sqrt{5}$.
  $(\sqrt{5})^2 = 5$ and $(-\sqrt{5})^2 = 5$ are true statements.
What are the two solutions of $x^2 = 17$?

$\sqrt{17}, -\sqrt{17}$

5
Find the two solutions of $x^2 = 3$.

$\sqrt{3}, -\sqrt{3}$

EQUATION SOLVING 317

6
Every quadratic equation of the form $x^2 = k$ can be solved using two linear equations:

$$x = \sqrt{k} \qquad x = -\sqrt{k}$$

Applying this fact to the equation $(x + 2)^2 = 25$ gives:

$$x + 2 = 5 \qquad x + 2 = -5$$

What are the solutions of the two linear equations?

3, -7

7
If x is replaced by either 3 or -7, will $(x + 2)^2 = 25$ become a true statement?

Yes, $(3 + 2)^2 = 25$ is true.
$(-7 + 2)^2 = 25$ is true.

8
The quadratic equation $(x - 3)^2 = 49$ may be used to write two linear equations.

$$x - 3 = 7 \qquad x - 3 = -7$$

What are the solutions of the two linear equations?

10, -4

9
If x is replaced by either 10 or -4, will $(x - 3)^2 = 49$ become a true statement?

Yes, $(7)^2 = 49$ is true.
$(-7)^2 = 49$ is true.

10
The quadratic equation $(x + 7)^2 = 17$ may be used to write two linear equations.

$$x + 7 = \sqrt{17} \qquad x + 7 = -\sqrt{17}$$

What are the solutions of the two linear equations?

$-7 + \sqrt{17}$, $-7 - \sqrt{17}$

11
If x is replaced by $-7 + \sqrt{17}$ in $(x + 7)^2 = 17$, the following statement is obtained.

$$([-7 + \sqrt{17}\,] + 7)^2 = 17$$

Is the statement true?

Yes, $(\sqrt{17}\,)^2 = 17$ is true

12

The quadratic equation $(x - 1)^2 = 5$ may be used to write two linear equations.

$$x - 1 = \sqrt{5} \qquad x - 1 = -\sqrt{5}$$

What are the solutions of the two linear equations?

$1 + \sqrt{5}, 1 - \sqrt{5}$

13

Every quadratic equation of the form $(x + c)^2 = k$ can be solved using the two linear equations shown below.

$$x + c = \sqrt{k} \qquad x + c = -\sqrt{k}$$

What are the solutions of the two linear equations?

$-c + \sqrt{k}, -c - \sqrt{k}$

14

Every quadratic equation of the form $(x + c)^2 = k$ is solvable using the two linear equations shown below.

$$x + c = \sqrt{k} \qquad x + c = -\sqrt{k}$$

Is the equation $x^2 - 5x - 7 = 0$ in the proper form for solving it by this approach?

No

15

To solve $x^2 - 5x - 7 = 0$, one method would be to generate an equivalent equation in the form $(x + c)^2 = k$. If an equivalent equation is generated its solutions _____ (will, will not) also be solutions of $x^2 - 5x - 7 = 0$.

will

16

The square of any binomial, $(x + c)^2$, gives a perfect square trinomial: $x^2 + 2cx + c^2$

$(x - 8)^2 = x^2 - 2 \cdot 8x + 64$
$(x + 6)^2 =$ _____

$x^2 + 2 \cdot 6x + 36$
$x^2 + 12x + 36$

17

$(x + 7)^2 = x^2 + 2 \cdot 7x + 49$
$(x + 2)^2 =$ _____

$x^2 + 2 \cdot 2x + 4$
$x^2 + 4x + 4$

EQUATION SOLVING   319

18
$(x - 4)^2 = x^2 - 2 \cdot 4x + 16$
$(x - 3)^2 =$ _____

$x^2 - 2 \cdot 3x + 9$
$x^2 - 6x + 9$

19
$(x + 5)^2 = x^2 + 2 \cdot 5x + 25$
$(x + 10)^2 =$ _____

$x^2 + 2 \cdot 10x + 100$
$x^2 + 20x + 100$

20
$(x + \frac{3}{2})^2 = x^2 + 2 \cdot \frac{3}{2}x + \frac{9}{4} = x^2 + 3x + \frac{9}{4}$

$(x + \frac{7}{2})^2 =$ _____

$x^2 + 2 \cdot \frac{7}{2}x + \frac{49}{4}$

$x^2 + 7x + \frac{49}{4}$

21
Any perfect square trinomial $x^2 + 2cx + c^2$ can be written as the square of a binomial, $(x + c)^2$.

$x^2 - 16x + 64 = x^2 - 2 \cdot 8x + 64 = (x - 8)^2$
$x^2 - 6x + 9 = x^2 - 2 \cdot 3x + 9 =$ _____

$(x - 3)^2$

22
$x^2 + 14x + 49 = x^2 + 2 \cdot 7x + 49 = (x + 7)^2$
$x^2 + 4x + 4 = x^2 + 2 \cdot 2x + 4 =$ _____

$(x + 2)^2$

23
$x^2 - 8x + 16 = x^2 - 2 \cdot 4x + 16 = (x - 4)^2$
$x^2 - 18x + 81 = x^2 - 2 \cdot 9x + 81 =$ _____

$(x - 9)^2$

24
$x^2 + 10x + 25 = x^2 + 2 \cdot 5x + 25 = (x + 5)^2$
$x^2 + 20x + 100 = x^2 + 2 \cdot 10x + 100 =$ _____

$(x + 10)^2$

25
$x^2 + 7x + \frac{49}{4} = x^2 + 2 \cdot \frac{7}{2}x + \frac{49}{4} = (x + \frac{7}{2})^2$

$x^2 + 3x + \frac{9}{4} = x^2 + 2 \cdot \frac{3}{2}x + \frac{9}{4} =$ _____

$(x + \frac{3}{2})^2$

26

The expression $x^2 + 6x$ is a binomial. To add a 3rd term and make it into a perfect square trinomial, begin by writing the 6x term using 2 as a factor.

$$x^2 + 6x = x^2 + 2 \cdot 3x$$

Is $6x = 2 \cdot 3x$ an identity?   Yes

27

Three examples of writing identities using 2 as a factor are shown below.

$$x^2 + 10x = x^2 + 2 \cdot 5x$$
$$x^2 + 14x = x^2 + 2 \cdot 7x$$
$$x^2 + 2x = x^2 + 2 \cdot 1x$$

Is $x^2 + 18x = x^2 + 2 \cdot 9x$ an identity?   Yes

28

Three more examples of writing identities using 2 as a factor are shown below.

$$x^2 + 5x = x^2 + 2 \cdot \frac{5}{2} x$$

$$x^2 + 13x = x^2 + 2 \cdot \frac{13}{2} x$$

$$x^2 + x = x^2 + 2 \cdot \frac{1}{2} x$$

Is $x^2 + 3x = x^2 + 2 \cdot \frac{3}{2} x$ an identity?   Yes

29

Three final examples of writing identities using 2 as a factor are shown below. These examples begin with coefficients of x that are not integers.

$$x^2 + \frac{3}{4} x = x^2 + 2 \cdot \frac{3}{8} x$$

$$x^2 + \frac{4}{7} x = x^2 + 2 \cdot \frac{2}{7} x$$

$$x^2 - \frac{5}{9} x = x^2 - 2 \cdot \frac{5}{18} x$$

Is $x^2 + \frac{5}{6} x = x^2 + 2 \cdot \frac{5}{12} x$ an identity?   Yes

30

To add a 3rd term to $x^2 + 8x$ which will make it a perfect square trinomial, the following steps are used:

    a. Write the 2nd term of $x^2 + 8x$ using a factor of 2.
        $x^2 + 8x = x^2 + 2 \cdot 4x$

    b. Square the coefficient of x.
        $(4)^2 = 16$

Is $x^2 + 8x + 16$ a perfect square trinomial?   Yes, $(x + 4)^2$

31

To add a 3rd term to $x^2 - 16x$ which will make it a perfect square trinomial, the following steps are used:

    a. Write the 2nd term of $x^2 - 16x$ using a factor of 2.
        $x^2 - 16x = x^2 - 2 \cdot 8x$

    b. Square the coefficient of x.
        $(8)^2 = 64$

Is $x^2 - 16x + 64$ a perfect square trinomial?   Yes, $(x - 8)^2$

32

To add a 3rd term to $x^2 + 5x$ which will make it a perfect square trinomial, the following steps are used:

    a. Write the 2nd term of $x^2 + 5x$ using a factor of 2.
        $x^2 + 5x = x^2 + 2 \cdot \frac{5}{2} x$

    b. Square the coefficient of x.
        $\left(\frac{5}{2}\right)^2 = \frac{25}{4}$

Is $x^2 + 5x + \frac{25}{4}$ a perfect square trinomial?   Yes, $\left(x + \frac{5}{2}\right)^2$

33

To add a 3rd term to $x^2 - \frac{3}{2}x$ which will make it a perfect square trinomial, the following steps are used:

    a. Write the 2nd term of $x^2 - \frac{3}{2}x$ using a factor of 2.

$$x^2 - \frac{3}{2}x = x^2 - 2 \cdot \frac{3}{4}x$$

    b. Square the coefficient of x.

$$\left(\frac{3}{4}\right)^2 = \frac{9}{16}$$

Is $x^2 - \frac{3}{2}x + \frac{9}{16}$ a perfect square trinomial?   Yes, $\left(x - \frac{3}{4}\right)^2$

34

The process for adding a 3rd term to $x^2 + 12x$ which will make it a perfect square trinomial is called **completing the square**. Complete the square for $x^2 + 12x$. $x^2 + 12x + 36 = (x + 6)^2$

35

Complete the square for $x^2 - 18x$. $x^2 - 18x + 81 = (x - 9)^2$

36

The completing the square process is applied to a binomial and requires adding a 3rd term which makes the polynomial a perfect square trinomial.

$$x^2 - 9x = x^2 - 2 \cdot \frac{9}{2}x$$

Complete the square for $x^2 - 9x$. $x^2 - 9x + \frac{81}{4} = \left(x - \frac{9}{2}\right)^2$

37

$$x^2 + \frac{5}{3}x = x^2 + 2 \cdot \frac{5}{6}x$$

Complete the square for $x^2 + \frac{5}{3}x$. $x^2 + \frac{5}{3}x + \frac{25}{36} = \left(x + \frac{5}{6}\right)^2$

38

Complete the square for $x^2 - 10x$. $x^2 - 10x + 25 = (x - 5)^2$

39
Complete the square for $x^2 + 2x$.

$x^2 + 2x + 1 = (x + 1)^2$

40
Complete the square for $x^2 + 6x$.

$x^2 + 6x + 9 = (x + 3)^2$

41
Complete the square for $x^2 - 14x$.

$x^2 - 14x + 49 = (x - 7)^2$

42
Complete the square for $x^2 + 5x$.

$x^2 + 5x + \frac{25}{4} = \left(x + \frac{5}{2}\right)^2$

43
Complete the square for $x^2 + \frac{7}{5}x$.

$x^2 + \frac{7}{5}x + \frac{49}{100} = \left(x + \frac{7}{10}\right)^2$

44
Complete the square for $x^2 - 3x$.

$x^2 - 3x + \frac{9}{4} = \left(x - \frac{3}{2}\right)^2$

45
Complete the square for $x^2 + \frac{4}{3}x$.

$x^2 + \frac{4}{3}x + \frac{4}{9} = \left(x + \frac{2}{3}\right)^2$

46
Complete the square for $x^2 + \frac{1}{7}x$.

$x^2 + \frac{1}{7}x + \frac{1}{196} = \left(x + \frac{1}{14}\right)^2$

FEEDBACK UNIT 4

This quiz reviews the preceding unit. Answers are at the back of the book.

Complete the square for each of the following binomials.

1. $x^2 + 12x$

2. $x^2 - 10x$

3. $x^2 + 7x$

4. $x^2 - 8x$

5. $x^2 - 9x$

6. $x^2 + 15x$

7. $x^2 + \frac{8}{5}x$

8. $x^2 - \frac{7}{2}x$

Unit 5: Polynomial Equations with Complex Number Solutions

The following mathematical term is crucial to an understanding of this unit.

Quadratic formula

1
Every quadratic equation of the form $(x + c)^2 = k$ can be solved using the two linear equations shown below.

$$x + c = \sqrt{k} \qquad x + c = -\sqrt{k}$$

What are the linear equations that should be solved for the quadratic equation $(x + 7)^2 = 13$?

$x + 7 = \sqrt{13}$, $x + 7 = -\sqrt{13}$

2
To solve $x^2 - 4x - 7 = 0$ the completing the square process can be used to write it in the form $(x + c)^2 = k$.
Add 7 to each side of $x^2 - 4x - 7 = 0$.

$x^2 - 4x = 7$

3
The first step in solving $x^2 - 4x - 7 = 0$ was to add 7 to both sides of the equation. This was done to make the left side of the equation a binomial, $x^2 - 4x$. Complete the square for this binomial.

$x^2 - 4x + 4$

4
For the equation $x^2 - 4x = 7$, if 4 is added to the left side what must be added to the right side?

4

5
$$\begin{aligned} x^2 - 4x - 7 &= 0 \\ x^2 - 4x &= 7 \\ x^2 - 4x + 4 &= 7 + 4 \end{aligned}$$

Factor the perfect square trinomial.

$(x - 2)^2$

EQUATION SOLVING   325

6

$$x^2 - 4x - 7 = 0$$
$$x^2 - 4x = 7$$
$$x^2 - 4x + 4 = 7 + 4$$
$$(x - 2)^2 = 11$$

Write the two linear equations used to solve $(x - 2)^2 = 11$.

$x - 2 = \sqrt{11},\ x - 2 = -\sqrt{11}$

7

$$x^2 - 4x - 7 = 0$$
$$x^2 - 4x = 7$$
$$x^2 - 4x + 4 = 7 + 4$$
$$(x - 2)^2 = 11$$
$$x - 2 = \sqrt{11} \qquad x - 2 = -\sqrt{11}$$

Find the solutions of $x^2 - 4x - 7 = 0$ by solving the two linear equations.

$2 + \sqrt{11},\ 2 - \sqrt{11}$

8
Any quadratic equation can be solved by the process used in frames 1-7. In fact, $ax^2 + bx + c = 0$ can be solved to derive a formula for solving any quadratic equation. Begin the solution of $ax^2 + bx + c = 0$ by adding -c to both sides of the equation. What new equation is generated?

$ax^2 + bx = -c$

9

$$ax^2 + bx + c = 0$$
$$ax^2 + bx = -c$$

Multiply each term of the equation by $\frac{1}{a}$ so the coefficient of x^2 will be 1.

$x^2 + \frac{b}{a}x = \frac{-c}{a}$

10

$$ax^2 + bx + c = 0$$
$$ax^2 + bx = -c$$
$$x^2 + \frac{b}{a}x = \frac{-c}{a}$$

In preparation for completing the square, write the term $\frac{b}{a}x$ so it has 2 as a factor.

$2 \cdot \frac{b}{2a} x$

11

$$ax^2 + bx + c = 0$$
$$ax^2 + bx = -c$$
$$x^2 + \frac{b}{a}x = \frac{-c}{a}$$
$$x^2 + 2 \cdot \frac{b}{2a}x = \frac{-c}{a}$$

Complete the square on the left side of the equation and add the same value to the right side.

$$x^2 + \frac{b}{a}x + \frac{b^2}{4a^2} = \frac{b^2}{4a^2} - \frac{c}{a}$$

12

$$ax^2 + bx + c = 0$$
$$ax^2 + bx = -c$$
$$x^2 + \frac{b}{a}x = \frac{-c}{a}$$
$$x^2 + \frac{b}{a}x + \frac{b^2}{4a^2} = \frac{b^2}{4a^2} - \frac{c}{a}$$

Factor the perfect square trinomial on the left side of the equation and add the fractions on the right side.

$$\left(x + \frac{b}{2a}\right)^2 = \frac{b^2 - 4ac}{4a^2}$$

13

$$ax^2 + bx + c = 0$$
$$ax^2 + bx = -c$$
$$x^2 + \frac{b}{a}x = \frac{-c}{a}$$
$$x^2 + \frac{b}{a}x + \frac{b^2}{4a^2} = \frac{b^2}{4a^2} - \frac{c}{a}$$
$$\left(x + \frac{b}{2a}\right)^2 = \frac{b^2 - 4ac}{4a^2}$$

Write the two linear equations that need to be solved.

$$x + \frac{b}{2a} = \sqrt{\frac{b^2 - 4ac}{4a^2}}$$

$$x + \frac{b}{2a} = -\sqrt{\frac{b^2 - 4ac}{4a^2}}$$

14

The two solutions of $ax^2 + bx + c = 0$ are $\frac{-b}{2a} + \frac{\sqrt{b^2 - 4ac}}{2a}$ and $\frac{-b}{2a} - \frac{\sqrt{b^2 - 4ac}}{2a}$. The **quadratic formula** is usually written as: $x = \frac{-b \pm \sqrt{b^2 - 4ac}}{2a}$

Will this formula provide solutions for any equation of the form $ax^2 + bx + c = 0$? Yes

15

In the quadratic formula, $x = \frac{-b \pm \sqrt{b^2 - 4ac}}{2a}$, the $\pm$ symbol means that two solutions can be found, one using the + sign and the other using the – sign. Can $3x^2 + 5x + 7 = 0$ be solved using the quadratic formula? Yes

16

To solve $3x^2 + 5x + 7 = 0$ using the quadratic formula, the replacements for a, b, and c must be specified. For this equation, a = 3, b = 5, and c = 7.

$$ax^2 + bx + c = 0$$
$$3x^2 + 5x + 7 = 0$$

For the equation $2x^2 + 4x + 9 = 0$, a = 2, b = 4, and c = _____. 9

17

Find values for a, b, and c to be used in the quadratic formula when solving $7x^2 + 3x - 5 = 0$. a = 7, b = 3, and c = -5

18

To solve $7x^2 + 3x - 5 = 0$, the following steps are used.

$$x = \frac{-(3) \pm \sqrt{(3)^2 - 4(7)(-5)}}{2(7)} = \frac{-3 \pm \sqrt{9 + 140}}{14} = \frac{-3 \pm \sqrt{149}}{14}$$

One solution is $\frac{-3 + \sqrt{149}}{14}$. The other is _____. $\frac{-3 - \sqrt{149}}{14}$

328 CHAPTER 8

19

To solve $2x^2 - 7x + 3 = 0$, the following steps are used.

$$x = \frac{-(-7) \pm \sqrt{(-7)^2 - 4(2)(3)}}{2(2)} = \frac{7 \pm \sqrt{49 - 24}}{4} = \frac{7 \pm \sqrt{25}}{4}$$

Can the radical expression be simplified?

Yes, $\sqrt{25} = 5$

20

The solutions of $2x^2 - 7x + 3 = 0$ are $\frac{7 \pm 5}{4}$.

Using the plus sign, one solution is $\frac{7+5}{4} = \frac{12}{4} = 3$

Using the minus sign, the other solution is _____.

$\frac{7-5}{4} = \frac{2}{4} = \frac{1}{2}$

21

Find the two solutions for $3x^2 - 5x + 2 = 0$ by simplifying the radical expression obtained from the formula.

$$x = \frac{-(-5) \pm \sqrt{(-5)^2 - 4(3)(2)}}{2(3)} = \frac{5 \pm \sqrt{25 - 24}}{6} = \underline{\qquad}$$

$\frac{5+1}{6} = 1, \quad \frac{5-1}{6} = \frac{2}{3}$

22

Complete the solution for $4x^2 - 2x - 1 = 0$.

$$x = \frac{-(-2) \pm \sqrt{(-2)^2 - 4(4)(-1)}}{2(4)} = \frac{2 \pm \sqrt{4 + 16}}{8} = \underline{\qquad}$$

$\frac{2 \pm \sqrt{20}}{8} = \frac{2 \pm 2\sqrt{5}}{8} = \frac{1 \pm \sqrt{5}}{4}$

23

Complete the solution for $x^2 - 5x + 2 = 0$.

$$x = \frac{-(-5) \pm \sqrt{(-5)^2 - 4(1)(2)}}{2(1)} = \underline{\qquad}$$

$\frac{5 \pm \sqrt{25 - 8}}{2} = \frac{5 \pm \sqrt{17}}{2}$

24

To solve $x^2 - 2x = 2$, the equation must first be written as _____.

$ax^2 + bx + c = 0$
$x^2 - 2x - 2 = 0$

25

Use the quadratic formula to solve $x^2 - 2x - 2 = 0$.

$\frac{2 \pm \sqrt{12}}{2} = \frac{2 \pm 2\sqrt{3}}{2} = 1 \pm \sqrt{3}$

EQUATION SOLVING 329

26
Use the quadratic formula to solve $x^2 + 5x = -3$.

$$\frac{-5 \pm \sqrt{13}}{2}$$

27
Use the quadratic formula to solve $x^2 + 5 = 7x$.

$$\frac{7 \pm \sqrt{29}}{2}$$

28
Use the quadratic formula to solve $4x^2 - 5x + 1 = 0$.

$$1, \frac{1}{4}$$

29
Use the quadratic formula to solve $x^2 - 4x - 6 = 0$.

$$2 \pm \sqrt{10}$$

30
Any quadratic equation can be solved with the quadratic formula whether the roots are rational, irrational, or complex. For example, the calculation below shows the solution of $x^2 + x + 2 = 0$.

$$\frac{-(1) \pm \sqrt{(1)^2 - 4(1)(2)}}{2(1)} = \frac{-1 \pm \sqrt{1-8}}{2} = \frac{-1 \pm \sqrt{-7}}{2} = \frac{-1 \pm i\sqrt{7}}{2}$$

Are the roots of $x^2 + x + 2 = 0$ real numbers?

No, complex

31
Find the two complex roots of $x^2 - 2x + 5 = 0$.

$$1 \pm 2i$$

32
Find the two complex roots of $3x^2 + 2x + 1 = 0$.

$$\frac{-2 \pm 2i\sqrt{2}}{6} = \frac{-1 \pm i\sqrt{2}}{3}$$

33
Solve. $x^2 + 3x + 7 = 0$

$$\frac{-3 \pm i\sqrt{19}}{2}$$

34
Solve. $x^2 + x + 1 = 0$

$$\frac{-1 \pm i\sqrt{3}}{2}$$

35
Solve. $x^2 + 4x - 3 = 0$

$$-2 \pm \sqrt{7}$$

36

$8x^3 + 27 = 0$ is a third degree (cubic) equation. It can be solved by factoring $8x^3 + 27$, setting each of the factors equal to 0, and solving the equations.

$$8x^3 + 27 = 0$$
$$(2x + 3)(4x^2 - 6x + 9) = 0$$
$$2x + 3 = 0 \qquad 4x^2 - 6x + 9 = 0$$

The root of $2x + 3 = 0$ is $\frac{-3}{2}$. The two roots of $4x^2 - 6x + 9 = 0$ are _____.

$\dfrac{6 \pm 6i\sqrt{3}}{8} = \dfrac{3 \pm 3i\sqrt{3}}{4}$

37

$6x^3 - 8x^2 - 14x = 0$ is a cubic equation that can be solved. Factor $6x^3 - 8x^2 - 14x$ as a first step in the process.

$2x(3x^2 - 4x - 7)$

38

$$6x^3 - 8x^2 - 14x = 0$$
$$2x(3x^2 - 4x - 7) = 0$$

What are the two equations, one linear and the other quadratic, that can now be written?

$2x = 0,\ 3x^2 - 4x - 7 = 0$

39

Complete the solution of $6x^3 - 8x^2 - 14x = 0$ and find its three roots.

$$6x^3 - 8x^2 - 14x = 0$$
$$2x(3x^2 - 4x - 7) = 0$$
$$2x = 0 \qquad 3x^2 - 4x - 7 = 0$$
_____ _____

$0,\ -1,\ \dfrac{7}{3}$

40

Complete the solution of $2x^3 - x^2 - 2x + 1 = 0$ and find its three roots.

$$2x^3 - x^2 - 2x + 1 = 0$$
$$x^2(2x - 1) - 1(2x - 1) = 0$$
$$2x - 1 = 0 \qquad x^2 - 1 = 0$$
_____ _____

$\dfrac{1}{2},\ 1,\ -1$

41

Find the 3 roots of $x^3 - 5x^2 - x = 0$ by first factoring $x^3 - 5x^2 - x$ and then setting each factor equal to zero.

$x = 0,\ x^2 - 5x - 1 = 0$

$0,\ \dfrac{5 \pm \sqrt{29}}{2}$

42
Solve $4x^3 - 5x^2 - 8x + 10 = 0$ by first factoring by grouping.

$4x - 5 = 0, x^2 - 2 = 0$

$\frac{5}{4}, \pm\sqrt{2}$

43
Solve. $x^3 - 125 = 0$

$5, \frac{-5 \pm 5i\sqrt{3}}{2}$

44
Solve. $x^3 + 1 = 0$

$-1, \frac{1 \pm i\sqrt{3}}{2}$

FEEDBACK UNIT 5

This quiz reviews the preceding unit. Answers are at the back of the book.

Solve each of the following.

1. $x^2 + 7x - 2 = 0$

2. $x^2 - 43 = 0$

3. $2x^2 - 5x + 6 = 0$

4. $3x^2 - 4x - 5 = 0$

5. $x^2 + 5x + 3 = 0$

6. $2x^2 - 8x + 9 = 0$

7. $x^3 + 125 = 0$

8. $x^3 - 27 = 0$

UNIT 6: SOLVING EQUATIONS WITH TWO VARIABLES

The following mathematical terms are crucial to an understanding of this unit.

    Ordered pair (x, y)   Ordered pair solution
    xy-plane   Circle
    Parabola   Ellipse
    Hyperbola

1
The preceding units of this chapter have been limited to equations with one variable. In this unit equations with 2 variables will be solved. Is $5x - 7 = 37$ the type of equation solved in this unit?

    No, it has only 1 variable.

2
Equations like $3x - 5y = 8$, $y = x^2 + 7x - 2$, and $x^2 + y^2 = 11$ have 2 variables, x and y. Each solution for these equations is an **ordered pair (x, y)**. Might (-5, -12) be a solution of one of these equations?

    Yes

3
Equations like $3x - 5y = 8$, $y = x^2 + 7x - 2$, and $x^2 + y^2 = 11$ have ordered pairs (x, y) as their solutions. In this unit, the set of real numbers will be used as replacements for x and y. Might ($\sqrt{2}$, $7\sqrt{2}$) be an acceptable solution of one of these equations?

    Yes, both $\sqrt{2}$ and $7\sqrt{2}$ are real numbers.

4
Every solution of $4x - y = 9$ is an ordered pair of real numbers. Could 6 be a solution of $4x - y = 9$?

    No, it is not an ordered pair.

5

The ordered pair (3, 4) is a **solution** for x + y = 7 because when x = 3 and y = 4 the equation becomes the true statement 3 + 4 = 7. Is (2, 5) a solution for 2x + y = 9?

Yes, 4 + 5 = 9 is true.

6

(5, 1) is a solution for 2x − y = 9 because 10 − 1 = 9 is true. Is (6, 3) also a solution for 2x − y = 9?

Yes, 12 − 3 = 9 is true.

7

To determine whether (2, 5) is a solution for 4x − y = 1, replace x by 2 and y by 5. Is 8 − 5 = 1 true?

No, 8 − 5 ≠ 1
(≠ means not equal)

8

Is (4, 1) a solution for 2x − y = 7?

Yes

9

Is (-1, 3) a solution for 5x + y = 7?

No

10

Is (4, -2) a solution for x − 3y = 10?

Yes

11

(8, 3), (4, 7), (12, -1), and (-5, 16) are some of the solutions for x + y = 11. Is (20, -9) also a solution for x + y = 11?

Yes

12

To complete the ordered pair (4, _____) as a solution for 2x − y = 7, the following steps are used:

$$\begin{aligned} 2x - y &= 7 \quad (4, \_\_) \\ 2 \cdot 4 - y &= 7 \\ 8 - y &= 7 \\ -y &= -1 \\ y &= 1 \end{aligned}$$

Is (4, 1) a solution for 2x − y = 7?

Yes, 8 − 1 = 7 is true.

13
To complete the solution (5, _____) for 2x − 3y = 6, solve 10 − 3y = 6.

$$\begin{align} 2x - 3y &= 6 \quad (5, \_\_) \\ 2 \cdot 5 - 3y &= 6 \\ 10 - 3y &= 6 \\ -3y &= -4 \\ y &= \tfrac{4}{3} \end{align}$$

Is $\left(5, \tfrac{4}{3}\right)$ a solution for 2x − 3y = 6? Yes

14
    (6, _____) is a solution for 2x − y = 7.
[Note: Solve 12 − y = 7.] (6, 5)

15
    (2, _____) is a solution for 3x + 2y = 13.
[Note: Solve 6 + 2y = 13.] $\left(2, \tfrac{7}{2}\right)$

16
    (_____ , -2) is a solution for 5x + y = 17. $\left(\tfrac{19}{5}, -2\right)$

17
Use the equation x + 2y = 18 to complete each of the following 10 solutions. Check each solution as you complete it.

    a. (2, _____) (2, 8)
    b. (_____ , 5) (8, 5)
    c. (6, _____) (6, 6)
    d. (_____ , 9) (0, 9)
    e. (14, _____) (14, 2)
    f. (_____ , -3) (24, -3)
    g. (-2, _____) (-2, 10)
    h. (_____ , 1) (16, 1)
    i. (-4, _____) (-4, 11)
    j. (_____ , 0) (18, 0)

18

There are an infinite number of solutions for x + 2y = 18.
Is it possible to list all of the solutions? No

19

One way to show all the solutions of x + 2y = 18 is
to show them on an xy-plane. Because x + 2y = 18
is a first degree(linear) equation, all of its solutions
lie on the same straight _____. line

20

Find 2 or more solutions
of x – y = 2, plot them
on the **xy-plane** shown,
and draw the straight line
that contains them.

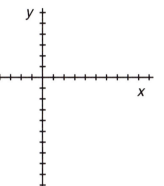

21

The line of x – y = 2 which is shown as the answer to the
preceding frame represents all the solutions for the equation.
Points not on the line _____ (are, are not) solutions of the
equation. are not

22
The line of $2x + y = 3$ is shown. It was graphed by finding some of its solutions, plotting them on the xy-plane, and drawing the straight line that contains them.

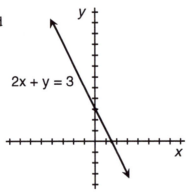

The line is **the graph of** $2x + y = 3$. Is every solution of $2x + y = 3$ on the line?

Yes

23
The figure shows the line of $2x - y = 6$.

Is every ordered-pair solution of $2x - y = 6$ represented by a point on the line?

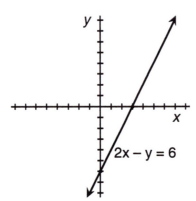

Yes

24
The ordered pair (-2, 11) is a **solution** for $y = x^2 + 7$ because when $x = -2$ and $y = 11$ the equation becomes the true statement $11 = 4 + 7$. Is (5, 32) a solution for $y = x^2 + 7$?

Yes, $32 = 25 + 7$ is true.

25
Determine if (2, 5) is a solution for $3x^2 - xy + y^2 = 23$ by replacing x by 2 and y by 5.

No, $12 - 10 + 25 = 23$ is false.

26
Is (4, 1) a solution for $x^2 - y^2 = 7$?

No, $16 - 1 = 7$ is false.

27
Is (-1, 2) a solution for $x^2 + 5x + 6 = y$? Yes

28
Is (-4, -3) a solution for $x^2 + y^2 = 25$? Yes

29
To complete the ordered pair (4, ____) as a solution for $y = x^2 - 5x - 2$, replace x by 4 and solve the equation for y.

$$y = x^2 - 5x - 2 \qquad (4, \underline{\ \ })$$
$$y = (4)^2 - 5(4) - 2$$
$$y = 16 - 20 - 2$$
$$y = -6$$

Complete (-2, ___) as a solution for $y = x^2 - 5x - 2$. (-2, 12)

30
To complete the ordered pair (-3, ___) as a solution for $x^2 - y^2 = 7$, replace x by -3 and solve the equation for y.

$$x^2 - y^2 = 7 \qquad (-3, \underline{\ \ })$$
$$(-3)^2 - y^2 = 7$$
$$9 - y^2 = 7$$
$$-y^2 = -2$$
$$y^2 = 2$$
$$y = \sqrt{2} \qquad y = -\sqrt{2}$$

Are both $(-3, \sqrt{2})$ and $(-3, -\sqrt{2})$ solutions of $x^2 - y^2 = 7$? Yes

31
To complete the ordered pair (6, ___) as a solution for $x^2 + y^2 = 25$, replace x by 6 and solve the equation for y.

$$x^2 + y^2 = 25 \qquad (6, \underline{\ \ })$$
$$(6)^2 + y^2 = 25$$
$$36 + y^2 = 25$$
$$y^2 = -11$$
$$y = i\sqrt{11} \qquad y = -i\sqrt{11}$$

Is there any ordered pair of real numbers (6, ___) that is a solution of $x^2 + y^2 = 25$?

No, $i\sqrt{11}$ is not a real number.

338 CHAPTER 8

32
   (-5, _____) is a solution for $y = 2x^2 + 7x - 4$.   (-5, 11)

33
   (-2, _____) is a solution for $2x^2 + 3x + y^2 = 8$. $(-2, \sqrt{6}), (-2, -\sqrt{6})$

34
   (5, _____) is a solution for $x^2 + y^2 + 3y = 30$. $\left(5, \dfrac{-3+\sqrt{29}}{2}\right), \left(5, \dfrac{-3-\sqrt{29}}{2}\right)$

35
   (-4, _____) is a solution of $y^2 - 3 = x$.
   [Note: Replacements for x and y are restricted to
   real numbers in this unit.]   There are no such
                                                                            solutions. i is not
                                                                            a real number.

36
The graph of a first degree equation with 2 variables
always is a straight line. The graph of a 2nd degree
equation with 2 variables always is a curve. Will the
graph of $2x^2 + 3x + y^2 = 8$ be a straight line?   No, it is a curve.

37
The figure at the right
shows the graph of
$x^2 + y^2 = 36$. It is
a **circle**. Two points
(-3, ____) are marked
on the circle.

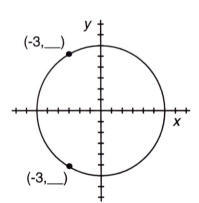

Complete (-3, ____)
as a solution of
$x^2 + y^2 = 36$. $(-3, 3\sqrt{3}), (-3, -3\sqrt{3})$

38

The figure at the right shows the graph of $y = x^2 - 6$. It is a **parabola**. The point $(2, \_\_\_)$ is marked on the parabola.

Complete $(2, \_\_\_)$ as a solution of $y = x^2 - 6$.

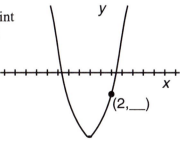

$(2, -2)$

39

The figure at the right shows the graph of $4x^2 + 9y^2 = 36$. It is an **ellipse**. Two points $(1, \_\_\_)$ are marked on the ellipse.

Complete $(1, \_\_\_)$ as a solution of $4x^2 + 9y^2 = 36$.

$\left(1, \frac{4\sqrt{2}}{3}\right), \left(1, \frac{-4\sqrt{2}}{3}\right)$

40

The figure at the right shows the graph of $x^2 - 9y^2 = 9$. It is a **hyperbola** and has two branches to its curve. Two points $(6, \_\_\_)$ are marked on the hyperbola.

Complete $(6, \_\_\_)$ as a solution of $x^2 - 9y^2 = 9$.

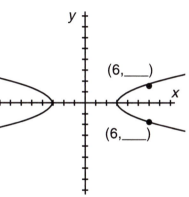

$(6, \sqrt{3}), (6, -\sqrt{3})$

FEEDBACK UNIT 6

This quiz reviews the preceding unit. Answers are at the back of the book.

1. (5, _____) is a solution for $2x - y = 13$.

2. (-5, _____) is a solution for $3x + 2y = -6$.

3. (-3, _____) is a solution for $y = \frac{4x}{3} + 5$.

4. (-4, _____) is a solution for $5x + 2y = 1$.

5. (-1, _____) is a solution for $x + 5y = 6$.

6. (5, _____) is a solution for $y = \frac{3x}{5} - 2$.

7. (2, _____) is a solution for $y = x^2 - 5x - 3$.

8. (3, _____) is a solution for $9x^2 + 16y^2 = 64$.

9. (2, _____) is a solution for $x^2 + y^2 = 49$.

10. (-5, _____) is a solution for $x^2 - y^2 = 16$.

Unit 7: Applications

In this Applications Section, the format of the text has been altered. Answers for the word problems will appear beneath the problems rather than in the right-hand column. The emphasis with these word problems is to select the correct procedure to follow with the numbers.

1

The 3-step process learned earlier for solving a word problem is:

a. Construct a table of the necessary translations of words, phrases, and sentences.
b. Solve any equation(s) obtained in the table.
c. Check answers in the original statement of the word problem.

Apply this process to the following problem.

The product of a number and 7 increased by 6 is 34.
What is the number?

Answer:

a.

word/phrase/sentence	translation
number	N
product of a number and 7	7N
product of a number and 7 increased by 6	7N + 6
The product of a number and 7 increased by 6 is 34.	7N + 6 = 34

b. $7N + 6 = 34$ is equivalent to $7N = 28$ or $N = 4$
c. It is true that the product of 4 and 7, which is 28, increased by 6 is 34.

2

The crucial step in the 3-step process used in frame 1 is the construction of a table of translations. The table allows the problem to be broken down into small parts. Even when the problems are longer and more complex, each line of the table will deal with a small part of the larger problem. Is the table of translations an important part of the process for solving word problems?

Answer: Yes

3

Some word problems require more than one number for their answer. This is the case with the following.

> A first number increased by a second number is 15. The first number is 7 more than the second number.

How many numbers are required for a complete answer to this problem?

Answer:
The problem requires 2 answers: one for the first number and the other for the second number.

4

> A first number increased by a second number is 15. The first number is 7 more than the second number. Find the numbers.

Because the problem above requires 2 numbers in its answer, its table of translations involves 2 variables.

word/phrase/sentence	translation
first number	F
second number	S
first number increased by a second number	$F + S$
A first number increased by a second number is 15.	$F + S = 15$
7 more than the second number	$S + 7$
The first number is 7 more than the second number.	$F = S + 7$

Does each line of the table of translations have a matching word, phrase, or sentence in the problem?

Answer: Yes

EQUATION SOLVING   343

5
Find the 2 equations in the table of translations of frame 4.

>Answer: F + S = 15 and F = S + 7

6
To solve the pair of equations, F + S = 15 and F = S + 7, the 2nd equation indicates that (S + 7) can replace F. Making this substitution in the first equation gives:

  F + S = 15 becomes (S + 7) + S = 15 when (S + 7) is substituted for F.
  (S + 7) + S = 15 is equivalent to 2S + 7 = 15 or 2S = 8 or S = 4.

When S = 4, then F = _____.

>Answer:
>Either equation, F + S = 15 and F = S + 7, can be used to replace S by 4. In either case, the result will be F = 11. The numbers 4 and 11 should be checked in the original problem statement.

7
Begin the 3-step process for solving the problem below by constructing a table of translations for all the necessary words, phrases, and sentences.

  Two times a first number decreased by 13 is a second number. Three times the first number increased by two times the second number is 44.

>Answer:
>
>| word/phrase/sentence | translation |
>|---|---|
>| first number | F |
>| second number | S |
>| 2 times a first number | 2F |
>| 2 times a first number decreased by 13 | 2F − 13 |
>| 2 times a first number decreased by 13 is a second number. | 2F − 13 = S |
>| 3 times a first number | 3F |
>| 2 times the second number | 2S |
>| 3 times a 1st number increased by 2 times the 2nd number | 3F + 2S |
>| 3 times a 1st number increased by 2 times the 2nd number is 44. | 3F + 2S = 44 |

8

Two times a first number decreased by 13 is a second number.
Three times the first number increased by two times the second number is 44.

Does each line of frame 7's table of translations have a matching word, phrase, or sentence in the problem?

Answer: Yes

9

The equations $2F - 13 = S$ and $3F + 2S = 44$ are in the table of frame 7. Solve them and find the solutions for the following problem.

Two times a first number decreased by 13 is a second number. Three times the first number increased by two times the second number is 44. Find the numbers.

Answer:
$3F + 2S = 44$ becomes $3F + 2(2F - 13) = 44$ when $(2F - 13)$ is substituted for S.
$3F + 2(2F - 13) = 44$ is equivalent to $7F = 70$ or $F = 10$.
When F is replaced by 10 in $2F - 13 = S$ then $S = 7$.
The numbers 10 and 7 should be checked in the original problem statement.

10

Begin the 3-step process for solving the problem below by constructing a table of translations for all the necessary words, phrases, and sentences.

The sum of two numbers is 51. The second number is 13 more than the first number.

Answer:

word/phrase/sentence	translation
first number	F
second number	S
sum of two numbers	F + S
The sum of two numbers is 51.	F + S = 51
13 more than the first number	F + 13
The second number is 13 more than the first number.	S = F + 13

11

Finish the 3-step process for solving the problem below by solving the pair of equations in frame 10 and checking the results in the problem itself.

The sum of two numbers is 51. The second number is 13 more than the first number. What are the two numbers?

> Answer:
> b. F = 19 and S = 32
> c. It is true that the sum of 19 and 32 is 51.
>   It is also true that 32 is 13 more than 19.

12

Begin the 3-step process for solving the problem below by constructing a table of translations for all the necessary words, phrases, and sentences.

The difference between two numbers is 17. Their sum is 41.

> Answer:
>
> | word/phrase/sentence | translation |
> |---|---|
> | first number | F |
> | second number | S |
> | the difference between two numbers | F − S |
> | The difference between two numbers is 17. | F − S = 17 |
> | their sum (of the two numbers) | F + S |
> | Their sum is 41. | F + S = 41 |

13

Finish the 3-step process for solving the problem below by solving the pair of equations in frame 12 and checking the results in the problem itself. Notice that the pair of equations may be easily solved by addition.

> The difference between two numbers is 17. Their sum is 41.
> What are the two numbers?

>> Answer:
>> b. The pair of equations may be added to eliminate the S's.
>> $F - S = 17$
>> $F + S = 41$
>> $2F = 58$ so $F = 29$
>> When $F = 29$ then $S = 12$.
>> c. It is true that 29 and 12 have a difference of 17.
>> It is also true that 29 and 12 have a sum of 41.

14

Begin the 3-step process for solving the problem below by constructing a table of translations for all the necessary words, phrases, and sentences.

> One number is 2 times another number. The sum of the two numbers is 21.

>> Answer:
>>
>> | word/phrase/sentence | translation |
>> |---|---|
>> | first number | F |
>> | second number | S |
>> | One number is 2 times another number. | $F = 2S$ |
>> | sum of the two numbers | $F + S$ |
>> | The sum of the two numbers is 21 | $F + S = 21$ |

15

Finish the 3-step process for solving the problem below by solving the pair of equations in frame 14 and checking the results in the problem itself.

> One number is 2 times another number. The sum of the two numbers is 21. What are the numbers?

Answer:
b. The pair of equations is best solved by substitution.   Since F = 2S then F + S = 21
    When S = 7 then F = 14.   2S + S = 21 or 3S = 21
c. It is true that 14 is 2 times 7.
    It is also true that the sum of 14 and 7 is 21.

FEEDBACK UNIT 7 FOR APPLICATIONS

This quiz reviews the preceding unit. Answers are at the back of the book.

Show the 3-step process and solve each of the following.

1. The sum of twice the first number and the second number is 4. Three times the first number diminished by the second number is 1. What are the numbers?

2. The sum of twice the first number and three times the second number is 13. Three times the first number increased by the second number is 9. What are the numbers?

3. The sum of twice a first number and the second number is 16. The second number subtracted from three times the first number is 14. What are the numbers?

4. A second number is equal to five times a first number. Three times the first number subtracted from two times the second number is 14. What are the numbers?

5. A first number is 3 less than a second. The sum of the numbers is 17. What are the numbers?

6. A first number increased by a second number is 12. The first number decreased by the second number is 2. What are the numbers?

Summary for Chapter 8

The following mathematical terms are crucial to an understanding of this chapter.

Statement	Linear equation
Complex numbers	Replacement set
Solve an equation	Truth set
Quadratic equation	Solution
Root	Equivalent equations
Additive Property for Equivalent Equations	Parabola
Multiplicative Property for Equivalent Equations	Hyperbola
Opposite	Reciprocal
Solve	Identity
Least common multiple (LCM)	Second degree equation
First degree equation	Third degree equation
Quadratic equation	Fourth degree equation
Cubic equation	Multiplicity
Polynomial equation	Quadratic formula
Completing the square	Ordered pair solution
Ordered pair (x, y)	Circle
xy-plane	Ellipse

Two equations are equivalent if they have exactly the same solution(s) (the same truth set). There are two, and only two, methods for generating equivalent linear equations. The first is the Additive Property of Equality which allows any number to be added to both sides of an equation. The second is the Multiplicative Property of Equality which allows any number, except 0, to be multiplied by both sides of an equation.

In solving equations with a variable in a denominator or where the variable is part of a radicand, procedures are used which may not generate an equivalent equation. In these cases, a check of each possible solution is a necessity.

Equations of degree 2 or higher may be solved by factoring, the use of the quadratic formula, or a combination of those processes.

Two variable equations have solutions (x, y) and replacements for x and y are generally restricted to the set of real numbers. For most equations there will be an infinite number of solutions. Often these solutions are shown by graphing.

Chapter 8 Mastery Test

The following questions test the objectives of Chapter 8. Answers are at the back of the book. The number in parentheses which follows each problem indicates the unit in which it can be learned.

Solve each of the following.

1. $5x - 3 = 9x + 16$ (1)
2. $6x - \sqrt{3} = 5\sqrt{3}$ (1)
3. $\frac{5}{3}x = \frac{2}{5}$ (1)
4. $11 = -4x$ (1)
5. $\sqrt{5} \cdot x = 10$ (1)
6. $\sqrt{3} \cdot x + \sqrt{7} = 2$ (1)
7. $(2 + i)x + (-3 + i) = (-1 + 4i)$ (1)
8. $3(x + 2) - 4(x - 1) = x + (5x - 3)$ (2)
9. $\frac{2}{3}x - \frac{1}{8} = \frac{3}{4}$ (2)
10. $\frac{9}{4x} - \frac{1}{3} = 5$ (2)
11. $\sqrt{x - 4} = 9$ (2)
12. $x^2 - 2x - 63 = 0$ (3)
13. $x^2 - 64 = 0$ (3)
14. $x^2 - 14x - 72 = 0$ (3)
15. $x^2 - 11x + 28 = 0$ (3)
16. $25x^2 - 49 = 0$ (3)
17. $x^2 - 9x + \frac{81}{4} = 0$ (3)
18. $\frac{x}{3} + \frac{1}{2x} = \frac{1}{2} + \frac{5}{6x}$ (3)
19. $\sqrt{3x} - 2 = \sqrt{x + 4}$ (3)
20. $x^2 + 7x + 3 = 0$ (5)
21. $x^2 - 9x - 3 = 0$ (5)
22. $x^2 + 4x - 8 = 0$ (5)
23. $x^2 + x + 5 = 0$ (5)
24. $x^3 - 27 = 0$ (5)
25. (3, ___) is a solution for $3x - y = 7$. (6)
26. (-2, ___) is a solution for $2x + 3y = -4$. (6)
27. (5, ___) is a solution for $y = \frac{3x}{5} + 6$. (6)
28. (-3, ___) is a solution for $y = x^2 + 6x - 2$. (6)
29. (1, ___) is a solution for $x^2 + 4y^2 = 4$. (6)
30. (-5, ___) is a solution for $x^2 + y^2 = 64$. (6)
31. Show the 3-step process and solve: The sum of twice the first number and the second number is 24. Three times the first number diminished by the second number is 11. What are the numbers? (7)
32. Show the 3-step process and solve: The second number is five times the first number. Nine times the first number diminished by the second number is 8. What are the numbers? (7)

CHAPTER 9 OBJECTIVES

The following problems illustrate the objectives of this chapter. At this time you are not expected to know how to do these problems. However, if all these problems are thoroughly understood, proceed directly to the Chapter Mastery Test. The number in parentheses which follows each problem indicates the unit in which it can be learned.

For problems 1-3 graph the solutions on the number line.

1. $x > -3$ (1)

2. $x > 5$ and $x \geq -4$ (1)

3. $x \geq 1$ or $x < -3$ (1)

4. Simplify. $7x - 9 > 10$ (2)
5. Simplify. $-8x > 24$ (2)
6. Simplify. $x - 9 \geq 5x - 4$ (2)
7. Simplify. $4x + 3 < 9x - 10$ (2)

For problems 8-10 find 2 ways to make the statement true by inserting symbols of inequality ($<, >, \leq, \geq$) to replace the question mark.

8. $(2, -5)$ is a solution for $3x - y \,?\, 7$. (3)
9. $(2, -9)$ is a solution for $y \,?\, x^2 - 2x + 5$. (3)
10. $(6, -6)$ is a solution for $x^2 + y^2 \,?\, 64$. (3)

For problems 11-14, find all solutions for each equation.

11. $|x| = 11$ (4)
12. $|3x - 4| = 8$ (4)
13. $|5x + 2| = 4$ (4)
14. $|3x - 7| = -3$ (4)

For problems 15-17, draw an interpretation of the inequality on the number line.

15. $|x - 2| \geq 5$ (5)

16. $|x + 5| < 2$ (5)

17. $|x + 1| \leq 4$ (5)

For problems 18-20, describe solutions using inequalities where necessary.

18. $|5x - 1| < 4$ (6)
19. $|2x + 1| \geq -3$ (6)
20. $|3x - 4| > 7$ (6)

Chapter 9

Inequalities and Absolute Values

Unit 1: Graphing Number Line Inequalities

The following mathematical terms are crucial to an understanding of this unit.

>   Symbols of inequality Less than
>   Less than or equal to Greater than
>   Greater than or equal to "Or" connective
>   Inequality Graph (of number line inequality)
>   Open dot Closed dot
>   "And" connective

1

The concept of size is best understood within the real number system. For that reason the only numbers considered in this chapter are real numbers.
The four **symbols of inequality** and their meanings are:

> is greater than
< is less than
≥ is greater than or equal to
≤ is less than or equal to

Is the following statement true?   10 > 3

Yes, 10 is greater than 3.

2

The symbol "<" means "**is less than.**"
Is the following statement true?   5 < 13

Yes, 5 is less than 13.

3

The symbol "≤" is a combination of (=) and (<).
The symbol "≤" means "**is less than or equal to.**"
Is the following statement true?   4 ≤ 1

No, 4 is neither less than 1 nor equal to 1.

4

The symbol ">" means "**is greater than.**"
Is the following statement true?   10 > 3

Yes, 10 is greater than 3.

5

The symbol "≥" is a combination of (=) and (>).
The symbol "≥" means "**is greater than or equal to.**"
Is the following statement true?   7 ≥ 2

Yes, 7 is greater than 2.

6

The number line is normally shown with the larger numbers to the right and the smaller numbers to the left. 10 is to the right of 3. Therefore, 10 is greater than 3. 10 > 3

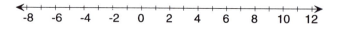

Since 1 is to the right of -5 on the number line, which of the following statements is true?

-5 > 1   1 > -5

1 > -5

INEQUALITIES AND ABSOLUTE VALUES   353

7
    2 is to the right of -5. Therefore, 2 > -5 and -5 < 2 are true.

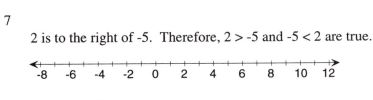

Which of the following statements is true?
        -6 < 10   10 < -6   -6 < 10

8
    -3 is to the right of -12. That makes -3 > -12 and -12 < -3 true statements. Write 2 true statements that may be made because 3 is to the right of -4.   3 > -4, -4 < 3

9
    ">" is a symbol for "greater than." "<" is a symbol for "less than." Which of the following statements are true?
(a) 6 > -1
(b) 5 < 2
(c) -6 > -10
(d) 4 < 15   (a), (c), (d)

10
    ">" means "greater than." "<" means "less than." Which of the following statements are true?
(a) 10 < 11
(b) 5 < 0
(c) 0 < -1
(d) -5 < 0   (a), (d)

11
The four symbols of inequality are: >, ≥, <, ≤.
9 ≥ 7 is true because 9 is greater than 7.
18 ≥ 18 is also true because 18 is equal to 18.
True or false?   7 ≥ 2   True

12
    "≥" means "greater than or equal to," which is an "**or**" condition and is true whenever "greater than" is true or "equal to" is true. True or false?   15 ≥ 15   True

13

$1 \geq 5$ is false because $1 > 5$ is false and $1 = 5$ is false.

$12 \geq 9$ is true because $12 > 9$ is true.

$7 \geq 7$ is _____ (true, false) because _____ is true.   true, $7 = 7$

14

Which of the following statements are true?
(a) $5 \geq 3$
(b) $18 \geq 21$
(c) $13 \geq 13$
(d) $6 \geq 5$ (a), (c), (d)

15

Which of the following statements are true?
(a) $-6 \geq 2$
(b) $-5 \geq -9$
(c) $-2 \geq -2$
(d) $7 \geq -6$ (b), (c), (d)

16

"$\leq$" is a symbol for "less than or equal to" which is an "or" condition. $4 \leq 9$ is true because $4 < 9$.

$-6 \leq -6$ is true because $-6 = -6$.

$4 \leq -1$ is _____ (true, false).   False. Both $4 < -1$ and $4 = -1$ are false.

17

Which of the following statements are true?
(a) $-9 \leq 10$
(b) $4 \leq -2$
(c) $0 \leq 9$
(d) $-3 \leq -3$ (a), (c), (d)

18

Which of the following statements are true?
(a) $5 \leq 2$
(b) $-6 \leq 3$
(c) $5 \leq -5$
(d) $11 \leq 11$ (b), (d)

19
Which of the following statements are true?
(a) $-4 \geq -3$
(b) $7 < -9$
(c) $-4 < 1$
(d) $15 \geq 15$

(c), (d)

20
The **inequality** $x > -3$ is true for any replacement of x by a number greater than (to the right of) -3. The **graph** of $x > -3$ is shown below.

The **open dot** at -3 means that -3 will not make $x > -3$ true.
On the number line below, graph $x > 7$.

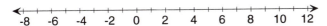   

21
The graph of $x < -2$ is all points to the left of -2 on the number line shown below.

An open dot at -2 means that -2 does not make $x < -2$ true.
On the number line below, graph $x < -6$.

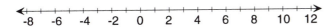   

22
The graph of $x \leq -1$ consists of -1 and all points to its left as shown on the number line below.

The **closed dot** at -1 means that -1 will make $x \leq -1$ true.
On the number line below, graph $x \leq 1$.

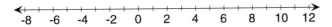   

23

The graph of x ≥ 3 includes 3 and all the points to its right as shown on the number line below.

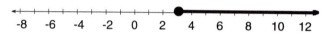

On the number line below, graph x ≥ -2.

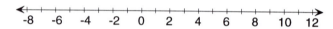 

24

Graph. x < -5

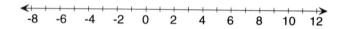 

25

Graph. x > -7

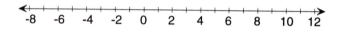 

26

Graph. x < 7

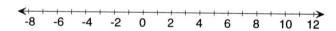 

27

Graph. x ≥ 0

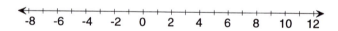 

28

Graph. x ≤ 6

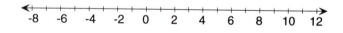 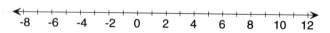

29

Graph. x > -4

30
When 2 inequalities are connected by "or," the entire sentence becomes true when at least one of its parts is true.

$$x > 2 \text{ or } x \leq -1$$

When x is replaced by 4 the "or" sentence becomes:

$$4 > 2 \text{ or } 4 \leq -1$$

Is the "or" sentence true?

Yes, 4 > 2 is true.

31

$$x > 2 \text{ or } x \leq -1$$

The "or" sentence above becomes true when either of its inequalities is true. If x is replaced by 0, is the sentence true?

No, neither 0 > 2 nor 0 ≤ -1 is true.

32
The graph of "x ≥ 2 or x < -5" is shown below.

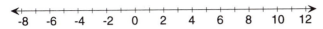

Notice: the graph contains all points obtained by graphing x ≥ 2. Does it also contain all points obtained by graphing x < -5?

Yes

33
To graph "x > 4 or x < -3," the 2 inequalities are graphed separately on the same number line.
Graph. x > 4 or x < -3

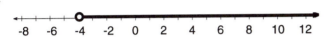

34

The figure above shows the graph of x > -4. Graph x ≤ 6 on the same number line and show the graph of x > -4 or x ≤ 6.

Each point on the number line is covered by at least one of the inequalities.

35

Graph x > 5 or x ≤ 10 by graphing the inequalities separately on the same number line.

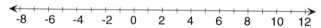 

36

The graph of x > 5 or x ≤ 10 covers each point on the number line. Each real number makes at least one of the inequalities true.

Graph. x ≥ 1 or x < -3

37

Graph. x < 1 or x ≥ 5

38

Graph. x < 2 or x ≤ -1

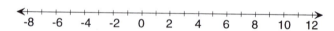 

39

When 2 inequalities are **connected by** "and," the entire sentence becomes true only when both of its parts are true.

x > -4 and x ≤ -1

When x is replaced by 4 the "and" sentence becomes:

4 > -4 and 4 ≤ -1

Is the "and" sentence true?   No, 4 ≤ -1 is false.

40
To graph x > -4 and x ≤ -1 each inequality is graphed separately and only those points that overlap are included in the final graph.

If -2 is used as a replacement for x will x > -4 and x ≤ -1 become true?

Yes

41
To graph an "and" sentence like x > -4 and x ≤ 5:
    1. Graph each inequality separately.
    2. Accept only the points where the graphs overlap.
Graph. x > -4 and x ≤ 5

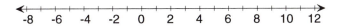 

42
The graph of x < 8 and x ≥ 1 consists of the real numbers that make both inequalities true. Graph. x < 8 and x ≥ 1

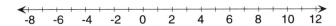 

43
To be on the graph of x > -6 and x < 0, a real number must make each inequality true. Graph. x > -6 and x < 0

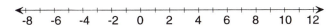 

44
The graph of x < 6 and x ≥ 8 has no points. There is no real number that is less than 6 and greater than or equal to 8.

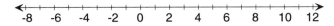

Graph. x ≤ -2 and x > 1

no points

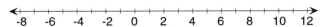 

45
Graph. $x \geq 2$ and $x < 0$

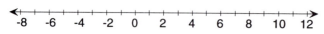

no points

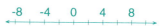

46
Graph. $x \geq -3$ and $x < 0$

FEEDBACK UNIT 1

This quiz reviews the preceding unit. Answers are at the back of the book.

Graph each of the following.

1. $x > -2$

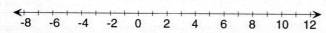

2. $x \leq 5$

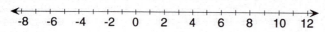

3. $x \geq 3$

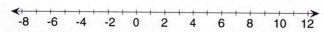

4. $x < 0$

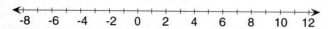

5. $x < 2$ or $x \geq 5$

6. $x \leq 3$ and $x > 0$

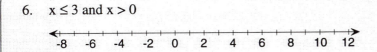

7. $x < 5$ and $x \geq 1$

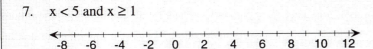

8. $x \geq 7$ or $x < 0$

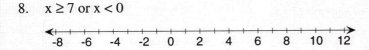

9. $x \geq -2$ and $x < 4$

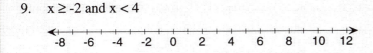

10. $x < -3$ or $x > 0$

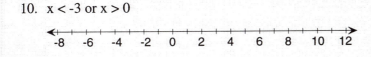

UNIT 2: EQUIVALENT INEQUALITIES

The following mathematical terms are crucial to an understanding of this unit.

Equivalent (inequalities)　　Reverse orientation
Additive Property of Inequality　　Multiplicative Property of Inequality

1

Two inequalities are **equivalent** if they are true for exactly the same real numbers.

$x \leq 9$ is equivalent to $9 \geq x$

Is $x > 5$ equivalent to $5 < x$?　　　　　　　　　　　Yes

2

The inequality symbol < has the **reverse orientation** of the symbol >. Similarly, the inequality symbol ≤ has the _____ _____ of the symbol ≥.

reverse orientation

3

An equivalent inequality to x > -3 can be found by interchanging the two sides and using the symbol of reverse orientation.

    x > -3 is equivalent to _____

-3 < x

4

Find an equivalent inequality for x ≤ 8 by interchanging the two sides and using the symbol of reverse orientation.

8 ≥ x

5

    7 ≤ x is equivalent to _____

x ≥ 7

6

Another method for generating an equivalent inequality is to add the same number to each side of the inequality. The example below shows this process.

$$\begin{array}{rcr} x - 7 & > & 2 \\ 7 & & 7 \\ \hline x & > & 9 \end{array}$$

Is x − 7 > 2 equivalent to x > 9?

Yes

7

The **Additive Property of Inequality** allows any real number to be added to both sides of an inequality to generate a new, equivalent inequality. Add 5 to both sides of x − 5 < 13. What equivalent inequality is generated?

$$\begin{array}{rcr} x - 5 & < & 13 \\ 5 & & 5 \\ \hline x & < & 18 \end{array}$$

8

    x − 5 < 13 and x < 18 are equivalent inequalities. What number can be added to both sides of x + 3 < 2 to generate x < -1?

-3

9
An **inequality is simplified** by isolating the variable on one side of the inequality. Simplify $x + 4 \leq 6$ by adding -4 to both sides of the inequality.

$$\begin{array}{rcr} x + 4 & \leq & 6 \\ -4 & & -4 \\ \hline x & \leq & 2 \end{array}$$

10
Simplify $x - 9 \leq 6$ by adding 9 to both sides of the inequality.

$x \leq 15$

11
Simplify $2x < x + 6$ by adding -x to both sides of the inequality.

$x < 6$

12
Simplify $5x \leq 9 + 4x$ by adding -4x to both sides of the inequality.

$x \leq 9$

13
Simplify. $2 + 4x \leq 5x$

$2 \leq x$ or $x \geq 2$

14
Simplify. $x - \frac{1}{4} \geq \frac{1}{2}$

$x \geq \frac{3}{4}$

15
Simplify. $x + \sqrt{2} > 3$

$x > 3 - \sqrt{2}$

16
Simplify. $\frac{1}{4} \leq \frac{1}{2} + x$

$\frac{-1}{4} \leq x$ or $x \geq \frac{-1}{4}$

17
The Additive Property works the same way with inequalities as with equations. This is not the case when multiplying both sides of an inequality by a negative number. Will $-3x > 12$ be equivalent to $3x > -12$?

No

18
The **Multiplicative Property of Inequality** has 2 parts:

1. If both sides of an inequality are multiplied by a **positive** number, an equivalent inequality with the same symbol is generated.
2. If both sides of an inequality are multiplied by a **negative** number, an equivalent inequality with the symbol of reverse orientation is generated.

Does the sign of the number make a difference when it is the multiplier times both sides of an inequality?   Yes

19
Equivalent inequalities can be generated by multiplying both sides by a positive number as in the example below.

$$7x > 35$$
$$\frac{1}{7} \cdot 7x > \frac{1}{7} \cdot 35$$
$$x > 5$$

Is $7x > 35$ equivalent to $x > 5$?   Yes, both sides were multiplied by a positive.

20
Equivalent inequalities can be generated by multiplying both sides by a negative number only when the symbol of inequality has its orientation reversed.

$$-3x > 12$$
$$\frac{-1}{3} \cdot -3x < \frac{-1}{3} \cdot 12$$
$$x < -4$$

Is $-3x > 12$ equivalent to $x < -4$?   Yes, the symbol of inequality was reversed.

21
Is $8x > 72$ equivalent to $\frac{1}{8} \cdot 8x > \frac{1}{8} \cdot 72$?   Yes, both sides were multiplied by a positive.

22
Simplify. $8x > 72$ $x > 9$

23
Is $-5x \leq -20$ equivalent to $\frac{-1}{5} \cdot -5x \geq \frac{-1}{5} \cdot -20$?

Yes, the symbol of inequality was reversed.

24
Simplify. $-5x \leq -20$

$x \geq 4$

25
Is $4x \geq 7$ equivalent to $\frac{1}{4} \cdot 4x \geq \frac{1}{4} \cdot 7$?

Yes

26
Simplify. $4x \geq 7$

$x \geq \frac{7}{4}$

27
Is $-7x < 42$ equivalent to $\frac{-1}{7} \cdot -7x < \frac{-1}{7} \cdot 42$?

No

28
Simplify. $-7x < 42$

$\frac{-1}{7} \cdot -7x > \frac{-1}{7} \cdot 42$, $x > -6$

29
Is $-2x \geq -4$ equivalent to $\frac{-1}{2} \cdot -2x \geq \frac{-1}{2} \cdot -4$?

No

30
Simplify. $-2x \geq -4$

$\frac{-1}{2} \cdot -2x \leq \frac{-1}{2} \cdot -4$, $x \leq 2$

31
Is $-5x < 10$ equivalent to $\frac{-1}{5} \cdot -5x > \frac{-1}{5} \cdot 10$?

Yes

32
Simplify. $-5x < 10$

$x > -2$

33
Simplify. $9x \geq -36$

$x \geq -4$

34
Simplify. $-6x \leq -54$

$x \geq 9$

35
Simplify. $-8x > 56$

$x < -7$

36
Simplify. $4x \leq -44$ $x \leq -11$

37
An inequality like $5x - 3 > 6$ is simplified by adding opposites and multiplying by reciprocals.

$$5x - 3 > 6$$
$$\underline{+3 +3}$$
$$5x > 9$$
$$\tfrac{1}{5} \cdot 5x > \tfrac{1}{5} \cdot 9$$
$$x > \tfrac{9}{5}$$

Simplify $8x - 7 \leq 9$ by adding opposites and multiplying by reciprocals. $8x \leq 16$
   $x \leq 2$

38
An inequality like $4 - 7x \geq 10$ is simplified by adding opposites and multiplying by reciprocals.

$$4 - 7x \geq 10$$
$$\underline{-4 -4}$$
$$-7x \geq 6$$
$$\tfrac{-1}{7} \cdot -7x \leq \tfrac{-1}{7} \cdot 6$$
$$x \leq \tfrac{-6}{7}$$

Simplify $6 - 5x > -9$ by adding opposites and multiplying by reciprocals. $-5x > -15$
   $x < 3$

39
Simplify. $9x - 7 > 3$ $x > \tfrac{10}{9}$

40
Simplify. $6x + 4 \leq 1$ $x \leq \tfrac{-1}{2}$

41
Simplify. $3 - 2x \leq 7$ $x \geq -2$

42
Simplify. $6 > 5 - x$ $x > -1$

43
Simplify. $3x + 5 \geq 17$ $x \geq 4$

44
Simplify. $5 - 3x < -4$ $x > 3$

45
To simplify $3x + 5 \geq 7x - 2$ the following steps are used.

$$\begin{array}{rcr} 3x + 5 & \geq & 7x - 2 \\ -5 & & -5 \\ \hline 3x & \geq & 7x - 7 \\ -7x & & -7x \\ \hline -4x & \geq & -7 \\ \frac{-1}{4} \cdot -4x & \leq & \frac{-1}{4} \cdot -7 \\ x & \leq & \frac{7}{4} \end{array}$$

Simplify. $7x + 3 \leq 2x - 6$ $x \leq \frac{-9}{5}$

46
Simplify. $6x - 7 > 4x + 8$ $x > \frac{15}{2}$

47
Simplify. $x + 3 \leq 4x + 9$ $x \geq -2$

48
Simplify. $13 - 4x < x - 2$ $x > 3$

49
Simplify. $12 - 3x < 4x + 5$ $x > 1$

Feedback Unit 2

This quiz reviews the preceding unit. Answers are at the back of the book.

Simplify each of the following inequalities.

1. $x - 4 \leq 7$
2. $3x + 2 > x - 5$
3. $-2x \geq 16$
4. $4 - 5x \geq 2x - 1$
5. $2 - (3x - 1) < 2 + 3(2x + 5)$
6. $-3 \geq -2x + 5$
7. $5(2x - 3) \leq 0$
8. $3x + 2 > x + 2$

Unit 3: Inequalities with Two Variables

The following mathematical terms are crucial to an understanding of this unit.

Inequality with two variables
xy-plane
Parabola
Hyperbola

Solution (inequality with 2 variables)
Circle
Ellipse

1

$x + y < 7$ is an **inequality with two variables**, x and y. Is $5x - 3y \geq 4$ an inequality with two variables?

Yes

2

The ordered pair (2, 3) is a **solution** for $x + y < 7$ because $2 + 3 < 7$ is true. Is (2, -5) a solution for $2x + y > 9$?

No, $4 - 5 > 9$ is false.

3

(5, -3) is a solution for $2x - y > 8$ because $10 + 3 > 8$ is true. Is (6, 3) a solution for $2x - y > 8$?

Yes

4

Is (4, -1) a solution for $x - 3y \leq 7$?

Yes, $4 + 3 = 7$

5

Is (-1, 3) a solution for $3x + y > 2$?

No

6

Is (4, -2) a solution for $2x - 3y \geq 10$?

Yes

7

(7, -4) is a solution for which of the following?

$x + y < 11$ $\qquad$ $x + y = 11$ $\qquad$ $x + y > 11$

$x + y < 11$

8

(6, 5) is a solution for which of the following?

$2x - 3y < -6$ $\qquad$ $2x - 3y = -6$ $\qquad$ $2x - 3y > -6$

$2x - 3y > -6$

9

Find 2 symbols of inequality ($<, \leq, >, \geq$) that will make the following statement true.

(5, 3) is a solution for $2x - 3y$? 6

$<, \leq$

10

Find 2 symbols of inequality ($<, \leq, >, \geq$) that will make the following statement true.

(4, -1) is a solution for $2x - y$? 7

$>, \geq$

11

Find 2 symbols of inequality ($<, \leq, >, \geq$) that will make the following statement true.

(-1, -3) is a solution for $2x - 3y$? 7

$\leq, \geq$

12

(4, -2) is a solution for $5x + y$ ($<, \leq, >, \geq$) 17

$>, \geq$

13

(-4, 1) is a solution for 2x + y (<, ≤, >, ≥) 9.   <, ≤

14

There are an infinite number of solutions for x + 2y < 8. Is it possible to list all of the solutions?   No

15

One way to visualize the solutions of x + 2y < 8 is to show them on an **xy-plane** like the one shown at the right. The points on the line labeled x + 2y = 8 are solutions of the equation. Points not on the line of x + 2y = 8 _____ (are, are not) solutions of the equation.   are not

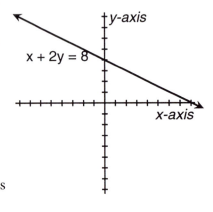

16

The graph of x – y = 2 is the straight line shown at the right. This line separates the xy-plane into 3 parts:
1. the line itself,
2. the area to one side of the line,
3. the area to the other side of the line.

Each ordered pair on the straight line is a solution of _____.   x – y = 2

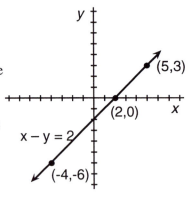

17

The figure at the right shows the line of $x - y = 2$. Five ordered pairs in the area to the upper left of the line are labeled. Are the ordered pairs in this area of the xy-plane solutions of $x - y > 2$?

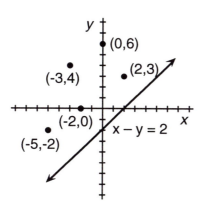

No, all of the labeled ordered pairs are, in fact, solutions of $x - y < 2$.

18

The figure at the right shows the line of $x - y = 2$ and the area to the upper left of the line where all of the ordered pairs are solutions of $x - y < 2$.

Test the ordered pairs labeled in the figure. Are they solutions of $x - y > 2$?

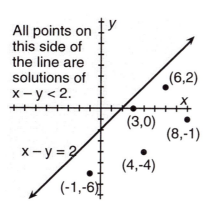

Yes, all the labeled ordered pairs are solutions of $x - y > 2$.

19

The graph of an equation like $2x + y = 3$ separates the xy-plane into 3 parts:

1. the line itself which contains the solutions of $2x + y = 3$,
2. the area to one side of the line which contains the solutions of $2x + y < 3$,
3. the area to the other side of the line which contains the solutions of _____.

$2x + y > 3$

20

The line of $2x + y = 3$ is shown. Test any ordered pair that is not on the line. Determine if its area is $2x + y < 3$ or $2x + y > 3$. Label the areas by their correct inequalities.

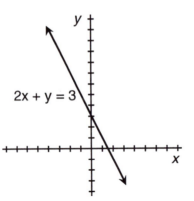

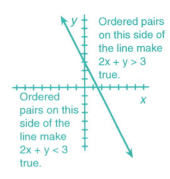

21

The line of $2x - y = 6$ is shown. Test any ordered pair that is not on the line. Determine if an area contains solutions of $2x - y < 6$ or solutions of $2x - y > 6$. Label each area by its correct inequality.

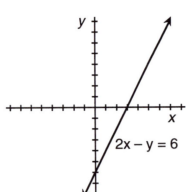

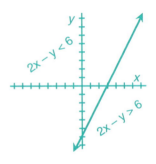

22

The line of x + 5y = -3 is shown. Test any ordered pair that is not on the line. Determine the area of x + 5y < -3 and the area of x + 5y > -3. Label each area by its correct inequality.

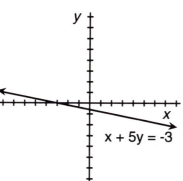

23

The line of 7x − 2y = 2 is shown. Test any ordered pair that is not on the line. Determine the area of 7x − 2y < 2 and the area of 7x − 2y > 2. Label each area by its correct inequality.

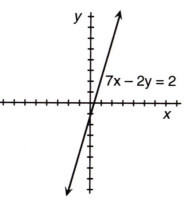

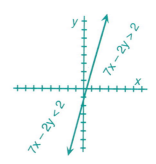

24
The ordered pair (-2, 5) is a solution for $y < x^2 + 7$ because $5 < 4 + 7$ is a true statement. Is (4, 25) a solution for $y < x^2 + 7$?

No, $25 < 16 + 7$ is false.

25
Determine if (2, 4) is a solution for $3x^2 - xy + y^2 > 18$ by replacing x by 2 and y by 4.

Yes, $12 - 8 + 16 > 18$ is true.

26
Is (4, 1) a solution for $x^2 - y^2 \leq 13$?

No, $16 - 1 \leq 13$ is false.

27
Is (-1, 3) a solution for $x^2 + 5x + 6 < y$?

Yes, $2 < 3$ is true.

28
Is (-4, 3) a solution for $x^2 + y^2 \geq 25$?

Yes, $25 = 25$ is true.

29
Find 2 symbols of inequality ($<, \leq, >, \geq$) that will make the following statement true.

   (4, -5) is a solution for y ? $x^2 - 5x - 2$

$>, \geq$

30
Find 2 symbols of inequality ($<, \leq, >, \geq$) that will make the following statement true.

   (-3, -1) is a solution for y ? $x^2 + 4x + 5$

$<, \leq$

31
Find 2 symbols of inequality ($<, \leq, >, \geq$) that will make the following statement true.

   (-4, 4) is a solution for $x^2 + y^2$? 25

$>, \geq$

32
   (-5, -14) is a solution for y ($<, \leq, >, \geq$) $2x^2 + 7x - 4$.

$\leq, <$

33
   (-2, 2) is a solution for $2x^2 + 3x + y^2$ ($<, \leq, >, \geq$) 8.

$<, \leq$

34

(5, -3) is a solution for $x^2 + y^2 + 3y$ (<, ≤, >, ≥) 30.   <, ≤

35

(-4, -7) is a solution of $y^2 - 3$ (<, ≤, >, ≥) x.   >, ≥

36
The graph of $2x^2 + 3x + y^2 = 8$ is a curve which separates the xy-plane into 3 parts:

1. the curve itself which contains the solutions of $2x^2 + 3x + y^2 = 8$,
2. one area which contains the solutions of $2x^2 + 3x + y^2 < 8$,
3. another area which contains the solutions of _____.

$2x^2 + 3x + y^2 > 8$

37
The figure at the right shows the graph of $x^2 + y^2 = 36$. Each ordered pair on the curve is a solution of $x^2 + y^2 = 36$. The curve is a **circle**. Choose any ordered pair inside the circle. Is it a solution of $x^2 + y^2 < 36$.

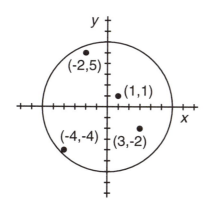

Yes

38
The figure at the right shows the circle of $x^2 + y^2 = 36$. Each ordered pair inside the circle is a solution of $x^2 + y^2 < 36$. Each ordered pair outside the circle is a solution of _____.

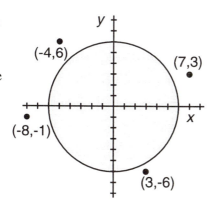

$x^2 + y^2 > 36$

39

The figure at the right shows the graph of $y = x^2 - 6$. It is a **parabola**. Each ordered pair on the curve is a solution of $y = x^2 - 6$. Three ordered pairs "inside" the parabola are labeled. Are they solutions of $y > x^2 - 6$?

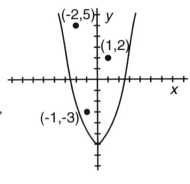

Yes

40

The figure at the right shows the parabola of $y = x^2 - 6$. Each ordered pair "inside" is a solution of $y > x^2 - 6$. Ordered pairs "outside" the parabola are solutions of _____.

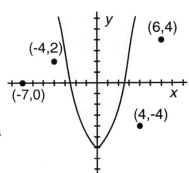

$y < x^2 - 6$

41

The figure at the right shows the graph of $9x^2 + 25y^2 = 225$. It is an **ellipse**. Choose any ordered pair inside the ellipse. Is it a solution of $9x^2 + 25y^2 < 225$.

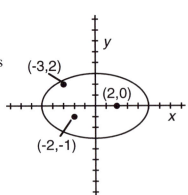

Yes

INEQUALITIES AND ABSOLUTE VALUES   377

42

The figure at the right shows the ellipse of $9x^2 + 25y^2 = 225$. Each ordered pair inside the ellipse is a solution of $9x^2 + 25y^2 < 225$. Each ordered pair outside the ellipse is a solution of

_____.

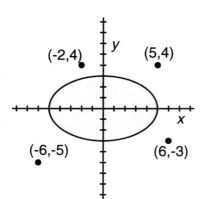

$9x^2 + 25y^2 > 225$

43

The figure at the right shows the graph of $9x^2 - 16y^2 = 144$. It is a **hyperbola** and has two branches to its curve. Four ordered pairs "between" the branches are labeled. Are they solutions of $9x^2 - 16y^2 > 144$?

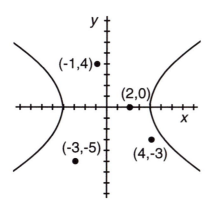

No, all are solutions of $9x^2 - 16y^2 < 144$.

44

The figure at the right shows the hyperbola of $9x^2 - 16y^2 = 144$. Each ordered pair "between" the branches is a solution of $9x^2 - 16y^2 < 144$. Ordered pairs along the x-axis to the far right and far left of the xy-plane are solutions of

_____.

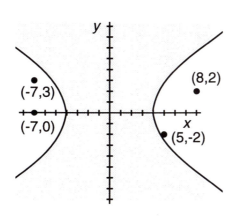

$9x^2 - 16y^2 > 144$

FEEDBACK UNIT 3

This quiz reviews the preceding unit. Answers are at the back of the book.

In each of the following problems find 2 ways to make the statement true by inserting symbols of inequality ($<, >, \leq, \geq$) to replace the question mark.

1. $(5, -7)$ is a solution for $2x - y$? 13.
2. $(-5, 9)$ is a solution for $3x + 2y$? 3.
3. $(-3, -1)$ is a solution for y ? $\frac{4x}{3} + 5$.
4. $(-4, 12)$ is a solution for $5x + 2y$? 1.
5. $(-1, 2)$ is a solution for $x + 5y$? 6.
6. $(5, 0)$ is a solution for y ? $\frac{3x}{5} - 2$.
7. $(2, -9)$ is a solution for y ? $x^2 - 5x - 3$.
8. $(2, -1)$ is a solution for $9x^2 + 16y^2$? 64.
9. $(2, 5)$ is a solution for $x^2 + y^2$? 49.
10. $(-5, 1)$ is a solution for $x^2 - y^2$? 16.

UNIT 4: SOLVING ABSOLUTE VALUE EQUATIONS

The following mathematical terms are crucial to an understanding of this unit.

Absolute value   Symbol for absolute value
Solve (absolute value equation)

1
The **absolute value** of a number is the distance that it is from 0. Since -8 is 8 from 0, the absolute value of -8 is 8. What is the absolute value of 3?

3

2

Since the distance from -7 to 0 is 7, the absolute value of -7 is 7. What is the absolute value of -13?

13

3

The **symbol for absolute value** is a pair of vertical parallel lines, "| |." To show "absolute value of -3" write | -3 |. Use the symbol, | |, to write the absolute value of -6.

| -6 |

4

| -6 | is the absolute value of -6. What is the distance between -6 and 0?

6

5

The absolute value of a number is always greater than or equal to 0.

| -9 | = 9 | -7 | = _____

7

6

| -11 | = 11 | -18 | = _____

18

7

| 5 | = 5 | 1 | = _____

1

8

| -27 | = _____

27

9

The absolute value of a number is always greater than or equal to 0. | 41 | = _____.

41

10

| 0 | = _____.

0

11

The solutions of | x | = 2 are 2 and -2 because they are numbers with a distance 2 from 0.

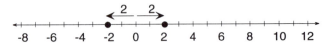

Find the 2 solutions of | x | = 6 by finding 2 numbers that are 6 units from 0.

6, -6

12

The solutions of | x | = 6 are 6 and -6. Both | 6 | = 6 and | -6 | = 6 are true statements. Find the solutions of | x | = 11.

11, -11

13

There are 2 solutions of | x | = 3 because there are 2 points on the number line that are 3 units from 0.

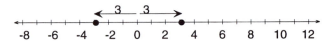

Find the solutions of | x | = 3.

3, -3

14

The solutions of | x | = 5 are those numbers which have a distance of 5 from zero. Solutions of | x | = 5 are 5 and -5.

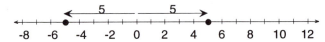

What are the solutions of | x | = 13?

13, -13

15

There is only one solution for | x | = 0 because only one number has a distance 0 from 0.
Find the solution of | x | = 0.

0

16

The distance between 2 distinct points on the number line is always positive. Is | -2 | = -2 a true statement?

No, (| -2 | = 2)

17
Is there any replacement for x that will make | x | = -2 a true statement?

No

18
There is no solution of | x | = -5 because there is no number that has a distance of -5 from 0. Distance is never negative. Find all solutions of | x | = -9.

There are none.

19
The absolute value of a positive number is the number itself.
If x ≥ 0, then | x | = x. | 3 | = 3 | 5 | = _____

5

20
The absolute value of a negative number is its opposite.
If x < 0, | x | = -x. | -7 | = -(-7) = 7 | -4 | = _____

-(-4) = 4

21
-8 is a negative number. | -8 | = -(-8) = _____

8

22
10 is a positive number. | 10 | = _____

10

23
-11 is a negative number. | -11 | = _____

11

24
If x ≥ 0, | x | = x. If x < 0, | x | = _____.

-x

25
True or false? | -6 | = -(6)

False

26
True or false? | -3 | = -(-3)

True

27
If (x – 4) represents a positive, then | x – 4 | = x – 4.
If (x – 4) represents a negative, then | x – 4 | = _____.

-(x – 4) or -x + 4

28

To **solve** | x – 5 | = 8 there are 2 possibilities:

```
              | x – 5 | = 8
             /             \
If (x – 5) is positive      If (x – 5) is negative
| x – 5 |  =  8             | x – 5 |  =  8
  x – 5   =  8              -(x – 5)   =  8
     x    =  13             -x + 5     =  8
                               -x      =  3
                                x      =  -3
```

If (x – 5) is positive, 13 is a solution. | 13 – 5 | = 8 is true.
If (x – 5) is negative, ___ is a solution. | ___ – 5 | = 8 is true.   -3, | -3 – 5 | = 8 is true

29

To solve | x + 4 | = 2 there are 2 possibilities:

```
              | x + 4 | = 2
             /             \
If (x + 4) is positive      If (x + 4) is negative
| x + 4 |  =  2             | x + 4 |  =  2
  x + 4   =  2              -(x + 4)   =  2
_____            _____
```

Solve the linear equations and find 2 solutions of | x + 4 | = 2.   -2, -6
                                                                                                                                                    | -2 + 4 | = 2 is true.
                                                                                                                                                    | -6 + 4 | = 2 is true.

30

Write 2 linear equations needed to solve | x – 3 | = 7.   x – 3 = 7   -(x – 3) = 7

31

Write 2 linear equations needed to solve | x + 6 | = 5.   x + 6 = 5   -(x + 6) = 5

32

Write 2 linear equations needed to solve | x – 9 | = 4.   x – 9 = 4   -(x – 9) = 4

33

Find the two roots of | x + 3 | = 5 by solving its 2 linear
equations.   x + 3 = 5   -(x + 3) = 5
                                                                                                                                                                   2, -8

INEQUALITIES AND ABSOLUTE VALUES   383

34
Find the two roots of | x − 5 | = 4.   9, 1

35
Solve (find the roots of) | x + 7 | = 2.   -5, -9

36
Solve. | x + 3 | = 10   7, -13

37
The equation | x − 2 | = -3 has no roots because an absolute
value cannot equal a negative. Solve. | x + 5 | = 1   -4, -6

38
Solve. | x − 5 | = -1   No roots

39
Solve. | x + 10 | = 12   2, -22

40
To solve | 5x − 2 | = 6 there are 2 possibilities:

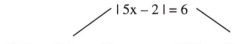

If (5x − 2) is positive	If (5x − 2) is negative
| 5x − 2 | = 6	| 5x − 2 | = 6
5x − 2 = 6	-(5x − 2) = 6
5x = 8	-5x + 2 = 6
x = $\frac{8}{5}$	-5x = 4
	x = $\frac{-4}{5}$

If (5x − 2) > 0, $\frac{8}{5}$ is a solution. | 5 • $\frac{8}{5}$ − 2 | = 6 is true.

If (5x − 2) < 0, ___ is a solution. | 5 • ___ − 2 | = 6 is true. $\frac{-4}{5}$

| 5 • $\frac{-4}{5}$ − 2 | = 6 is true.

41
To solve | 4x – 3 | = 9 write its 2 linear equations.

| 4x – 3 | = 9

If (4x – 3) is positive
| 4x – 3 | = 9
4x – 3 = 9

If (4x – 3) is negative
| 4x – 3 | = 9
-(4x – 3) = 9

Solve both equations and find 2 solutions of | 4x – 3 | = 9.

$3, \frac{-3}{2}$

| 4 • 3 – 3 | = 9 is true.

| 4 • $\frac{-3}{2}$ – 3 | = 9 is true.

42
Write 2 linear equations needed to solve | 3x + 5 | = 1.

3x + 5 = 1   -(3x + 5) = 1

43
Solve. | 3x + 5 | = 1

$\frac{-4}{3}, -2$

44
Write 2 linear equations needed to solve | 4x – 3 | = 11.

4x – 3 = 11
-(4x – 3) = 11

45
Solve. | 4x – 3 | = 11

$\frac{7}{2}, -2$

46
Solve. | 2x + 5 | = 5

0, -5

47
Solve. | 3x – 5 | = 7

$4, \frac{-2}{3}$

48
Solve. | 5x + 6 | = 4

$\frac{-2}{5}, -2$

49
Solve. | 4x + 9 | = 1

$-2, \frac{-5}{2}$

50
Solve. $|3 + 2x| = 5$ 1, -4

51
The equation $|3x + 2| = -5$ has no solutions because the absolute value of a real number cannot be negative.
Solve. $|5x - 3| = -4$ No solution

52
Solve. $|6x - 2| = -3$ No solution

53
Solve. $|6x - 7| = 5$ $2, \frac{1}{3}$

FEEDBACK UNIT 4

This quiz reviews the preceding unit. Answers are at the back of the book.

Solve each of the following.

1. $|x| = 5$
2. $|x + 2| = 6$
3. $|x - 5| = 1$
4. $|x + 2| = 7$
5. $|x + 8| = -3$
6. $|x - 12| = 0$
7. $|x + 3| = -2$
8. $|x - 7| = 7$
9. $|x - 21| = -21$
10. $|x + 4| = 7$
11. $|3x - 2| = 8$
12. $|2x + 5| = 7$

Unit 5: Interpreting Absolute Value as Distance

The following mathematical terms are crucial to an understanding of this unit.

> Distance between 2 numbers   Line segment
> Rays

1
Every absolute value expression can be interpreted as a distance. For example, the absolute value of a difference is always the **distance between the two numbers**.

$|2 - 9|$ is the distance between 2 and 9.
$|-3 - 5|$ is the distance between ___ and ___.   -3, 5

2
The absolute value of a difference is always the distance between the two numbers.

$|8 + 5| = |8 - (-5)|$ is the distance between 8 and -5.
$|4 + 3|$ is the distance between ___ and ___.   4, -3

3
Write an absolute value expression for the distance between 8 and 13. $|8 - 13|$ or $|13 - 8|$

4
Write an absolute value expression for the distance between -7 and 9. $|-7 - 9|$ or $|9 + 7|$

5
Write an absolute value expression for the distance between -6 and -4. $|-6 + 4|$ or $|-4 + 6|$

6

An absolute value equation, like $|x - 5| = 8$, can be interpreted as a distance.

$$\underbrace{|x-5|}_{\text{The distance between x and 5}} \underbrace{=}_{\text{is equal to}} \underbrace{8}_{8}$$

What two numbers are 8 away from 5?   13, -3

7

A number line interpretation of $|x - 5| = 8$ is shown below. Notice that 13 and -3 are the two points at distance 8 from 5.

Draw an interpretation of $|x - 3| = 4$ on the number line.

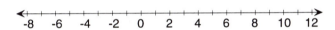   

8

A number line interpretation of $|x + 7| = 3$ is shown below.

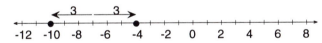

Draw an interpretation of $|x + 1| = 6$ on the number line.

9

Why is it not possible to draw a number line representation of $|x - 8| = -5$?   Distance cannot be -5.

10

An absolute value inequality, like $|x - 4| < 6$, can be interpreted as a distance.

The distance between x and 4   is less than   6

What two numbers are 6 away from 4?   10, -2

11

A number line interpretation of $|x - 4| < 6$ is shown below. Notice that 10 and -2 are shown as open dots.

Draw an interpretation of $|x - 3| < 4$ on the number line.

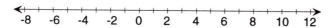

12

A number line interpretation of $|x + 2| \leq 5$ is shown below. Notice that 3 and -7 are shown as closed dots.

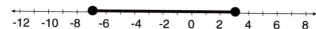

Draw an interpretation of $|x + 1| \leq 7$ on the number line.

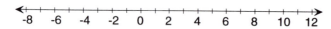

13

Why is it not possible to draw a number line representation of $|x + 3| < -8$?   Distance cannot be less than -8.

14

An absolute value inequality, like $|x + 3| > 2$, can be interpreted as a distance.

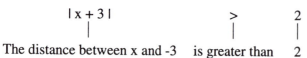

The distance between x and -3   is greater than   2

What two numbers are 2 away from -3?   -1, -5

15

A number line interpretation of $|x + 3| > 2$ is shown below. Notice that -1 and -5 are shown as open dots.

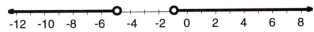

Draw an interpretation of $|x - 3| > 4$ on the number line.

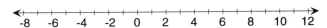

16

A number line interpretation of $|x - 4| \geq 3$ is shown below. Notice that 7 and 1 are shown as closed dots.

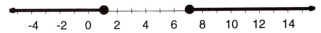

Draw an interpretation of $|x + 1| \geq 5$ on the number line.

17

Why is the number line representation of $|x + 3| > -6$ going to cover all points on the line?

Any distance is greater than -6.

18

Any equation of the form $|x - c| = d$ where $d > 0$ can be represented on the number line as 2 points.

Draw an interpretation of $|x - 6| = 2$ on the number line.

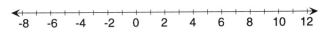

19

Draw an interpretation of $|x + 3| = 5$ on the number line.

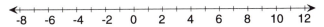

20

Any inequality of the form $|x - c| < d$ where $d > 0$ can be represented on the number line as a **line segment** joining the two roots of $|x - c| = d$.

Draw an interpretation of $|x + 2| < 3$ on the number line.

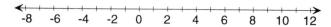   

21

Draw an interpretation of $|x - 1| \leq 4$ on the number line.

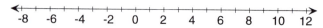   

22

Any inequality of the form $|x - c| > d$ where $d > 0$ can be represented on the number line as two **rays** whose endpoints are the two roots of $|x - c| = d$.

Draw an interpretation of $|x - 1| > 3$ on the number line.

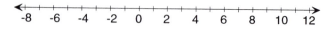   

23

Draw an interpretation of $|x + 4| \geq 2$ on the number line.

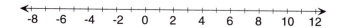   

24

The inequality $|x - 5| > -4$ is true for every point on the number line because an absolute value is always greater than -4. Is $|x - 3| > -8$ true for every point on the number line?

Yes

25

The inequality | x − 7 | < -5 is false for any point on the number line because an absolute value is never greater than -5. Is | x − 2 | < -6 false for every point on the number line? Yes

---

FEEDBACK UNIT 5

This quiz reviews the preceding unit. Answers are at the back of the book.

For each problem draw an interpretation on the number line.

1. | x − 1 | > 3

   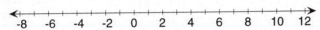

2. | x − 4 | < 6

   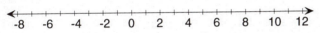

3. | x − 5 | = 2

   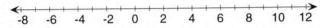

4. | x + 3 | < -2

   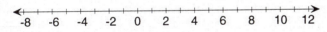

5. | x − 1 | ≤ 7

6. | x + 2 | > 3

7. | x − 2 | ≥ 6

8. | x + 4 | < 3

   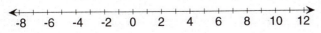

Unit 6: Absolute Value Inequalities

1
The number line interpretation of | x − 1 | > 3 is 2 rays.

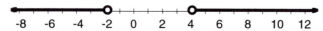

Is the number line interpretation of | x − 7 | > -2 also 2 rays?

No, it is the entire number line. Any distance is greater than -2.

2
The number line interpretation of | x − 4 | < 6 is a line segment.

Is the number line interpretation of | x + 3 | < -2 also a line segment?

No, it has no points. A distance is never less than -2.

3
Watch for the situations where an absolute value inequality results in the entire line or no points. Except for those instances, absolute value inequalities will be either 2 rays or a line segment. What shape is the number line interpretation of | x − 4 | > 7?

2 rays

4
The points on 2 rays may always be described by an "or" inequality. For example, the 2 rays of | x − 4 | > 7 are described by: x > 11 ___ x < -3

or

5
The points on a line segment may always be described by an "and" inequality. For example, the line segment of | x − 5 | < 3 is described by: x < 8 _____ x > 2

and

6

| x + 5 | > 2 has an equivalent "or" sentence.
| x + 6 | < 5 has an equivalent "____" sentence.   and

7
To write | x – 5 | > 7 as an "or" sentence, the following steps are used.

$$|x - 5| > 7$$

If (x – 5) is positive		If (x – 5) is negative
| x – 5 | > 7		| x – 5 | > 7
x – 5 > 7	or	-(x – 5) > 7
x > 12	or	-x + 5 > 7
x > 12	or	-x > 2
x > 12	or	x < -2

Does | x – 5 | > 7 describe the same numbers as the "or" sentence x > 12 or x < -2?   Yes

8
Any inequality of the form | x – c | > d where d > 0 can be described by an "or" sentence containing two linear inequalities. Can | x – 3 | > 2 be described by an "or" sentence?   Yes

9
To find the "or" sentence describing | x – 3 | > 2 the following steps are used.

$$|x - 3| > 2$$

If (x – 3) is positive		If (x – 3) is negative
x – 3 > 2	or	-(x – 3) > 2
x > 5	or	-x + 3 > 2
x > 5	or	-x > -1
x > 5	or	x < 1

The "or" sentence for | x – 3 | > 2 is: _____   x > 5 or x < 1

10

Complete the process for an "or" sentence for $|x - 2| > 8$.

$$|x - 2| > 8$$

If $(x - 2)$ is positive		If $(x - 2)$ is negative
$x - 2 > 8$	or	$-(x - 2) > 8$
$x > 10$	or	$-x + 2 > 8$
_____	__	_____

$x > 10$ or $x < -6$

11

Describe $|x + 8| > 1$ by a pair of linear inequalities.

$x > -7$ or $x < -9$

12

To describe $|5x - 3| > 4$ by a pair of linear inequalities use the following steps.

$$|5x - 3| > 4$$

If $(5x - 3)$ is positive		If $(5x - 3)$ is negative
$5x - 3 > 4$	or	$-(5x - 3) > 4$
$5x > 7$	or	$-5x + 3 > 4$
$x > \frac{7}{5}$	or	$-5x > 1$
$x > \frac{7}{5}$	or	x _____

$< \frac{-1}{5}$

13

The solutions of $|5x - 3| > 4$ may be written as a pair of linear inequalities:

$$x > \frac{7}{5} \text{ or } x < \frac{-1}{5}$$

Describe the solutions of $|3x - 1| > 2$ by a pair of linear inequalities.

$x > 1$ or $x < \frac{-1}{3}$

14

Describe the solutions of $|x + 3| \geq 7$ by a pair of linear inequalities.

$x \geq 4$ or $x \leq -10$

15

Describe the solutions of $|x - 5| \geq 3$ by a pair of linear inequalities.

$x \geq 8$ or $x \leq 2$

INEQUALITIES AND ABSOLUTE VALUES   395

16
The points on 2 rays may always be described by an "or" inequality. The points on a line segment may always be described by an "_____" inequality.

and

17
$|x + 13| > 5$ has an equivalent "or" sentence.
$|x - 7| < 9$ has an equivalent "____" sentence.

and

18
To write $|x - 5| < 7$ as an "and" sentence, the following steps are used.

$$|x - 5| < 7$$

If $(x - 5)$ is positive		If $(x - 5)$ is negative
$|x - 5| < 7$		$|x - 5| < 7$
$x - 5 < 7$	and	$-(x - 5) < 7$
$x < 12$	and	$-x + 5 < 7$
$x < 12$	and	$-x < 2$
$x < 12$	and	$x > -2$

Does $|x - 5| < 7$ describe the same numbers as the "and" sentence $x < 12$ and $x > -2$?

Yes

19
Any inequality of the form $|x - c| < d$ where $d > 0$ can be described by an "and" sentence containing two linear inequalities. Can $|x - 3| < 2$ be described by an "and" sentence?

Yes

20
To find the "and" sentence describing $|x - 3| < 2$ the following steps are used.

$$|x - 3| < 2$$

If $(x - 3)$ is positive		If $(x - 3)$ is negative
$x - 3 < 2$	and	$-(x - 3) < 2$
$x < 5$	and	$-x + 3 < 2$
$x < 5$	and	$-x < -1$
$x < 5$	and	$x > 1$

The "and" sentence for $|x - 3| < 2$ is: _____

$x < 5$ and $x > 1$

21
Complete the process for an "and" sentence for $|x - 2| < 8$.

$$|x - 2| < 8$$

If $(x - 2)$ is positive		If $(x - 2)$ is negative
$x - 2 < 8$	and	$-(x - 2) < 8$
$x < 10$	and	$-x + 2 < 8$
_____	_	_____

$x < 10$ and $x > -6$

22
Describe $|x + 8| < 1$ by a pair of linear inequalities.

$x < -7$ and $x > -9$

23
Complete the following to describe $|5x - 3| < 4$ by a pair of linear inequalities.

$$|5x - 3| < 4$$

If $(5x - 3)$ is positive		If $(5x - 3)$ is negative
$5x - 3 < 4$	and	$-(5x - 3) < 4$
$5x < 7$	and	$-5x + 3 < 4$
$x < \frac{7}{5}$	and	$-5x < 1$

$x < \frac{7}{5}$ and $x > \frac{-1}{5}$

24
Describe the solutions of $|x + 3| \leq 7$ by a pair of linear inequalities.

$x \leq 4$ and $x \geq -10$

25
Describe the solutions of $|x - 5| \leq 3$ by a pair of linear inequalities.

$x \leq 8$ and $x \geq 2$

26
To describe $|4x - 3| > 7$ by a pair of linear inequalities,
1. Determine whether an "or" or "and" sentence is needed.
2. Simplify the inequalities in the sentence.

Does $|4x - 3| > 7$ require an "and" sentence?

No, it is an "or."

27
Describe $|4x - 3| > 7$ by a pair of linear inequalities. $x > \frac{5}{2}$ or $x < -1$

28
To describe $|5x + 2| \leq 8$ by a pair of linear inequalities,

1. Determine whether an "or" or "and" sentence is needed.
2. Simplify the inequalities in the sentence.

Does $|5x + 2| \leq 8$ require an "and" sentence?   Yes

29
Describe $|5x + 2| \leq 8$ by a pair of linear inequalities. $x \leq \frac{6}{5}$ and $x \geq -2$

30
Describe $|7x - 2| \geq 3$ by a pair of linear inequalities. $x \geq \frac{5}{7}$ or $x \leq \frac{-1}{7}$

31
Describe $|x + 3| > 7$ by a pair of linear inequalities. $x > 4$ or $x < -10$

32
Describe $|4x - 7| < 2$ by a pair of linear inequalities. $x < \frac{9}{4}$ and $x > \frac{5}{4}$

33
Are inequalities needed to describe the solutions of $|4x + 7| > -5$?   No, it is true for every real number.

34
Are inequalities needed to describe the solutions of $|6x - 5| \leq -1$?   No, it is not true for any real number.

Feedback Unit 6

This quiz reviews the preceding unit. Answers are at the back of the book.

Describe solutions for each of the following using inequalities where necessary.

1. $|x - 5| > 7$

2. $|x + 4| < -2$

3. $|x + 9| \geq 4$

4. $|x - 6| \leq 1$

5. $|3x - 2| > 7$

6. $|2x + 3| \leq 4$

7. $|6x - 1| > -6$

8. $|4x + 7| \geq 3$

9. $|9x - 4| < 8$

10. $|3x + 10| \leq 4$

Unit 7: Applications

In this Applications Section, the format of the text has been altered. Answers for the problems appear beneath them rather than in the right-hand column. Your studying emphasis should be on learning the best procedures to follow with word problems.

1

In a base 10 numeration system, the numeral 57 has 5 as its tens digit and 7 as its ones digit. The number it represents is $10 \cdot 5 + 7$. If a numeral has T as its tens digit and U as its units (ones) digit, what number does it represent?

Answer: $10T + U$

2

The digits of 83 are 8 and 3. The sum of the digits is 11. Find the sum of the digits of 97.

Answer: 9 + 7 = 16

3

If a two-digit numeral has K as its tens digit and S as its ones digit, the sum of the digits is K + S. The number it represents is 10K + S. If M is a tens digit and N is a units (ones) digit then the sum of the digits is _____ and the number it represents is _____.

Answer:
M + N is the sum of the digits.
10M + N is the number the digits represent.

4

The 3-step process learned earlier for solving a word problem is:

a. Construct a table of the necessary translations of words, phrases, and sentences.
b. Solve any equation(s) obtained in the table.
c. Check answers in the original statement of the word problem.

For the digit problem shown below, begin the 3-step process by constructing a table of translations. Since neither digit is known, let T represent the tens digit and U represent the ones digit.

A 2-digit numeral has 10 as the sum of its digits and the tens digit is 4 more than the ones digit.

Answer:

word/phrase/sentence	translation
tens digit	T
units (ones) digit	U
sum of its digits	T + U
10 as the sum of its digits	10 = T + U
4 more than the ones digit	U + 4
tens digit is 4 more than the ones digit	T = U + 4

5

The table of translations for the following problem is shown in frame 4.

    Find the two-digit numeral if the sum of the two digits is 10 and the tens digit is 4 more than the ones digit.

Complete the last 2 steps of the problem solving process by solving the equations and checking the result in the original word problem.

    Answer:
    b. To solve $T + U = 10$ and $T = U + 4$ replace T by $(U + 4)$ in the 1st equation.
         $T + U = 10$ becomes $(U + 4) + U = 10$ or $2U = 6$. So $U = 3$.
         When $U = 3$, then $T = 7$.
         The 2-digit numeral is $10 \cdot 7 + 3$ or 73.

    c. The sum of the digits of 73 is $7 + 3 = 10$.
         The tens digit, 7, is 4 more than the ones digit, 3.

6

Use the 3-step process to solve the following word problem. Let T represent the tens digit and U represent the ones digit in the table of translations.

    Find the 2-digit numeral for which the tens digit subtracted from the ones digit is 5 and the sum of the digits is 11.

    Answer:
    a.

word/phrase/sentence	translation
tens digit	T
ones digit	U
tens digit subtracted from the ones digit	$U - T$
tens digit subtracted from the ones digit is 5	$U - T = 5$
sum of the digits	$T + U$
sum of the digits is 11	$T + U = 11$

    b. The best way to solve $U - T = 5$ and $T + U = 11$ is to add the equations.
         This will eliminate the Ts and give $2U = 16$ or $U = 8$.
         When $U = 8$ then $T = 3$.
         The 2-digit numeral is $10 \cdot 3 + 8$ or 38.

    c. The tens digit, 3, subtracted from the ones digit, 8, is 5.
         The sum of the digits, $3 + 8$, is 11.

7

Use the 3-step process to solve the following word problem. Let T represent the tens digit and U represent the ones digit in the table of translations.

Find the two-digit numeral in which the tens digit is 8 more than the ones digit and the tens digit is 4 more than 5 times the ones digit.

Answer:

a.

word/phrase/sentence	translation
tens digit	T
ones digit	U
8 more than the ones digit	U + 8
tens digit is 8 more than the ones digit	T = U + 8
5 times the ones digit	5U
4 more than 5 times the ones digit	5U + 4
tens digit is 4 more than 5 times the ones digit	T = 5U + 4

b. Since T = U + 8 and T = 5U + 4, substitution gives U + 8 = 5U + 4. Solving the equation gives U = 1 and either equation will lead to T = 9. The 2-digit numeral is 10 • 9 + 1 or 91.

c. The tens digit, 9, is 8 more than the ones digit, 1.
The tens digit, 9, is 4 more than 5 times 1, the ones digit.

8

The problem below involves both a number and its digits.

Find the two-digit numeral in which the tens digit is 1 more than the ones digit and the number is 6 times the sum of its digits.

Construct a table of translations for the problem representing the tens digit by T and the ones digit by U. Notice that this means the number is 10T + U.

Answer:

a.

word/phrase/sentence	translation
tens digit	T
ones digit	U
1 more than the ones digit	U + 1
tens digit is 1 more than the ones digit	T = U + 1
sum of its digits	T + U
6 times the sum of its digits	6(T + U)
the number	10T + U
the number is 6 times the sum of its digits	10T + U = 6(T + U)

9

Find the two-digit numeral in which the tens digit is 1 more than the ones digit and the number is 6 times the sum of its digits.

Complete the 3-step process for the problem above by solving the pair of equations from frame 8 and checking the results in the original word problem.

Answer:
b. First simplify $10T + U = 6(T + U)$ to $4T = 5U$ then use $T = U + 1$ to substitute. This gives $4(U + 1) = 5U$ or $U = 4$. Hence $T = 5$. The number is $10 \cdot 5 + 4$ or 54.

c. The tens digit, 5, is 1 more than the ones digit, 4.
The number, 54, is 6 times the sum of the digits, $5 + 4$, because $6 \cdot 9 = 54$.

10
The problem below involves both a number and its digits.

Find the two-digit numeral for which the following is true. The tens digit is 1 more than twice the ones digit. The number is 7 times the tens digit increased by 8 times the ones digit.

Construct a table of translations for the problem representing the tens digit by T and the ones digit by U. Notice that this means the number is $10T + U$.

Answer:
a.

word/phrase/sentence	translation
tens digit	T
ones digit	U
twice the ones digit	2U
1 more than twice the ones digit	$2U + 1$
tens digit is 1 more than twice the ones digit	$T = 2U + 1$
7 times the tens digit	7T
8 times the ones digit	8U
7 times the tens digit increased by 8 times the ones digit	$7T + 8U$
the number	$10T + U$
the number is 7 times the tens digit increased by 8 times the ones digit	$10T + U = 7T + 8U$

11

Find the two-digit numeral for which the following is true. The tens digit is 1 more than twice the ones digit. The number is 7 times the tens digit increased by 8 times the ones digit.

Complete the 3-step process for the problem above by solving the pair of equations from frame 10 and checking the results in the original word problem.

Answer:
b. First simplify $10T + U = 7T + 8U$ to $3T = 7U$.
Then use $T = 2U + 1$ to substitute. This gives $3(2U + 1) = 7U$ or $U = 3$.
Hence $T = 7$. The number is $10 \cdot 7 + 3$ or 73.

c. The tens digit, 7, is 1 more than twice the ones digit, 3.
The number, 73, is 7 times the tens digit increased by 8 times the ones digit. $73 = 7 \cdot 7 + 8 \cdot 3$

12

The numerals 47 and 74 are alike in that they have the same digits, but 47 and 74 are different numbers because $10 \cdot 4 + 7$ is not equal to $10 \cdot 7 + 4$. If the digits of a numeral are reversed, as with 47 and 74, will the numbers they represent be equal?

Answer: No. $10T + U$ is not equal to $10U + T$ except when $T = U$.

13

Reversing the digits of 81 gives 18. The numbers are different because $81 = 10 \cdot 8 + 1$ while $18 = 10 \cdot 1 + 8$. If M and N represent digits, what are the 2 different numbers that can be represented by their 2-digit numerals?

Answer: $10M + N$ is one number and $10N + M$ is the number with its digits reversed.

14

If $10K + L$ is the number of a 2-digit numeral, what is the number when the digits are reversed?

Answer: $10L + K$

15

The problem below involves both a number and the number with its digits reversed.

> The sum of the digits of a two-digit number is 10. If the digits are reversed, the new number is 36 less than the original number. Find the two-digit number.

Construct a table of translations for the problem representing the tens digit by T and the ones digit by U. Notice that this means the number is $10T + U$ and the number with its digits reversed is $10U + T$.

Answer:

a.

word/phrase/sentence	translation
tens digit	T
ones digit	U
the sum of the digits	$T + U$
the sum of the digits is 10	$T + U = 10$
the number	$10T + U$
the number with its digits reversed	$10U + T$
36 less than the original number	$(10T + U) - 36$
the new number is 36 less than the original number	$10U + T = (10T + U) - 36$

16

The sum of the digits of a two-digit number is 10. If the digits are reversed, the new number is 36 less than the original number. Find the two-digit number.

Complete the 3-step process for the problem above by solving the pair of equations from frame 15 and checking the results in the original word problem.

Answer:
b. First simplify $10U + T = (10T + U) - 36$ to $9U - 9T = -36$, and dividing each term by 9 gives $U - T = -4$. This may be added to $T + U = 10$ to give $2U = 6$ or $U = 3$. Hence, $T = 7$. The number is $10 \cdot 7 + 3$ or 73.

c. The sum of the digits, 7 and 3, is 10. The number with its digits reversed, 37, is 36 less than the original number, 73.

17

The problem below involves both a number and the number with its digits reversed.

> If the digits of a two-digit number are reversed, the new number is 18 more than the original number. The sum of the digits is 6. Find the original number.

Complete the 3-step process for solving the problem.

Answer:

a.

word/phrase/sentence	translation
tens digit	T
ones digit	U
the original number	$10T + U$
18 more than the original number	$(10T + U) + 18$
the number with its digits reversed	$10U + T$
the new number is 18 more than the original number	$10U + T = (10T + U) + 18$
the sum of the digits	$T + U$
the sum of the digits is 6	$T + U = 6$

b. First simplify $10U + T = (10T + U) + 18$ to $9U - 9T = 18$, and dividing each term by 9 gives $U - T = 2$. This may be added to $T + U = 6$ to give $2U = 8$ or $U = 4$. Hence, $T = 2$. The number is $10 \cdot 2 + 4$ or 24.

c. The sum of the digits, 4 and 2, is 6.
The number with its digits reversed, 42, is 18 more than the original number, 24.

Feedback Unit 8 for Applications

Show the 3-step process for solving each of the following.

1. The sum of the digits of a two-digit numeral is 10. If the ones digit is 4 less than the tens digit, what is the numeral?

2. Find the two-digit numeral whose tens digit is twice the ones digit and the sum of the digits is 12.

3. Find the two-digit numeral which has 3 as the sum of its digits and the number itself is 6 times the ones digit.

4. Find the two-digit numeral whose ones digit is 3 more than the tens digit and the number itself is 4 times the sum of its digits.

5. The ones digit of a two-digit number is twice the tens digit. If the digits are reversed, the new number is 36 more than the original number. What is the original number?

6. The ones digit of a two-digit number is three times the tens digit. If the digits are reversed, the new number is 54 more than the original number. What is the original number?

Summary for Chapter 9

The following mathematical terms are crucial to an understanding of this chapter.

Symbols of inequality	Less than
Less than or equal to	Greater than
Greater than or equal to	"Or" connective
Inequality	Graph (of number line inequality)
Open dot	Closed dot
"And" connective	Reverse orientation
Equivalent (inequalities)	Multiplicative Property of Inequality
Additive Property of Inequality	Solution (inequality with 2 variables)
Inequality with two variables	Circle
xy-plane	Ellipse
Parabola	Symbol for absolute value
Hyperbola	Line segment
Absolute value	Solve (absolute value equation)
Distance between 2 numbers	Rays

To simplify an inequality, use the Additive Property of Inequality and/or the Multiplicative Property of Inequality to generate the equivalent inequality, $x > k$, $x < k$, $x \geq k$, or $x \leq k$.

The Additive Property of Inequality states:

if $x + a > b$, then $x + a - a > b - a$ and $x > b - a$.

The Multiplicative Property of Inequality states:

1. If a is positive ($a > 0$) and $ax > b$, then $\frac{1}{a} \cdot ax > \frac{1}{a} \cdot b$ and $x > \frac{b}{a}$.

2. If a is negative ($a < 0$) and $ax > b$, then $\frac{1}{a} \cdot ax < \frac{1}{a} \cdot b$ and $x < \frac{b}{a}$.

Each of the absolute value sentences below can be solved (simplified) by first writing the equivalent compound sentence on the right.

$\lvert x \rvert = a$	$x = a$ or $-x = a$
$\lvert x + a \rvert = b$	$x + a = b$ or $-(x + a) = b$
$\lvert x + a \rvert > b$	$x + a > b$ or $-(x + a) > b$
$\lvert x + a \rvert < b$	$x + a < b$ and $-(x + a) < b$

Chapter 9 Mastery Test

The following questions test the objectives of Chapter 9. Answers are at the back of the book. The number in parentheses which follows each problem indicates the unit in which it can be learned.

For problems 1-3 graph the solutions on the number line.

1. $x < -1$ (1)

2. $x > 4$ and $x \leq 7$ (1)

3. $x > 3$ or $x < -4$ (1)

4. Simplify. $6x - 7 > 5$ (2)
5. Simplify. $-5x > -20$ (2)
6. Simplify. $7 - x \leq 3x + 4$ (2)
7. Simplify. $5x - 3 < x + 7$ (2)

For problems 8-10 find 2 ways to make the statement true by inserting symbols of inequality ($<, >, \leq, \geq$) to replace the question mark.

8. $(3, -4)$ is a solution for $x - 3y \: ? \: 11$. (3)
9. $(5, -1)$ is a solution for $y \: ? \: x^2 - 3x - 5$. (3)
10. $(7, -4)$ is a solution for $x^2 - y^2 \: ? \: 25$. (3)

For problems 11-14, find all solutions for each equation.

11. $|x| = 8$ (4)
12. $|4x - 3| = 6$ (4)
13. $|3x - 10| = -8$ (4)
14. $|2x - 5| = 7$ (4)

For problems 15-17, draw an interpretation of the inequality on the number line.

15. $|x - 5| \geq 3$ (5)

16. $|x + 2| < 5$ (5)

17. $|x + 1| \leq 3$ (5)

For problems 18-20, describe solutions using inequalities where necessary.

18. $|2x - 5| < 9$ (6)
19. $|x + 1| \geq 4$ (6)
20. $|4x + 9| > -3$ (6)

21. Show the 3-step process and solve: Find the two-digit numeral whose tens digit is 1 less than twice the ones digit and the sum of the digits is 11. (7)

22. Show the 3-step process and solve: The ones digit of a two-digit number is 1 more twice the tens digit. If the digits are reversed, the new number is 27 more than the original number. What is the original number? (7)

Chapter 10 Objectives

The following problems illustrate the objectives of this chapter. At this time you are not expected to know how to do these problems. However, if all these problems are thoroughly understood, proceed directly to the Chapter Mastery Test. The number in parentheses which follows each problem indicates the unit in which it can be learned.

1. List the domain elements of $\{(2, 4), (3, 6), (4, 8), (5, -11)\}$. (1)

2. Is $\{(4, 0), (2, -2), (0, -4), (-2, 15)\}$ a function? (1)

3. What number is not in the domain of $\{(x, y) \mid y = \frac{4x + 7}{3x - 2}\}$? (1)

4. Is $\{(x, y) \mid y = -5x + 7\}$ a linear function? (1)

5. Graph $2x + 3y = 9$. (2)

6. Graph $y = 3x - 5$. (2)

7. Find the slope through the points (5, -3) and (-2, 5). (3)

8. Find the slope through the points (6, 1) and (4, 1). (3)

9. Find the y-intercept of $y = 7x - 1$. (4)

10. Find the y-intercept of $4x - 7y = 10$. (4)

11. Find the slope of $f(x) = \frac{-2}{5}x + 7$. (4)

12. Find the slope of $3x - 6y = 13$. (4)

13. Find the slope-intercept equation of the line with slope $\frac{4}{5}$ and y-intercept (0, -5). (5)

14. Find the slope-intercept equation of the line with slope -3 through (-2, -3). (5)

15. Find the slope-intercept equation of the line through (5, 4) and (-2, -1). (5)

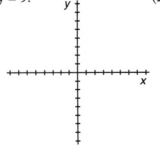

Chapter 10
Linear Functions

Unit 1: Relations and Functions

The following mathematical terms are crucial to an understanding of this unit.

- Set
- Ordered pairs
- Relation
- Domain
- Linear function

- Element
- Set selector method
- First component
- Function

1
Any collection of numbers, objects, or ideas may be called a **set**. The collection of numbers 6, 13, 19, 27, and 41 is a set and may be shown as {6, 13, 19, 27, 41}. Use braces to show the set containing 9, 51, 96, and 103.

{9, 51, 96, 103}

2
Each number in the set {6, 13, 19, 27, 41} is an **element** of the set. 19 is an element of the set {6, 13, 19, 27, 41}, and 21 _____ (is, is not) an element of the set {6, 13, 19, 27, 41}.

is not

3
The symbol for "is an element of" is "$\in$."
$43 \in \{6, 19, 43, 81\}$ is a true statement. Is the following statement true? $65 \in \{18, 51, 60, 65, 91, 106\}$

Yes

4
True or false? $43 \in \{19, 27, 43, 89\}$

True

5
The elements of {(2, 7), (8, 5), (9, 7)} are **ordered pairs**. $(2, 7) \in \{(2, 7), (8, 5), (9, 7)\}$ is a true statement. $7 \in \{(2, 7), (8, 5), (9, 7)\}$ is a false statement because the only elements of the set are ordered pairs.
True or false? $(5, 3) \in \{(2, 1), (-5, 1), (5, 3)\}$

True

6
The set {6, 7, 12, 19} has four elements. The set {(5, 7), (6, 13)} has two elements, each of which is an ordered pair. Does {-2, 5, 6, -10} have the same number of elements as {(-2, 5), (6, -10)}?

No

7
The notation {x | x is a counting number} is read as:

    The set of all replacements for x that will make
        "x a counting number"
    a true statement.

Will every element of {1, 2, 3, . . . } belong to {x | x is a counting number}?

Yes

8
Which of the following is the condition for membership in {x | x is an integer}?
    a.   The replacement of x is a counting number.
    b.   The replacement of x is an integer.
    c.   The replacement of x is a rational number.
    d.   The replacement of x is an irrational number.   (b)

9
Which of the following is the condition for membership in {x | x is an odd counting number}?
    a.   A replacement of x is a counting number.
    b.   A replacement of x is an integer.
    c.   A replacement of x is an even counting number.
    d.   A replacement of x is an odd counting number.   (d)

10
True or false? $-7 \in \{x \mid x \text{ is an integer}\}$ True

11
True or false? $13 \in \{x \mid x \text{ is an odd counting number}\}$ True

12
$\{x \mid x + 9 = 11\}$ is a set described by the **set selector method**. To be an element of the set, a replacement for x must make $x + 9 = 11$ a true statement. What is the only element of $\{x \mid x + 9 = 11\}$?   2

13
$\{2\}$ and $\{x \mid x + 9 = 11\}$ are two methods for describing the same set. Describe $\{x \mid x + 4 = 19\}$ in a simpler way.   {15}

14
There are two elements in $\{x \mid (x - 6)(x + 2) = 0\}$ because either 6 or -2 make $(x - 6)(x + 2) = 0$ a true statement.

    $\{x \mid (x - 6)(x + 2) = 0\} = \{6, -2\}$.
    $\{x \mid (x + 9)(x - 3) = 0\} = $ _____.   {-9, 3}

15

$\{(x, y) \mid y = 2x - 3\}$ is a set of ordered pairs.
Is (5, 7) an element of the set?

Yes, $7 = 2 \cdot 5 - 3$ is true.

16

Is (4, -3) an element of $\{(x, y) \mid y = x - 7\}$?

Yes, $-3 = 4 - 7$ is true.

17

Is (-5, 25) an element of $\{(x, y) \mid x^2 = y\}$?

Yes

18

A set is a **relation** if every element is an ordered pair.
Is $\{(4, 3), (7, 2), (6, -2), (5, 4), (1, -3)\}$ a relation?

Yes

19

To be a relation, each element of a set must be an ordered pair. Is $\{6, 5, 4, (2, 1)\}$ a relation?

No

20

Every element of $\{(x, y) \mid x^2 - 5 = y\}$ is an ordered pair. Is the set a relation?

Yes

21

Is $\{x \mid x + 5 = 4\}$ a relation?

No

22

In the ordered pair (7, -3), the **first component** is 7.
What is the first component of (-6, 4)?

-6

23

The set of all first components of a relation is its **domain**.
The domain of $\{(3, 5), (4, 6), (11, 13), (17, 20)\}$ is the set of first components or $\{3, 4, 11, 17\}$. Find the domain of $\{(6, 5), (-2, 4), (5, 1), (4, 7)\}$.

$\{6, -2, 5, 4\}$

24

The domain of $\{(2, 5), (9, 1), (-6, 3)\}$ is $\{2, 9, -6\}$.
Find the domain of $\{(5, -2), (4, 1), (7, 3), (-5, 0)\}$.

$\{5, 4, 7, -5\}$

25
The domain of $\{(x, y) \mid 3x - 2 = y\}$ is the set of real numbers because any real number can replace x. Find the domain of $\{(x, y) \mid y = 4x + 7\}$.

Set of real numbers

26
In the open expression $\frac{3}{x-5}$, 5 cannot be used as a replacement for x because $\frac{3}{5-5}$ with 0 as a denominator is not acceptable. What replacement for x cannot be used in the open expression $\frac{7}{x-3}$?

3

27
Zero cannot be used for the variable x in the expression $\frac{10}{x}$. What number cannot be used to replace x in the open expression $\frac{-12}{x}$?

0

28
The domain of $\{(x, y) \mid \frac{7}{x+4} = y\}$ is all real numbers except -4. Find the domain of $\{(x, y) \mid \frac{6}{x+2} = y\}$.

all reals except -2

29
Find the domain of $\{(x, y) \mid \frac{5}{x-7} = y\}$.

all reals except 7

30
Find the domain of $\{(x, y) \mid \frac{x-3}{5} - 7 = y\}$.

all reals

31
A relation is a set of ordered pairs (x, y). Its domain is the set of real numbers that are acceptable replacements for x. Find the domain of $\{(x, y) \mid \frac{6}{x-3} + 13 = y\}$.

all reals except 3

32
Find the domain of $\{(x, y) \mid 4x - 3y = 5\}$.　　　　　　all reals

33
$\{(6, 3), (4, -1), (4, 9), (1, 3)\}$ has $\{6, 4, 1\}$ as its domain. Is every element of the domain used exactly once as a first component?　　　　　　No, 4 is used 2 times

34
In $\{(5, 7), (6, 3), (4, 7), (5, 3), (9, 6)\}$, what number is used as a first component in more than one ordered pair?　　　　　　5

35
In the relation $\{(6, 3), (4, -1), (4, 9), (1, 3)\}$, what element of the domain is used as a first component of more than one ordered pair?　　　　　　4

36
Some relations use each domain element exactly once as a first component. Is there any domain element used more than once in $\{(5, 9), (4, 7), (9, 3), (7, -2), (1, 4)\}$?　　　　　　No

37
Is there any domain element used more than once in $\{(5, 9), (4, -2), (8, 3), (4, 5), (1, -6)\}$?　　　　　　Yes, 4 is used twice

38
A relation is a **function** if each domain element is used exactly one time as a first component.
Is $\{(1, 6), (2, 14), (3, 6), (4, 5), (5, 6)\}$ a function?　　　　　　Yes

39
A function uses each domain element exactly one time as a first component. Is $\{(7, 9), (7, 3), (4, 5)\}$ a function?　　　　　　No, 7 is used twice.

40
Are both $(4, 2)$ and $(4, -2)$ elements of $\{(x, y) \mid x = y^2\}$?　　　　　　Yes

41

(4, 2) and (4, -2) are elements of $\{(x, y) \mid x = y^2\}$.
Is $\{(x, y) \mid x = y^2\}$ a function?

No, (4, 2) and (4, -2) have same first component.

42
A function is a relation in which each domain element is used exactly one time as a first component.
Is $\{(1, 3), (2, 5), (3, 7), (1, 8)\}$ a function?

No [(1, 3), (1, 8)]

43
For a relation to be a function it is necessary for each domain element to be used exactly _____ as a first component.

once

44
Is $\{(4, 7), (3, 6), (5, 12), (1, 2)\}$ a function?

Yes

45
Is $\{(6, 3), (5, 9), (4, 7), (9, 6)\}$ a function?

Yes

46
Is $\{(9, 3), (-2, 4), (5, 7)\}$ a function?

Yes

47
Is $\{(6, 5), (4, 7), (5, 6), (4, 9)\}$ a function?

No, 4 is used twice

48
When a relation has a domain element that is used more than once, the relation _____ (is, is not) a function.

is not

49

$(5, -3) \in \{(x, y) \mid y = x - 8\}$ because $-3 = 5 - 8$.
Is there any other ordered pair (5, ___) that is also an element of $\{(x, y) \mid y = x - 8\}$?
[Hint: Replace x by 5 in $y = x - 8$.]

No

50

$(-3, 7) \in \{(x, y) \mid 5x - y = -22\}$ because $5 \cdot -3 - 7 = -22$. Is there any other ordered pair (-3, ___) that is also an element of $\{(x, y) \mid 5x - y = -22\}$?

No

51

How many ordered pairs (4, ___) are elements of the relation $\{(x, y) \mid y = 5x - 2\}$?

(4, 18) is the one and only such element.

52

$\{(x, y) \mid y = 5x - 2\}$ is a function because for each replacement of x there will be exactly one replacement for y that will complete the solution. Complete the ordered pair (-3, _____) for $\{(x, y) \mid y = 5x - 2\}$.

(-3, -17)

53

(2, 10) is an element of $\{(x, y) \mid y = 3x + 4\}$; it is the only element of $\{(x, y) \mid y = 3x + 4\}$ that has 2 as its first component. How many ordered pairs (-1, ___) are elements of $\{(x, y) \mid y = 3x + 4\}$?

One, (-1, 1)

54

To decide if $\{(x, y) \mid y = 9x - 2\}$ is a function, it is necessary to decide whether any first component could be used in 2 or more elements of the set.
Is $\{(x, y) \mid y = 9x - 2\}$ a function?

Yes

55

$\{(x, y) \mid y = 4x - 2\}$ is a function because any choice of a first component determines exactly one second component. Every first component is used exactly once.
Is $\{(x, y) \mid y = 5x - 6\}$ a function?

Yes

56

$\{(x, y) \mid x = y^2\}$ is not a function because its elements include ordered pairs, like (4, 2) and (4, -2), which have the same first component. Find two elements of $\{(x, y) \mid x = y^2\}$ with a first component of 9.

(9, 3), (9, -3)

57

Any set of ordered pairs (x, y) that is described by an equation of the form y = mx + b, where m and b are real numbers is a function. For example, {(x, y) | y = 3x − 2} is a function.

Is $\{(x, y) \mid y = \frac{-1}{5}x + 7\}$ a function?

Yes, y = mx + b becomes

$y = \frac{-1}{5}x + 7$ when

$m = \frac{-1}{5}$ and b = 7

58

Relations that are described by equations of the form y = mx + b, where m and b are real numbers, are **linear functions**. Is {(x, y) | y = 4x + 11} a linear function?

Yes, m = 4 and b = 11

59

Equations of the form y = mx + b describe linear functions. Is {(x, y) | y = x² − 7} a linear function?

No

60

Is $\{(x, y) \mid y = \frac{x}{2} - 3\}$ a linear function?

Yes, $m = \frac{1}{2}$ and b = −3

61

Is {(x, y) | y = -7x} a linear function?

Yes, m = −7 and b = 0

62

Is {(x, y) | y = 5} a linear function?

Yes, m = 0 and b = 5

63

Is {(x, y) | y = 19x + 1} a linear function?

Yes

64

Is $\{(x, y) \mid y = \frac{15}{13}x + \frac{1}{4}\}$ a linear function?

Yes

65

Is {(x, y) | y = 6x² − 5} a linear function?

No

66
Is $\{(x, y) \mid y = \frac{3}{4} - x\}$ a linear function?   Yes

67
Every equation of the form $y = mx + b$ describes a linear function. Frequently, the equation is used without braces and is treated as its function. For example, $y = -5x - 13$ describes a linear function. Does $y = 3x + 8$ describe a linear function?   Yes

68
Is $y = \frac{4}{5}x + 2$ a linear function?   Yes

69
Is $y = -2x^2 + x - 5$ a linear function?   No, it has an x^2 term

70
Is $y = -2x + 5$ a linear function?   Yes

FEEDBACK UNIT 1

This quiz reviews the preceding unit. Answers are at the back of the book.

1. True or false? $\{(4, 7), (3, 16)\}$ has the same number of elements as $\{3, 4, 7, 16\}$.

2. True or false? $-7 \in \{x \mid x \text{ is an odd counting number}\}$.

3. Is every element of $\{x \mid x + 3x = 12\}$ an ordered pair?

4. What replacement for x cannot be used in the open expression $\frac{-3}{x} + 7$?

5. List the domain elements of $\{(3, 7), (-3, 9), (3, 14), (5, 8)\}$.

6. Is $\{(4, 8), (3, -7), (4, 5), (2, -6)\}$ a function?

7. What number is not in the domain of $\{(x, y) \mid \frac{3}{x} + 4 = y\}$?

8. Is $\{(x, y) \mid y = 4x + \frac{1}{2}\}$ a function?

9. Is $\{(x, y) \mid y = x^2\}$ a function?

10. Is $\{(3, 4), (6, 8), (-3, 4)\}$ a function?

11. Is $\{(x, y) \mid 3x - 10 = y\}$ a linear function?

12. Is $\{(x, y) \mid y = 3 - x^2\}$ a linear function?

Unit 2: Graphing Linear Functions

The following mathematical terms are crucial to an understanding of this unit.

    xy-plane   x-axis
    y-axis   Origin
    Graph of a linear function   Table of values

1
A linear function is a set of ordered pairs (x, y).
Is (3, -4) an element of $y = 2x - 10$?   Yes

2
Ordered pairs may be shown on an **xy-plane** such as the one shown at the right. The point where the **x-axis** intersects the **y-axis** is the **origin** and represents the ordered pair _____.

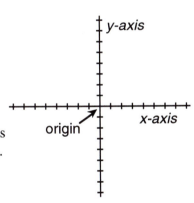

(0, 0)

3
The xy-plane at the right has the locations of two specific ordered pairs clearly indicated. The ordered pair (3, 4) is 3 units to the right of (0, 0) and 4 units above the origin. The ordered pair (-2, 5) is 2 units to the left of the origin and ____ units above (0, 0).

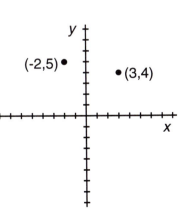

5

4

Plot the ordered pairs (7, -1), (2, 5), (-6, 3), (3, 0), (-4, -2), and (0, -1) on the xy-plane shown at the right.

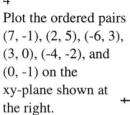

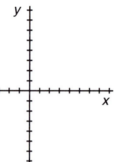

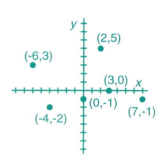

5

Plot the ordered pairs (2, -5), (-1, -6), (-4, 3), (-3, 0), (1, -2), and (0, 5) on the xy-plane shown at the right.

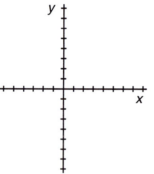

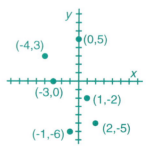

6

The **graph of a linear function** contains an infinite number of ordered pairs (points), but all the points are on the same straight line. Will all the ordered pairs of $y = 5x - 3$ be on the same straight line?

Yes

7

The graph of y = 5x + 2 is shown at the right. Because y = 5x + 2 is a linear function, its ordered pairs are all on the same _____ line.

                                                                                   straight

8

The graph of a linear function is a straight line that represents all its ordered pairs. If (4, 5) is an element of a linear function, the straight line of that function must include _____.

                                                                                   (4, 5)

9

To graph the straight line of a linear function, enough of its elements must be plotted to determine the position of all its ordered pairs. How many points determine a straight line?

                                                                                    Two

10

To graph a linear function, find two or more of its ordered pair elements. Plot them and then draw the _____ _____ that contains them.

                                                                                  straight line

11

Some ordered pairs of y = 3x − 2 can be found using a table such as the one shown at the right. The table shows (2, 4) and (−3, −11) as ordered pairs of the function. Complete the ordered pair (5, _____) by finding the value of y when x = 5.

x	y
2	4
−3	−11
5	___

                                                                                   13

12

A **table of values** is an easy way of tabulating ordered pairs of any relation (x, y). The table at the right is for the linear function $y = -2x + 5$. Complete the y-column of the table.

x	y
0	5
3	-1
7	___
-4	___
-8	___

-9
13
21

13

Complete the table of values shown at the right for the linear function $y = 4x - 3$.

x	y
3	___
0	___
-1	___

9
-3
-7

14

The table of values found in frame 13 contains the ordered pairs (3, 9), (0, -3), and (-1, -7). Plot these three points on the xy-plane.

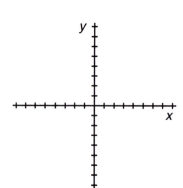

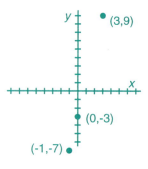

15
The three points plotted in frame 14 are elements of the linear function y = 4x – 3. Draw the straight line that contains the three points. This straight line is the graph of the linear function.

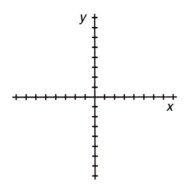

16
Complete the table of values shown at the right for the linear function
y = 5x + 1.

x	y
1	___
0	___
-2	___

6
1
-9

17

The table of values found in frame 16 contains the ordered pairs (1, 6), (0, 1), and (-2, -9). Plot these three points on the xy-plane.

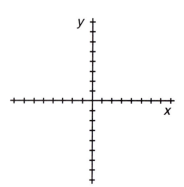

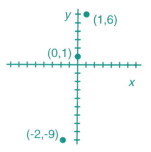

18

The three points plotted in frame 17 are elements of the linear function $y = 5x + 1$. Draw the straight line that contains the three points. This straight line is the graph of the linear function.

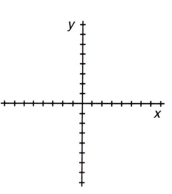

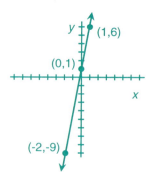

19
Two points are all that are necessary for determining a straight line, but tables for linear functions usually contain three pairs. This provides an easy check because all three points must lie on the same

_____ _____ .

straight line

20
Complete at least three rows in the table of values shown at the right for the linear function y = x − 6.

x	y
8	__
3	__
0	__
-2	__
-4	__

2
-3
-6
-8
-10

21
Plot the values found in the table of frame 20. Draw the line that contains them. The line is the graph of _____ .

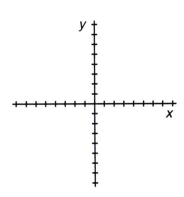

y = x − 6

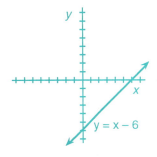

22

Construct a table of at least three elements of y = 2x + 5. Then draw the graph of the linear function.

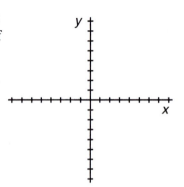

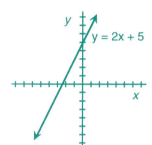

23

Construct a table of at least three elements of y = x + 4. Then draw the graph of the linear function.

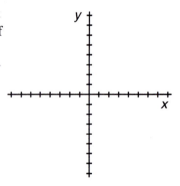

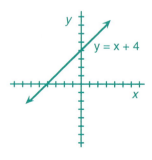

24
Construct a table of at least three elements of y = 3x − 7. Then draw the graph of the linear function.

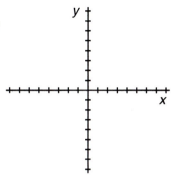

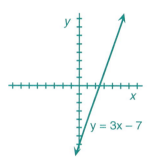
y = 3x − 7

Feedback Unit 2

This quiz reviews the preceding unit. Answers are at the back of the book.

1. Complete the table of values for $y = 3x - 2$.

x	y
6	___
2	___
0	___
-3	___
-5	___

2. Complete the table of values for $y = \frac{-1}{2}x + 3$.

x	y
8	___
6	___
3	___
0	___

3. Graph $y = 3x - 2$.

   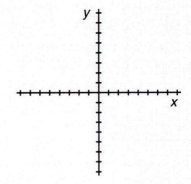

4. Graph $y = \frac{-1}{2}x + 3$.

   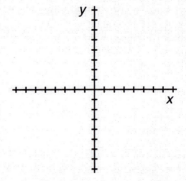

UNIT 3: THE SLOPE OF A LINEAR FUNCTION

The following mathematical terms are crucial to an understanding of this unit.

    Difference   Slope (through 2 points)
    Sub-one   Sub-two
    Formula for the slope   0 as a slope
    Slope undefined   Slope (of a linear function)

1

The figure at the right shows the line of $y = \frac{1}{2}x - 3$ with two of its solutions, (-2, -4) and (6, 0). Moving from (-2, -4) to (6, 0), the value of x increases by 8 (from -2 to 6). Moving from (-2, -4) to (6, 0), the value of y _____ (increases, decreases) by ____ .

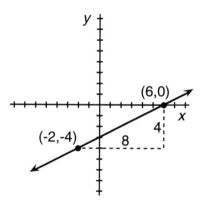

increases

4 (from -4 to 0)

2

The graph at the right shows the line of y = -2x + 5 with two of its solutions, (0, 5) and (3, -1). Moving from (0, 5) to (3, -1), the value of x increases by 3 (from 0 to 3). Moving from (0, 5) to (3, -1), the value of y _____ (increases, decreases) by _____.

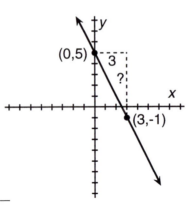

decreases

6 (from 5 to -1)

3

An increase in the value of y is shown by a positive number. A decrease in the value of y is shown by a negative number. If the value of y changes by 4, the change has been a(n) _____ (increase, decrease).

increase

4

A positive change in the value of y denotes an increase. A negative change in the value of y denotes a decrease. If the value of y changes -5, it denotes a(n) _____ (increase, decrease).

decrease

5

(2, 5) and (6, 13) are two of the solutions of y = 2x + 1.

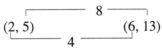

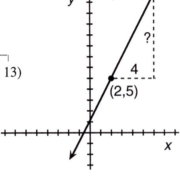

In the diagram above, it can be seen that when x changes by 4, y changes by _____.

8

6

(0, -3) and (3, -6) are two of the solutions of y = -x – 3.

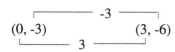

In the diagram above, it can be seen that when x changes by 3, y changes by _____.

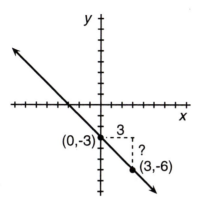

-3

7

(-4, 3) and (-2, -2) are two of the solutions of y = $\frac{-5}{2}$ x – 7.

In the diagram above, it can be seen that when x increases by 2, y _____ by _____.

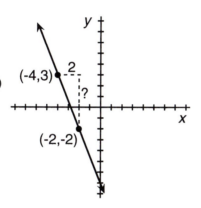

decreases
5

8

(-3, 2) and (5, 7) are two of the solutions of $y = \frac{5}{8}x + \frac{31}{8}$.

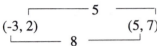

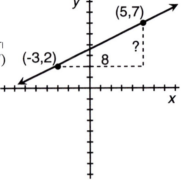

In the diagram above, it can be seen that when x increases by 8, y _____ by _____.

increases
5

9

(1, 6) and (2, 8) are solutions of y = 2x + 4.

(1, 6) (2, 8) — 2 — — 1 —

As the value of x increases by 1, the value of y _____ by _____.

increases, 2

10

(3, 17) and (4, 22) are solutions of y = 5x + 2.

(3, 17) (4, 22) — 5 — — 1 —

As the value of x increases by 1, the value of y _____ by _____.

increases, 5

11

(2, -6) and (3, -9) are solutions of y = -3x.

(2, -6) (3, -9) — -3 — — 1 —

As the value of x increases by 1, the value of y _____ by _____.

decreases, 3

LINEAR FUNCTIONS   435

12

(3, 7) and (6, 9) are solutions of $y = \frac{2}{3}x + 5$.

```
       ┌──────── ? ────────┐
    (3, 7)              (6, 9)
       └──────── 3 ────────┘
```

As the x component changes 3, the y component changes _____ .

2

13

(0, -1) and (4, 4) are solutions of $y = \frac{5}{4}x - 1$.

```
       ┌──────── ? ────────┐
    (0, -1)             (4, 4)
       └──────── 4 ────────┘
```

As x changes 4, y changes _____ .

5

14

(5, 0) and (10, -2) are solutions of $y = \frac{-2}{5}x + 2$.

```
       ┌──────── ? ────────┐
    (5, 0)             (10, -2)
       └──────── 5 ────────┘
```

As the x component changes 5, the y component changes _____ .

-2

15

$y = 3x + 2$ is equivalent to $y = \frac{3}{1}x + 2$.

(2, 8) and (3, 11) are solutions of $y = \frac{3}{1}x + 2$.

```
       ┌──────── ? ────────┐
    (2, 8)              (3, 11)
       └──────── 1 ────────┘
```

As x changes 1, y changes _____ .

3

16

$y = 6x - 1$ is equivalent to $y = \frac{6}{1}x - 1$.

$(0, -1)$ and $(1, 5)$ are solutions of $y = \frac{6}{1}x - 1$.

```
        ┌──────── ? ────────┐
       (0, -1)            (1, 5)
        └──────── 1 ────────┘
```

As x changes 1, y changes _____.   6

17

$y = -3x - 8$ is equivalent to $y = \frac{-3}{1}x - 8$.

$(-1, -5)$ and $(0, -8)$ are solutions of $y = \frac{-3}{1}x - 8$.

```
        ┌──────── ? ────────┐
       (-1, -5)           (0, -8)
        └──────── 1 ────────┘
```

As x changes 1, y changes _____.   -3

18
As a line goes from (1, 2) to (3, 7), the change in x is 2 and the change in y is _____.   5

19
As a line goes from (-1, 2) to (1, 4), the change in x is [1 – (-1)] or 2. The change in y is _____.   2

20
From (-2, 3) to (3, -7), the change in x is [3 – (-2)] or 5 and the change in y is [-7 – 3] or _____.   -10

21
The change between two numbers is their **difference**.
From (1, 0) to (3, 17) the change in x is [3 – 1] or 2 and the change in y is _____.   [17 – 0] or 17

22
From (2, 0) to (3, -1), the change in x is [3 – 2] or 1 and the change in y is _____.   [-1 – 0] or -1

LINEAR FUNCTIONS 437

23

The **slope** through two points is the change in y divided by the change in x. The slope through (2, 1) and (3, 7) is

$$\frac{7-1}{3-2} = \frac{6}{1}$$

The slope through (1, 2) and (5, 4) is $\frac{4-2}{5-1}$ = _____.

$\frac{2}{4}$ or $\frac{1}{2}$

24

The change in y divided by the change in x is the slope through two points.

The slope through (1, 3) and (2, 1) is $\frac{1-3}{2-1} = \frac{-2}{1}$

The slope through (3, 4) and (4, 7) is $\frac{7-4}{4-3}$ = _____.

$\frac{3}{1}$

25

The slope through two points is the change in y divided by the change in x.

The slope through (-2, 7) and (-5, 8) is $\frac{8-7}{-5-(-2)} = \frac{1}{-3} = \frac{-1}{3}$

Find the slope through (2, 1) and (-7, -3).

$$\frac{-3-1}{-7-2} = $$ _____.

$\frac{4}{9}$

26

The slope through two points is the change in y divided by the change in x.

The slope through (2, 5) and (0, 7) is $\frac{7-5}{0-2} = \frac{-1}{1}$

The slope through (2, 0) and (8, 6) is $\frac{6-0}{8-2}$ = _____

$\frac{1}{1}$

27

The slope through (7, -2) and (10, 5) is $\frac{5-(-2)}{10-7} = \frac{7}{3}$

The slope through (3, -1) and (5, 6) is $\frac{6-(-1)}{5-3}$ = _____.

$\frac{7}{2}$

CHAPTER 10

28

The slope through (6, -2) and (7, -10) is $\frac{-10-(-2)}{7-6} = \frac{-8}{1}$

The slope through (8, 1) and (13, -7) is $\frac{-7-1}{13-8} =$ _____ $\frac{-8}{5}$

29

The slope through (-3, -6) and (-4, 8) is $\frac{8-(-6)}{-4-(-3)} = \frac{14}{-1} = \frac{-14}{1}$

The slope through (5, 7) and (2, -8) is $\frac{-8-7}{2-5} =$ _____ $\frac{5}{1}$

30

The slope through (6, 2) and (10, 1) is $\frac{1-2}{10-6} =$ _____ $\frac{-1}{4}$

31

A formula for the slope through two points is shown below. If (x_1, y_1) and (x_2, y_2) are the two points, the slope m is

$$m = \frac{y_2 - y_1}{x_2 - x_1}$$

Find the slope through (7, 8) and (15, 3).

[Note: x_1 is "x **sub-one**" and x_2 is "x **sub-two**."]

$\frac{3-8}{15-7} = \frac{-5}{8}$

32

The slope, m, through (x_1, y_1) and (x_2, y_2) is $m = \frac{y_2-y_1}{x_2-x_1}$

Find the slope through (3, 5) and (6, -2). $\frac{-7}{3}$

33

The **formula for the slope** through two points is

$$m = \frac{y_2 - y_1}{x_2 - x_1}$$

Find the slope through (-6, 4) and (-1, 5). $\frac{1}{5}$

34

The slope through (7, -2) and (10, 6) is found by evaluating either one of the following.

$$\frac{6-(-2)}{10-7} = \frac{8}{3} \quad \text{or} \quad \frac{-2-6}{7-10} = \frac{8}{3}$$

Find the slope through (1, 5) and (2, 7). $\frac{2}{1}$

35

The slope through (-8, -3) and (0, 2) is found by evaluating either one of the following.

$$\frac{2-(-3)}{0-(-8)} \quad \text{or} \quad \frac{-3-2}{-8-0}$$

Find the slope through (0, 7) and (1, 8).

$\frac{1}{1}$

36

Find the slope through (2, 0) and (8, -3).

$\frac{-1}{2}$

37

Find the slope through (-3, 2) and (4, -7).

$\frac{-9}{7}$

38

Find the slope through (8, 2) and (-2, 12).

$\frac{-1}{1}$

39

Find the slope through (0, 1) and (3, 0).

$\frac{-1}{3}$

40

Find the slope through (-6, -2) and (-1, -8).

$\frac{-6}{5}$

41

Find the slope through (-2, -7) and (-3, -1).

$\frac{-6}{1}$

42

Find the slope through (0, -3) and (7, -8).

$\frac{-5}{7}$

43

Find the slope through (2, 7) and (-3, 0).

$\frac{7}{5}$

44

Find the slope through (-1, 4) and (3, -13).

$\frac{-17}{4}$

45
Graph the ordered pairs (2, 1) and (5, 1) and draw a straight line through the two points.

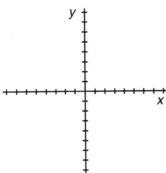

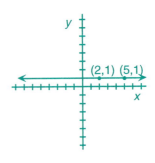

46
Find the slope through (2, 1) and (5, 1).

$\frac{0}{3} = 0$

47
When a **slope is 0**, its points lie on a horizontal line which is parallel to the _____ (x-axis, y-axis).

x-axis

48
Find the slope through (-1, -3) and (-8, -3).

$\frac{0}{7} = 0$

49
If a slope is $\frac{0}{1}$, its points lie on a horizontal line. Every pair of points with slope $\frac{0}{1}$ lie on a line parallel to the _____ -axis.

[Note: $\frac{0}{n} = \frac{0}{1}$ for any nonzero replacement of n.]

x

50
Find the slope through (2, 7) and (-3, 7).

[Note: $\frac{0}{n} = \frac{0}{1}$ when $n \neq 0$.]

$\frac{0}{1} = 0$

51

Graph the ordered pairs (-2, 3) and (-2, -5) and draw a straight line through the two points.

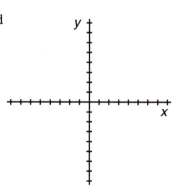

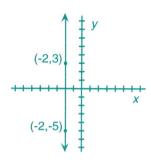

52

The **slope** through (-2, 3) and (-2, 5) is **undefined** because the formula leads to a 0 denominator.

$$\frac{5-3}{-2-(-2)} = \frac{2}{0}$$

Find the slope through (3, -2) and (3, 2).   undefined

53

When two different ordered pairs have the same first component, the line through the two points is a vertical line and is parallel to the _____ -axis.   y

54

When two points lie on a vertical line parallel to the y-axis, the slope is undefined. Since the line through (-2, 3) and (-2, 7) is a vertical line, the slope through the two points is _____.   undefined

55

If two points lie on a line parallel to the x-axis, the slope is 0. If two points lie on a line parallel to the y-axis, the slope is _____.   undefined

56
Find the slope through (-1, 1) and (0, 0).

$\frac{-1}{1}$

57
Find the slope through (3, 7) and (2, 1).

$\frac{6}{1}$

58
Find the slope through (-3, 2) and (-3, -6).

undefined

59
Find the slope through (-2, 3) and (-11, 3).

$\frac{0}{1} = 0$

60
    (-1, -1), (2, 8), (0, 2), and (-3, -7) are four solutions of y = 3x + 2. Find the slope through (-1, -1) and (2, 8). Compare it with the slope through (0, 2), and (-3, -7).

Both have 3 as the slope.

61
Every pair of solutions for the linear function y = -x + 5 will have -1 as their slope. Since (4, 1) and (-2, 7) are solutions of y = -x + 5, their slope is _____.

-1

62
The **slope of a linear function** is the slope through any two of its solutions. Will the slope of a linear function vary on which two solutions are used?

No

63
A linear function has a slope that is equal to the slope through any two of its solutions. If (4, 9) and (-2, 7) are solutions of a linear function, what is the slope for the linear function?

$\frac{1}{3}$

64
Find the slope of a linear function which contains (5, -3) and (7, 6) as two of its solutions.

$\frac{9}{2}$

FEEDBACK UNIT 3

This quiz reviews the preceding unit. Answers are at the back of the book.

Find the slope of a linear function with the following pairs of solutions.

1. (2, -3) and (2, 7)
2. (-1, 1) and (0, 4)
3. (8, -2) and (1, -2)
4. (-4, 2) and (-6, 1)
5. (0, 2) and (-1, -8)
6. (4, 0) and (0, 4)

UNIT 4: y-INTERCEPTS OF LINEAR FUNCTIONS

The following mathematical terms are crucial to an understanding of this unit.

y-intercept
Slope-intercept

Solved for y

1
If a point is on the y-axis, its ordered pair has the form (0, y). In other words, the first component of any point on the y-axis is _____.

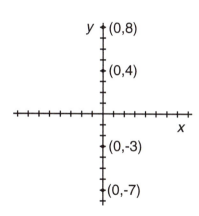

0

2
The ordered pair for any point on the y-axis has 0 as its first component. What is the value of x for each ordered pair on the y-axis?   0

3
Complete the ordered pair (0, ___) as a solution for y = 4x − 3.   (0, -3)

4
(0, -3) is a solution of y = 4x − 3. What point on the line of y = 4x − 3 is also on the y-axis?   (0, -3)

5
The **y-intercept** of y = 4x − 7 is the solution when x = 0.
(0, _____) is the y-intercept of y = 4x − 7.   (0, -7)

6
The y-intercept of $y = \frac{-3}{5}x - 1$ is the solution when x = 0.
What is the y-intercept of $y = \frac{-3}{5}x - 1$?   (0, -1)

7
When a linear function is written as y = mx + b, then (0, b) is the y-intercept. If x = 0, then y = mx + b becomes y = m • 0 + b. Simplify m • 0 + b.   b

8
The y-intercept of y = -3x + 6 is the solution when x = 0.
Find the y-intercept of y = -3x + 6.   (0, 6)

9
(0, b) is the y-intercept of y = mx + b.
Find the y-intercept of $y = -x + \frac{7}{5}$. $(0, \frac{7}{5})$

10
Find the y-intercept of y = 7x + 5.   (0, 5)

11
Find the y-intercept of $y = \frac{-2}{3}x - \frac{1}{4}$. $(0, \frac{-1}{4})$

12
Find the y-intercept of $y = \frac{-5}{9} x + 1$. $\qquad$ (0, 1)

13
Find the y-intercept of $y = -5x$. $\qquad$ (0, 0)

14
To find the y-intercept of $3x - 2y = 8$, the equation is **solved for y**.

$$\begin{aligned} 3x - 2y &= 8 \\ -2y &= -3x + 8 \\ \tfrac{-1}{2} \cdot -2y &= \tfrac{-1}{2} \cdot (-3x + 8) \\ y &= \tfrac{3}{2} x - 4 \end{aligned}$$

What is the y-intercept of $3x - 2y = 8$? $\qquad$ (0, -4)

15
To find the y-intercept of $4x + 3y = 5$, solve the equation for y.

$$\begin{aligned} 4x + 3y &= 5 \\ 3y &= -4x + 5 \\ \tfrac{1}{3} \cdot 3y &= \tfrac{1}{3} \cdot (-4x + 5) \\ y &= \tfrac{-4}{3} x + \tfrac{5}{3} \end{aligned}$$

What is the y-intercept of $4x + 3y = 5$? $\qquad$ $\left(0, \tfrac{5}{3}\right)$

16
Solve $2x - y = 5$ for y and find the y-intercept. $\qquad$ $y = 2x - 5$; (0, -5)

17
Solve $4x - 2y = 12$ for y and find its y-intercept. $\qquad$ $y = 2x - 6$; (0, -6)

18
Solve $x + 3y = 6$ for y and find its y-intercept. $\qquad$ $y = \tfrac{-x}{3} + 2$; (0, 2)

446 CHAPTER 10

19

When a linear function is in the form $y = mx + b$, its y-intercept is $(0, b)$. Equally important, the slope of the line is m, the coefficient of x. For example, $y = 3x - 2$ has slope $\frac{3}{1}$ because the coefficient of x is _____.

3

20

A linear function in the form $y = mx + b$ discloses its y-intercept as $(0, b)$ and its slope as m, the coefficient of x. The linear function $y = -x + 4$ has slope $m = \frac{-1}{1}$.
The linear function $y = \frac{x}{3} + 8$ has slope $m = $ ____.

$\frac{1}{3}$

21

The y-intercept of $y = mx + b$ is determined by b and its slope by m, the coefficient of x. For $y = \frac{-2}{3}x + \frac{5}{8}$ the y-intercept is $\left(0, \frac{5}{8}\right)$ and the slope $m = $ _____.

$\frac{-2}{3}$

22

For $y = mx + b$ the y-intercept is $(0, b)$ and the slope is m. For $y = \frac{4}{5}x - \frac{1}{3}$ the y-intercept is _____ and the slope $m = \frac{4}{5}$.

$\left(0, \frac{-1}{3}\right)$

23

For $y = 7x - 1$ the y-intercept is _____.

$(0, -1)$

24

For $y = 7x - 1$ the slope is _____.

$\frac{7}{1}$

25

For $y = \frac{-6}{11}x + 5$ the y-intercept is _____.

$(0, 5)$

26

For $y = \frac{-7}{9}x + 6$ the slope is _____.

$\frac{-7}{9}$

27

For $y = \frac{-x}{4} + 11$ the slope is _____.

$\frac{-1}{4}$

28
For y = 7 the slope is _____ .
[Note: y = 7 is equivalent to y = 0x + 7.]

$\frac{0}{1}$

29
For y = -x + 9 the slope is _____.

$\frac{-1}{1}$

30
For y = 5x – 7 the y-intercept is _____ .

(0, -7)

31
For y = $\frac{-8}{3}$ x the y-intercept is _____ .
[Note: y = $\frac{-8}{3}$ x is equivalent to y = $\frac{-8}{3}$ x + 0.]

(0, 0)

32
For y = -5x + 4 the y-intercept is _____ .

(0, 4)

33
For y = -5x + 11 the slope is _____ .

$\frac{-5}{1}$

34
For y = $\frac{2}{3}$ x the slope is _____ .

$\frac{2}{3}$

35
For y = $\frac{-5}{7}$ x the y-intercept is _____ .
[Note: y = $\frac{-5}{7}$ x is equivalent to y = $\frac{-5}{7}$ x + 0.]

(0, 0)

36
For y = -6 the y-intercept is _____ .
[Note: y = -6 is equivalent to y = 0 • x – 6.]

(0, -6)

37
For y = 4 the slope is _____.

$\frac{0}{1}$

38
For y = $\frac{5}{2}$ x – 3 the slope is _____.

$\frac{5}{2}$

39
For y = 4x + 3 the y-intercept is _____.

(0, 3)

40

An equation in the form y = mx + b is in its **slope-intercept** form. Solve 4x + y = 7 for y to find the slope-intercept form of the equation.

y = -4x + 7

41

    4x + y = 7 and y = -4x + 7 are equivalent equations; they have the same solutions. y = -4x + 7 is the slope-intercept form of the equation. Find the slope-intercept form of 3x – y = 5.

y = 3x – 5

42

Write 2x + y = -4 in its slope-intercept form and find its slope.

y = -2x – 4; m = $\frac{-2}{1}$

43

Write 5x – y = 6 in its slope-intercept form and find its slope.

y = 5x – 6; m = $\frac{5}{1}$

44

Write 4x – 3y = 12 in its slope-intercept form and find its y-intercept.

y = $\frac{4}{3}$x – 4; (0, -4)

45

Write 5x – 2y = 7 in its slope-intercept form and find the y-intercept.

y = $\frac{5}{2}$x – $\frac{7}{2}$; $\left(0, \frac{-7}{2}\right)$

46

Find the y-intercept of 2x – y = 13.

(0, -13)

47

Find the y-intercept of 7x – 3y = 4.

$\left(0, \frac{-4}{3}\right)$

48

Find the slope and y-intercept of 3y – 2x + 8 = 0.

m = $\frac{2}{3}$; $\left(0, \frac{-8}{3}\right)$

49

Find the slope of 2 = 3x – y.

m = $\frac{3}{1}$

50
Find the slope of $\frac{1}{2}y + x = 2$.

$m = \frac{-2}{1}$

51
Find the slope and y-intercept of $y = -13$.

$m = \frac{0}{1}$; $(0, -13)$

52
$\quad y = -13$ is a linear function because each ordered pair $(x, -13)$ has a different value for x. This is not the case with $x = 9$ which is not a linear function. Each ordered pair $(9, y)$ of $x = 9$ has 9 as its first component. A function uses each first component exactly _____ time as a first component.

1

53
$\quad x = 2$ is the vertical line through $(2, 0)$. The line has no slope because it cannot be written in the form $y = mx + b$. $y = 5$ is the horizontal line through $(0, 5)$. $y = 5$ is equivalent to $y = 0x + 5$ and has slope _____.

0

54
A vertical line has no slope and its associated equation is not a function. A horizontal line has slope _____.

0

55
Does $x = -2$ have a slope?

No, its slope is undefined.

56
$\quad x = 6$ is the vertical line through $(6, 0)$. $x = 6$ has neither a slope nor a y-intercept because it does not cross the y-axis. Does $x = -1$ have either a slope or a y-intercept?

No

57
$\quad y = 5$ is equivalent to $y = 0x + 5$ and has slope zero and y-intercept $(0, 5)$. What are the slope and y-intercept of $y = -4$?

0; $(0, -4)$

58

  y = -3 is equivalent to y = 0x + (-3) and has slope 0 and y-intercept (0, -3). Does x = 1 have either slope or a y-intercept?   No

---

FEEDBACK UNIT 4

This quiz reviews the preceding unit. Answers are at the back of the book.

Find the y-intercept and slope of each of the following.

1. $3x + y = 2$

2. $x - y = 1$

3. $y = \frac{-2}{5}x - \frac{7}{8}$

4. $y = mx + b$

5. $-2x - 3y = 15$

6. $y = 2x + 3$

7. $y = \frac{-5}{3}x - 1$

8. $x + y = 2$

9. $2x - 3y = 6$

10. $3x - 5y = 10$

UNIT 5: WRITING THE EQUATION OF A LINE

1
When the slope and y-intercept of a line are known, its equation can be written by replacing m and b. For example, to write the equation of the line with slope 3 and y-intercept (0, -2), replace m with 3 and b with -2 as shown below.

$$y = mx + b \text{ and } m = 3, b = -2$$
$$y = 3x - 2$$

Write the equation of the line with slope 5 and y-intercept (0, 6).

$y = mx + b$
$y = 5x + 6$

2
Write the equation of the line with slope 4 and y-intercept (0, -5).

$y = 4x - 5$

3
Write the equation of the line with slope $\frac{-2}{3}$ and y-intercept (0, 1).

$y = \frac{-2}{3}x + 1$

4
Write the equation of the line with slope -6 and y-intercept $\left(0, \frac{4}{5}\right)$.

$y = -6x + \frac{4}{5}$

5
Write the equation of the line with slope $\frac{2}{3}$ and y-intercept (0, 5).

$y = \frac{2}{3}x + 5$

6
If the slope and a point other than the intercept are given, the slope formula can be used to find an equation of the line.

The formula for slope is: $m = \frac{y_2 - y_1}{x_2 - x_1}$

Replace m by $\frac{3}{5}$ and (x_1, y_1) by (7, 4) in the formula. What equation is obtained?

$\frac{3}{5} = \frac{y_2 - 4}{x_2 - 7}$

452   CHAPTER 10

7

To find an equation of the line through (1, -6) with a slope of $\frac{1}{2}$, use the slope formula and replace m by $\frac{1}{2}$ and (x_1, y_1) by (1, -6).

$$m = \frac{y_2 - y_1}{x_2 - x_1} \qquad m = \frac{1}{2} \quad \text{and} \quad (x_1, y_1) = (1, -6)$$

What equation is obtained? $\frac{1}{2} = \frac{y_2 + 6}{x_2 - 1}$

8

To find an equation of the line through (5, 3) with a slope of $\frac{4}{7}$, use the slope formula.

$$m = \frac{y_2 - y_1}{x_2 - x_1} \qquad m = \frac{4}{7} \quad \text{and} \quad (x_1, y_1) = (5, 3)$$

What equation is obtained? $\frac{4}{7} = \frac{y_2 - 3}{x_2 - 5}$

9

To find an equation of the line through (-4, 3) with a slope of $\frac{-3}{5}$, use the slope formula.

$$m = \frac{y_2 - y_1}{x_2 - x_1} \qquad m = \frac{-3}{5} \quad \text{and} \quad (x_1, y_1) = (-4, 3)$$

What equation is obtained? $\frac{-3}{5} = \frac{y_2 - 3}{x_2 + 4}$

10

To find an equation of the line through (-8, -1) with a slope of $\frac{-5}{9}$, use the slope formula.

$$m = \frac{y_2 - y_1}{x_2 - x_1} \qquad m = \frac{-5}{9} \quad \text{and} \quad (x_1, y_1) = (-8, -1)$$

What equation is obtained? $\frac{-5}{9} = \frac{y + 1}{x + 8}$ (the sub-twos are not necessary)

LINEAR FUNCTIONS 453

11
To find an equation of the line through (2, 11) with a slope of 13, use the slope formula.

$$m = \frac{y_2 - y_1}{x_2 - x_1} \qquad m = 13 \text{ and } (x_1, y_1) = (2, 11)$$

What equation is obtained?

$\frac{13}{1} = \frac{y - 11}{x - 2}$ (the sub-twos are not necessary)

12
To find an equation of the line through (-7, 2) with a slope of 6, use the slope formula.

$$m = \frac{y_2 - y_1}{x_2 - x_1} \qquad m = 6 \text{ and } (x_1, y_1) = (-7, 2)$$

What equation is obtained?

$\frac{6}{1} = \frac{y - 2}{x + 7}$ (the sub-twos are not necessary)

13
To find an equation of the line through (-2, 3) with a slope of 1, use the slope formula.

$$m = \frac{y_2 - y_1}{x_2 - x_1} \qquad m = 1 \text{ and } (x_1, y_1) = (-2, 3)$$

What equation is obtained?

$\frac{1}{1} = \frac{y - 3}{x + 2}$

14
To find an equation of the line through (5, -3) with a slope of -4, use the slope formula.
$$m = -4 \text{ and } (x_1, y_1) = (5, -3)$$

What equation is obtained?

$\frac{-4}{1} = \frac{y + 3}{x - 5}$

15
To find an equation of the line through (4, -5) with a slope of $\frac{2}{3}$, use the slope formula.

$$m = \frac{y_2 - y_1}{x_2 - x_1} \qquad m = \frac{2}{3} \text{ and } (x_1, y_1) = (4, -5)$$

What equation is obtained?

$\frac{2}{3} = \frac{y + 5}{x - 4}$

16

To write the slope-intercept form of $\frac{2}{3} = \frac{y+5}{x-4}$ first multiply both sides of the equation by $3(x-4)$ to remove the denominators.

$$\frac{2}{3} = \frac{y+5}{x-4}$$

$$3(x-4) \cdot \frac{2}{3} = 3(x-4) \cdot \frac{y+5}{x-4}$$

Simplify each side of the equation. What equation is obtained? $2x - 8 = 3y + 15$

17

Complete the process below for writing the slope-intercept form of the equation with slope $\frac{2}{3}$ and the point $(4, -5)$.

$$\frac{2}{3} = \frac{y+5}{x-4}$$

$$3(x-4) \cdot \frac{2}{3} = 3(x-4) \cdot \frac{y+5}{x-4}$$

$$2x - 8 = 3y + 15$$

Solve the equation for y. $3y = 2x - 23$

$y = \frac{2}{3}x - \frac{23}{3}$

18

To write the slope-intercept equation of a line with slope $\frac{-4}{5}$ and the point $(2, 6)$ begin with the following steps.

$$\frac{-4}{5} = \frac{y-6}{x-2}$$

$$5(x-2) \cdot \frac{-4}{5} = 5(x-2) \cdot \frac{y-6}{x-2}$$

Simplify each side of the equation. What equation is obtained? $-4x + 8 = 5y - 30$

19
Complete the process below for writing the slope-intercept form of the equation with slope $\frac{-4}{5}$ and the point (2, 6).

$$\frac{-4}{5} = \frac{y-6}{x-2}$$

$$5(x-2) \cdot \frac{-4}{5} = 5(x-2) \cdot \frac{y-6}{x-2}$$

$$-4x + 8 = 5y - 30$$

Solve the equation for y.

$5y = -4x + 38$

$y = \frac{-4}{5}x + \frac{38}{5}$

20
Complete the process below. Write the slope-intercept form of the equation with slope -2 and the point (-3, -1).

$$\frac{-2}{1} = \frac{y+1}{x+3}$$

$$(x+3) \cdot \frac{-2}{1} = (x+3) \cdot \frac{y+1}{x+3}$$

_____

_____

$-2x - 6 = y + 1$

$y = -2x - 7$

21
Begin the process for writing the slope-intercept form of the line through (-3, 4) with slope -7. Use the slope formula to write an equation.

$\frac{-7}{1} = \frac{y-4}{x+3}$

22
The slope formula gives $\frac{-7}{1} = \frac{y-4}{x+3}$ as an equation for the line through (-3, 4) with slope -7. Multiply both sides of the equation by $1(x+3)$ to remove the denominators.

$-7x - 21 = y - 4$

23
Solve $-7x - 21 = y - 4$ for y to obtain the slope-intercept form of the equation.

$y = -7x - 17$

24

Begin the process for writing the slope-intercept form of the line through (6, -7) with slope $\frac{-3}{4}$. Use the slope formula to write an equation.

$\frac{-3}{4} = \frac{y+7}{x-6}$

25

The slope formula gives $\frac{-3}{4} = \frac{y+7}{x-6}$ as an equation for the line through (6, -7) with slope $\frac{-3}{4}$. Multiply both sides of the equation by $4(x-6)$ to remove the denominators.

$-3x + 18 = 4y + 28$

26

Solve $-3x + 18 = 4y + 28$ for y to obtain the slope-intercept form of the equation.

$4y = -3x - 10$

$y = \frac{-3}{4}x - \frac{5}{2}$

27

To write the slope-intercept form of the line through (-2, 3) with slope $\frac{4}{3}$ the following steps are used.

1. Use the slope formula to write an equation.

   $m = \frac{y_2 - y_1}{x_2 - x_1}$ becomes $\frac{4}{3} = \frac{y-3}{x+2}$

2. Multiply by the LCM of the denominators.

   $3(x+2) \cdot \frac{4}{3} = 3(x+2) \cdot \frac{y-3}{x+2}$

3. Simplify and solve for y.

   $4x + 8 = 3y - 9$

   $3y = 4x + 17$

   $y = \frac{4}{3}x + \frac{17}{3}$

Write the slope-intercept form of the line through (5, -1) with slope $\frac{3}{5}$.

$y = \frac{3}{5}x - 4$

28

Find the slope-intercept equation of the line through (4, -5) with slope $\frac{5}{7}$.

$\frac{5}{7} = \frac{y+5}{x-4}$

$y = \frac{5}{7}x - \frac{55}{7}$

29

Find the slope-intercept equation of the line through (-7, 2) with slope -5.

$\frac{-5}{1} = \frac{y-2}{x+7}$

$y = -5x - 33$

30

Find the slope-intercept equation of the line through (7, 3) with slope $\frac{-2}{9}$.

$-2x + 14 = 9y - 27$

$y = \frac{-2}{9}x + \frac{41}{9}$

31

Find the slope-intercept equation of the line through (-5, 5) with slope $\frac{-7}{3}$.

$-7x - 35 = 3y - 15$

$y = \frac{-7}{3}x - \frac{20}{3}$

32

Find the slope-intercept equation of the line through (10, 1) with slope 3.

$3x - 30 = y - 1$

$y = 3x - 29$

33

Find the slope-intercept equation of the line through (-6, -4) with slope $\frac{-3}{5}$.

$y = \frac{-3}{5}x - \frac{38}{5}$

34

Find the slope-intercept equation of the line through (-7, 3) with slope 4.

$y = 4x + 31$

35

Find the slope-intercept equation of the line through (4, -4) with slope $\frac{-2}{5}$.

$y = \frac{-2}{5}x - \frac{12}{5}$

36
Find the slope-intercept equation of the line through (-6, 1) with slope 4.

$y = 4x + 25$

37
To find the equation of a line through two points, begin by finding the slope. Find the slope of the line through (2, -4) and (3, -1).

$\frac{-1+4}{3-2} = 3$

38
The slope through (2, -4) and (3, -1) is 3. Find the slope-intercept equation of the line through (2, -4) with slope 3.

$\frac{y+4}{x-2} = 3$

$y = 3x - 10$

39
The slope through (2, -4) and (3, -1) is 3. Find the slope-intercept equation of the line through (3, -1) with slope 3.

$\frac{y+1}{x-3} = 3$

$y = 3x - 10$

40
The results of the last 2 frames show that either (2, -4) or (3, -1) can be used with the slope 3 to give the slope-intercept equation. Use the slope 1 and one of the points (3, 4) and (1, 2) to find the slope-intercept equation of the line that contains the two points.

$\frac{1}{1} = \frac{y-4}{x-3}$ or $\frac{1}{1} = \frac{y-2}{x-1}$

Either gives $y = x + 1$

41
To find the equation of the line through (3, 4) and (5, 7):

1. Find the slope. $\frac{7-4}{5-3} = \frac{3}{2}$

2. Use either one of the points to find an equation.

   If (3, 4), $\frac{3}{2} = \frac{y-4}{x-3}$ gives $3x - 9 = 2y - 8$

   If (5, 7), $\frac{3}{2} = \frac{y-7}{x-5}$ gives $3x - 15 = 2y - 14$

Change both equations to the slope-intercept form.

Both give $y = \frac{3}{2}x - \frac{1}{2}$

42
Find the slope-intercept equation of the line through (3, 2) and (7, 4) by first finding the slope and then using it with one of the points.

$\frac{1}{2} = \frac{y-4}{x-7}$ or $\frac{1}{2} = \frac{y-2}{x-3}$

$y = \frac{1}{2}x + \frac{1}{2}$

43
Find the slope-intercept equation of the line through (-1, -5) and (4, 9) by first finding the slope and then using it with one of the points.

$\frac{14}{5} = \frac{y+5}{x+1}$ or $\frac{14}{5} = \frac{y-9}{x-4}$

$y = \frac{14}{5}x - \frac{11}{5}$

44
Find the slope-intercept equation of the line through (3, 2) and (5, 6) by first finding the slope and then using it with one of the points.

$\frac{2}{1} = \frac{y-2}{x-3}$ or $\frac{2}{1} = \frac{y-6}{x-5}$

$y = 2x - 4$

45
Find the slope-intercept equation of the line through (4, -1) and (6, 5) by first finding the slope and then using it with one of the points.

$y = 3x - 13$

46
Find the slope-intercept equation of the line through (-2, -5) and (-6, -2) by first finding the slope and then using it with one of the points.

$y = \frac{-3}{4}x - \frac{13}{2}$

47
Find the slope-intercept equation of the line through (-6, 1) and (5, -4).

$y = \frac{-5}{11}x - \frac{19}{11}$

48
Find the slope-intercept equation of the line through (-5, -3) and (5, 10).

$y = \frac{13}{10}x + \frac{7}{2}$

49
Find the slope-intercept equation of the line through (-7, 1) and (-5, 5).

$y = 2x + 15$

50
Find the slope-intercept equation of the line through (-3, 2) and (7, 2).

$y = 2$

51
Find the slope-intercept equation of the line through (-2, -5) and (0, -9).

$y = -2x - 9$

52
Find the slope-intercept equation of the line through (-5, 3) and (6, 7).

$y = \frac{4}{11}x + \frac{53}{11}$

53
Find the slope-intercept equation of the line through (-1, 1) and (3, -3).

$y = -x$

FEEDBACK UNIT 5

This quiz reviews the preceding unit. Answers are at the back of the book.

Find the slope-intercept equation of the line with:

1. slope 2, point (0, 3)
2. slope -1, point (0, 7)
3. slope $\frac{-5}{2}$, point (0, 0)
4. slope 1, point (2, 1)
5. slope $\frac{3}{4}$, point (6, 0)
6. slope 0, point (5, -3)
7. slope 6, point (-2, -5)
8. points (-2, 1), (3, -4)
9. points (3, -2), (-6, 5)
10. points (2, -1), (3, -2)

Unit 6: Applications

In this Applications Section, the format of the text has been altered. Answers for the problems appear beneath them rather than in the right-hand column. Your studying emphasis should be on learning the best procedures to follow with word problems.

1
The value of a box of nickels is directly related to the number of nickels in the box. If a box contains 34 nickels, what is the value of the coins?

> Answer: The value of 34 nickels is found by multiplying 34 by .05 which gives $1.70.

2
If a box of quarters contains coins worth $5.75, how many quarters are in the box?

> Answer: The number of quarters is found by dividing $5.75 by .25 which gives 23.

3

The value of a box of dimes is related to the number of dimes by the first two rows of the following table. Complete the remaining entries for the table.

word/phrase/sentence	translation
number of dimes	d
value of d dimes	.10d
number of quarters	q
value of q quarters	_____
number of $3.50 movie tickets	t
value of t movie tickets	_____
number of $10 bills	B
value of B bills	_____

Answer:

word/phrase/sentence	translation
value of q quarters	.25q
value of t movie tickets	3.5t
value of B bills	10B

4

To solve the following problem, it is necessary to use the relationship between the number of dimes and the value of the dimes.

> A pile of 47 coins contains dimes and nickels. The total value of the coins is $4.00. Find the number of dimes and the number of nickels.

The 3-step process for solving the problem begins with the construction of the following table of translations.

word/phrase/sentence	translation
the number of dimes	d
the number of nickels	n
A pile of 47 coins contains dimes and nickels.	$d + n = 47$
the value of d dimes	_____
the value of n nickels	_____
The total value of the coins is $4.00.	_____

Complete the entries for the last 3 lines of this table.

Answer:

word/phrase/sentence	translation
the value of d dimes	.10d
the value of n nickels	.05n
The total value of the coins is $4.00	$.10d + .05n = 4.00$

5

> A pile of 47 coins contains dimes and nickels. The total value of the coins is $4.00. Find the number of dimes and the number of nickels.

Use the table of frame 4. Find 2 equations that may be used to solve the problem.

Answer: $d + n = 47$ and $.10d + .05n = 4.00$ are the 2 equations. The decimal numerals in the 2nd equation can be eliminated by multiplying each term by 100. This gives $10d + 5n = 400$.

6

Solve d + n = 47 and 10d + 5n = 400 by first solving d + n = 47 for n and then substituting in the other equation.

> Answer: d + n = 47 is equivalent to n = 47 − d. Substituting in 10d + 5n = 400 gives 10d + 5(47 − d) = 400 or 5d = 165. Hence d = 33 and n = 14.

7

A pile of 47 coins contains dimes and nickels. The total value of the coins is $4.00. Find the number of dimes and the number of nickels.

The work of frame 6 provides the answers to this problem. There were 33 dimes and 14 nickels. Show that these answers check in the original problem.

> Answer: A pile of 33 dimes and 14 nickels will have 47 coins in it. The value of 33 dimes is $3.30; the value of 14 nickels is $.70. The total value is $4.00.

8

The 3-step process learned earlier for solving a word problem is:

   a. Construct a table of the necessary translations of words, phrases, and sentences.
   b. Solve any equation(s) obtained in the table.
   c. Check answers in the original statement of the word problem.

For the problem shown below, begin the 3-step process by constructing a table of translations. The table must include the number of each coin and the value of the coins.

   A pile of nickels and dimes contains 19 coins. The value of the coins is $1.30. What is the number of nickels and the number of dimes in the pile?

Answer:

word/phrase/sentence	translation
the number of dimes	d
the number of nickels	n
A pile of nickels and dimes contains 19 coins.	d + n = 19
the value of d dimes	.10d
the value of n nickels	.05n
The value of the coins is $1.30.	.10d + .05n = 1.30

9

   A pile of nickels and dimes contains 19 coins. The value of the coins is $1.30. What is the number of nickels and the number of dimes in the pile?

Complete the 3-step process for solving the problem shown above. Use the table constructed in frame 8. Find and solve 2 equations and check the solutions in the original wording of the problem.

Answer:
   b. $d + n = 19$ and $10d + 5n = 130$ are solved by substitution. $10d + 5n = 130$ becomes $10d + 5(19 - d) = 130$ or $5d = 35$. Hence $d = 7$ and $n = 12$. There are 7 dimes and 12 nickels.

   c. A pile of 7 dimes and 12 nickels contains 19 coins. The value of 7 dimes and 12 nickels is: $.70 + .60 = \$1.30$

10

For the problem shown below, begin the 3-step process by constructing a table of translations. The table must include the number of each coin and the value of the coins.

> A box of coins has twice as many quarters as dimes. The value of the coins is $11.40. What is the number of quarters and dimes in the box?

Answer:

word/phrase/sentence	translation
the number of quarters	q
the number of dimes	d
A box of coins has twice as many quarters as dimes.	$q = 2d$
the value of q quarters	$.25q$
the value of d dimes	$.10d$
The value of the coins is $11.40.	$.25q + .10d = 11.40$

11

> A box of coins has twice as many quarters as dimes. The value of the coins is $11.40. What is the number of quarters and dimes in the box?

Complete the 3-step process for solving the problem shown above. Use the table constructed in frame 10. Find and solve 2 equations and check the solutions in the original wording of the problem.

Answer:
b. $q = 2d$ and $25q + 10d = 1140$ are solved by substitution. $25q + 10d = 1140$ becomes $25(2d) + 10d = 1140$ or $60d = 1140$. Hence $d = 19$ and $q = 38$. There are 19 dimes and 38 quarters.

c. A box of 38 quarters and 19 dimes has twice as many quarters as dimes. The value of 38 quarters and 19 dimes is: $9.50 + 1.90 = \$11.40$

12

Use the 3-step process to solve the following problem.

A piggy bank contains only dimes and quarters. Find the number of each if there are 15 coins and their total value is $2.85.

Answer:

a.
word/phrase/sentence	translation
the number of dimes	d
the number of quarters	q
there are 15 coins	d + q = 15
the value of d dimes	.10d
the value of q quarters	.25q
their total value is $2.85	.10d + .25q = 2.85

b. $d + q = 15$ and $10d + 25q = 285$ are solved by substitution. $d = 6$ and $q = 9$. There are 6 dimes, 9 quarters.

c. A collection of 6 dimes, 9 quarters is 15 coins.
The value of 6 dimes, 9 quarters is: $.60 + 2.25 = \$2.85$

FEEDBACK UNIT 6 FOR APPLICATIONS

Show the 3-step process for solving each of the following.

1. A piggy bank contains only dimes and quarters. It has 51 coins in all that have a total value of $7.20. How many of each coin are in the piggy bank?

2. John has 23 bills in his pocket that have a value of $51. If he only has $1 and $5 bills, how many of each kind does he have?

3. A greeting card dealer specializes in 2 types of cards. The less expensive type sells for $.75 each and the deluxe card sells for $1.25 each. Last week he sold $38,500 worth of cards. If he sold 3 times as many inexpensive cards as deluxe, how many of each kind did he sell?

4. If a jar contains seven times as many pennies as nickels, and their total value is $1.44, how many of each coin are in the jar?

Summary for Chapter 10

The following mathematical terms are crucial to an understanding of this chapter.

Set	Element	Ordered pairs
Set selector method	Relation	First component
Domain	Function	Linear function
Slope-intercept	xy-plane	x-axis
y-axis	Origin	Graph of a linear function
Table of values	Difference	Slope (through 2 points)
Sub-one	Sub-two	Formula for the slope
0 as a slope	Slope undefined	Slope (of a linear function)
y-intercept	Solved for y	

A set is defined as any collection of numbers, objects, or ideas. Elements of sets can be either real numbers or ordered pairs.

A relation was defined as a set of ordered pairs. The domain of a relation is the set of all first components of the ordered pairs that make up the relation.

A function was defined as a relation in which each domain element is used exactly one time as a first component.

Functions described by the form $y = mx + b$ are linear functions. This particular type of function was the focus of the chapter.

Graphs of linear functions were reviewed. Every linear function has a straight line as its graph. The graph can be found by constructing a table of values for the function, plotting at least three points, and drawing the straight line that contains them.

The slope of a linear function is the rate of change of y as compared to the rate of change of x. The slope of a function is positive if as x increases, y increases, and it is negative if y decreases as x increases.

The slope through two points is the rate of change of y as compared to the rate of change of x. The slope of a linear function is equal to the slope through any two of its solutions.

Linear functions of the form $y = mx + b$ show that the slope of the function is m and the y-intercept is (0, b). Linear functions with slope 0 are of the form $y = k$ and are associated with lines that are parallel to the x-axis.

Linear equations of the form $x = k$ are not functions. The use of the slope formula with these ordered pairs always results in a denominator of 0. These equations have no slope and are associated with lines that are parallel to the y-axis.

Chapter 10 Mastery Test

The following questions test the objectives of Chapter 10. Answers are at the back of the book. The number in parentheses which follows each problem indicates the unit in which it can be learned.

1. List the domain elements of $\{(4, 15), (-6, 7), (3, 3), (-1, 7)\}$. (1)

2. Is $\{(6, 3), (8, 4), (0, -4), (12, 6)\}$ a function? (1)

3. What number is not in the domain of $\{(x, y) \mid y = \frac{2x - 9}{5x + 4}\}$? (1)

4. Is $\{(x, y) \mid y = x^2 - 3x + 1\}$ a linear function? (1)

5. Graph $x + 3y = 9$. (2)

6. Graph $y = -3x - 5$. (2)

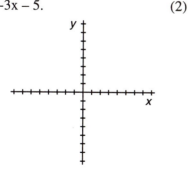

7. Find the slope through the points $(2, -7)$ and $(-2, 5)$. (3)

8. Find the slope through the points $(9, 0)$ and $(4, 1)$. (3)

9. Find the y-intercept of $y = 2x + 13$. (4)

10. Find the y-intercept of $3x - 8y = 18$. (4)

11. Find the slope of $y = 13x - 3$. (4)

12. Find the slope of $2x + 5y = 23$. (4)

13. Find the slope-intercept equation of the line with slope $\frac{-3}{5}$ and y-intercept $(0, 6)$. (5)

14. Find the slope-intercept equation of the line with slope -4 through $(6, -2)$. (5)

15. Find the slope-intercept equation of the line through $(-1, -3)$ and $(-2, 5)$. (5)

16. Show the 3-step process and solve: Joan has 19 bills in her purse that have a value of $43. If she only has $1 and $5 bills, how many of each kind does she have? (6)

17. Show the 3-step process and solve: A piggy bank contains dimes and quarters with 9 times as many dimes as quarters. If their total value is $7.65, how many of each coin are in the bank? (6)

Chapter 11 Objectives

The following problems illustrate the objectives of this chapter. At this time you are not expected to know how to do these problems. However, if all these problems are thoroughly understood, proceed directly to the Chapter Mastery Test. The number in parentheses which follows each problem indicates the unit in which it can be learned.

1. Which of the following functions is/are linear? (1)
   a. $f(x) = 3x - 9$
   b. $g(x) = x^2 + x - 3$
   c. $k(x) = (x - 3)^2 - 4$

2. Which of the following functions is/are quadratic? (1)
   a. $k(x) = \frac{2}{9}x - 5$
   b. $g(x) = x$
   c. $f(x) = 4 + x^2$

3. Identify the shape of the graph of each function. (2)
   a. $f(x) = x^2 + 4x - 11$
   b. $g(x) = 16 - x$

4. Complete a table of solutions for $f(x) = x^2 + 4x - 1$, place axes where convenient, and draw its graph. (3)

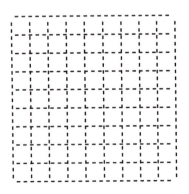

5. Complete a table of solutions for $f(x) = x^2 - 6x + 2$, place axes where convenient, and draw its graph. (3)

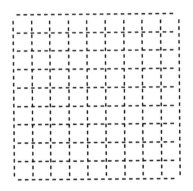

6. Find the vertex and axis of $f(x) = (x - 4)^2 - 1$. (4)

7. Find the vertex and axis of $g(x) = (x + 1)^2 + 3$. (4)

8. Graph. $f(x) = (x - 2)^2 - 5$ (4)

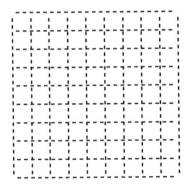

9. Graph. $g(x) = -(x+5)^2 + 1$ (4)

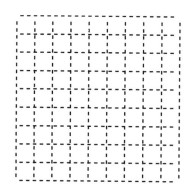

10. Graph. $y = x^2 + 10x + 17$ (4)

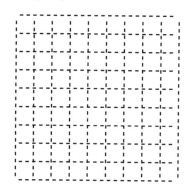

11. Graph. $x^2 + y^2 = 36$ (5)

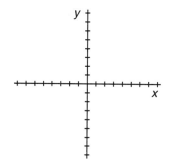

12. Graph. $(x-3)^2 + (y-1)^2 = 4$ (5)

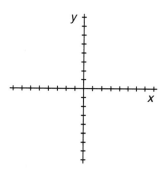

13. Graph. $x^2 + 4x + y^2 = 21$ (5)

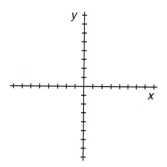

14. Write the equation of a circle with center (2, -4) and radius 4. (5)

15. Write the equation of a circle with center (0, 2) and radius 1. (5)

Chapter 11
Quadratic Functions and Relations

Unit 1: Introduction to Quadratic Functions

The following mathematical terms are crucial to an understanding of this unit.

        Linear function   f(x) notation
        Quadratic function

1
The **linear function** $y = 9x + 1$ is often written as $f(x) = 9x + 1$, where $f(x)$ replaces y. Write $y = 4x - 3$ using $f(x)$ in place of y.

$f(x) = 4x - 3$

2

f(x) is read as "f of x" or "f evaluated at x."
To find f(-2) when f(x) = 9x + 1 replace x by -2.
When f(x) = 9x + 1, then f(-2) = 9 • -2 + 1 or
f(-2) = _____   -17

3

If f(x) = 5x – 3, then f(7) is the value of 5x – 3 when x = 7.

f(x) = 5x – 3 and f(7) = 5(7) – 3 = 35 – 3 = 32

Find f(0) when f(x) = -3x + 13 by replacing x by 0.   f(0) = 13

4

Find g(5) when g(x) = -2x – 7 by replacing x by 5.   -17 (-2 • 5 – 7 = -17)

5

Find m(x) when m(x) = 5x – 6, and x = 3.   9

6

Find h(x) when h(x) = -4 + 2x, and x = -9.   -22

7

Find f(x) when f(x) = 8x, and x = 0.   0

8

Find g(x) when g(x) = -3x – 10, and x = -1.   -7

9

The function j(x) = 4 is a **constant function**. j(x) is constantly going to be 4 regardless of the replacement used for x.

j(9) = 4   j(-37) = 4   j(6) = 4   j(0) = 4

Find t(x) when t(x) = 7, and x = -5.   7

10

Find f(x) when f(x) = 5x + 4, and x = -2.   -6

11
Find f(x) when f(x) = x, and x = 10.

10

12
Equations of the form $y = ax^2 + bx + c$ or
$f(x) = ax^2 + bx + c$ where a, b, and c are real numbers
and $a \neq 0$, describe **quadratic functions**. Does
$y = 3x^2 + 7x - 2$ describe a quadratic function?

Yes, a = 3, b = 7, c = -2

13
$y = -2x^2 - 5x + 7$ is a quadratic function. Whenever x is replaced by a real number, then exactly one ordered pair (x, y) is determined. For $y = x^2 + 9x - 3$, when x is replaced by 2, how many ordered pairs (2, y) are solutions of the equation?

One, (2, 19)

14
Is $y = -3x^2 + 7x - 2$ a quadratic function?

Yes

15
Is $j(x) = -3x^2 - 9x - 2$ a quadratic function?

Yes

16
Is $m(x) = x^2 - x - 1$ a quadratic function?

Yes

17
Is $n(x) = 3x + 7$ a quadratic function?

No, it is a linear function

18
Is $p(x) = 2x^2 - 9$ a quadratic function?

Yes

19
Is $g(x) = 5x^2 + 7x$ a quadratic function?

Yes

20
Is $r(x) = 4x^3 - 2x^2 + 5x$ a quadratic function?

No

21
Is $h(x) = 7x^2$ a quadratic function?

Yes

22
Is $f(x) = 9x^2 + 7$ a quadratic function?   Yes

23
Is $t(x) = x^2$ a quadratic function?   Yes

24
Is $g(x) = 7$ a quadratic function?   No

25
Is $t(x) = 5x^2 - 7x + 3$ a quadratic function?   Yes

26
Is $g(x) = \frac{1}{2}x^2 - 3x + 8$ a quadratic function?   Yes

27
Is $d(x) = 5x - 4$ a quadratic function?   No

28
To find $f(-3)$ for $f(x) = 2x^2 - 9x - 1$, the following steps are used.

$$f(x) = 2x^2 - 9x - 1 \qquad f(-3) = 2 \cdot (-3)^2 - 9(-3) - 1$$
$$= 2 \cdot 9 + 27 - 1$$
$$= 18 + 27 - 1$$
$$= 44$$

Find $f(0)$ for $f(x) = 2x^2 - 9x - 1$. $2(0)^2 - 9(0) - 1 = -1$

29
Find $f(2)$ for $f(x) = 2x^2 - 9x - 1$. $2(2)^2 - 9(2) - 1 = -11$

30
Find $f(5)$ when $f(x) = 2x^2 - 9x - 1$.   4

31
Find $f(0)$ when $f(x) = 5x^2 - 7$.   -7

32
Find $f(2)$ when $f(x) = x^2 - 6x + 9$.   1

FEEDBACK UNIT 1

This quiz reviews the preceding unit. Answers are at the back of the book.

1. Which of the following are linear functions?
   a. $f(x) = 3x - 7$
   b. $g(x) = x^2 + 7$
   c. $h(x) = \frac{-2}{3}x + 5$
   d. $h(x) = -x$

2. Which of the following are quadratic functions?
   a. $f(x) = 2x^2 - 5x + 3$
   b. $g(x) = x^2$
   c. $k(x) = 3x + 4$
   d. $j(x) = \frac{-1}{2}x^2 - 5$

3. For the linear function $f(x) = 4x - 7$ find:
   a. $f(5)$
   b. $f(0)$
   c. $f(-2)$
   d. $f\left(\frac{-1}{2}\right)$

4. For the quadratic function $g(x) = x^2 - 5x - 2$ find:
   a. $g(-3)$
   b. $g(0)$
   c. $g(2)$
   d. $g\left(\frac{1}{5}\right)$

Unit 2: Graphs of Quadratic Functions of the Form $f(x) = ax^2 + bx + c$

The following mathematical term is crucial to an understanding of this unit.

parabola

1
The simplest quadratic function is $y = x^2$. Its graph, which is a **parabola**, is shown at the right. Every quadratic function has a parabola as its graph. Will $f(x) = x^2 - 19x + 12$ have a parabola as its graph?

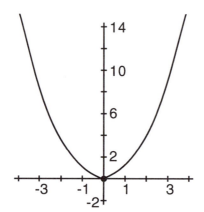

Yes

2
Another simple quadratic function is $y = -x^2$. Its graph, which is a parabola, is shown at the right. Every quadratic function has a parabola as its graph. Will $r(x) = 5x - 3$ have a parabola as its graph?

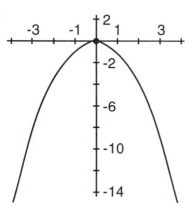

No, it is a linear function.

3

The graph of any quadratic function is a parabola. When the coefficient of x^2 is positive, as in the function $f(x) = x^2$, the parabola opens upward. Will the parabola of $f(x) = 3x^2 - 5x - 2$ open upward?

Yes, 3 is positive

4

The graph of any quadratic function is a parabola. When the coefficient of x^2 is negative, as in the function $f(x) = -x^2$, the parabola opens downward. Will the parabola of $f(x) = 5x^2 - 7x - 13$ open downward?

No, 5 is positive

5

Will the parabola of $f(x) = -2x^2$ open upward or downward?

Downward

6

Will the parabola of $f(x) = 3x^2 - x - 4$ open upward or downward?

Upward

7

Will the parabola of $f(x) = 5x^2 - 6$ open upward or downward?

Upward

8

Will the parabola of $f(x) = -15x^2 + 3x + 8$ open upward or downward?

Downward

9

$k(x) = 9x + 7$ is a linear function.
$j(x) = 5x^2 - 3x + 2$ _____ (is, is not) a linear function.

is not

10

$t(x) = 5x^2 - 3x + 2$ is a quadratic function.
$r(x) = 2x^2 - 4x - 11$ _____ (is, is not) a quadratic function.

is

11

Which of the following are linear functions?

    a. $f(x) = \frac{3}{4}x - 7$

    b. $g(x) = 4x^2 + 3$

    c. $j(x) = -x + 4$

    d. $k(x) = \frac{9}{5}x + \frac{2}{3}$ a, c, d

12

$g(x) = 4x^2 + 3$ is a quadratic function. Which of the following are quadratic functions?

    a. $r(x) = 9x^2 - 3x + 7$

    b. $t(x) = 4x + 9$

    c. $s(x) = x^2$

    d. $w(x) = \frac{3}{4}x^2 + 5$ a, c, d

13

A linear function always has a straight line as its graph. Which of the following has a straight-line as its graph?

    a. $f(x) = 3x - 4$

    b. $g(x) = 3x^2 - 5x - 7$ a

14

A quadratic function of the form $f(x) = ax^2 + bx + c$ always has a parabola as its graph. Which of the following has a parabola as its graph?

    a. $j(x) = x^2 - 9x - 11$

    b. $k(x) = \frac{3}{4}x - 7$ a

15

Which of the following has a straight-line as its graph?

    a. $f(x) = -8x^2 - 1$

    b. $g(x) = -8x - 1$ b

16
Which of the following has a parabola as its graph?

    a. $f(x) = -3x - 17$
    b. $g(x) = -3x^2 + 17$ b

17
The graph of $f(x) = 9x^2 + 5x - 2$ is a
_____ (straight line, parabola).   parabola

18
The graph of $g(x) = -2x^2 + 3x$ is a
_____ (straight line, parabola).   parabola

19
The graph of $h(x) = \frac{2}{5}x + 7$ is a
_____ (straight line, parabola).   straight line

20
The graph of $k(x) = x^2$ is a
_____ (straight line, parabola).   parabola

21
The graph of $j(x) = -x + \frac{3}{5}$ is a
_____ (straight line, parabola).   straight line

22
    $m(x) = x^2$ is a quadratic function. Its graph is
a _____ (straight line, parabola).   parabola

23
    $n(x) = (x - 5)^2$ is a quadratic function. Its graph
is a _____ (straight line, parabola).   parabola

Feedback Unit 2

This quiz reviews the preceding unit. Answers are at the back of the book.

Find the shape, straight line or parabola, for the graph of each of the following functions.

1. $f(x) = x^2 - 5x + 14$
2. $g(x) = 3x - 10$
3. $h(x) = (x - 7)^2 - 1$
4. $j(x) = -2x^2 - 5$
5. $k(x) = 5 - 3x$
6. $m(x) = 5x - x^2 + 3$

Unit 3: Graphing Quadratic Functions by Tables of Solutions

The following mathematical term is crucial to an understanding of this unit.

Table of solutions

1
A quadratic function is a set of ordered pairs (x, y).
Is (3, 9) an element of $f(x) = x^2$?   Yes, $9 = 3^2$

2
Ordered pairs of a quadratic function may be shown on the xy-plane. (3, 9), (-2, 4), and (1, 1) are three solutions of $y = x^2$. Plot the points on the xy-plane at the right.

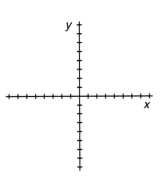

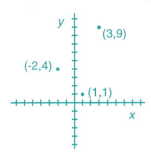

3
(3, 9), (-2, 4), and (1, 1) are three solutions of $y = x^2$. The location of these three points is shown on the xy-plane. Can a single straight line be drawn through all three points?

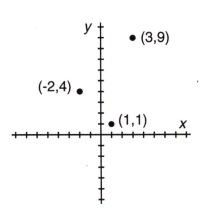

No

4
The solutions of any quadratic function $y = ax^2 + bx + c$ will never be on a straight line; they will always be on a parabola. The points of a linear function always will be on a _____ _____.

straight line

5
The solutions of a linear function always lie on a straight line, but the points of a quadratic function will _____ (sometimes, always, never) lie on a straight line.

never

6
The graph of a quadratic function contains an infinite number of ordered pairs (points), but these solutions lie on a parabola rather than a straight line. Will all the ordered pairs of $y = x^2 + 5x + 3$ be on the same straight line?

No

7
The graph of the quadratic function $y = x^2 + 4x + 2$ is shown at the right. (-5, 7) is a point on the curve. Is (-5, 7) also an ordered pair of the quadratic function?

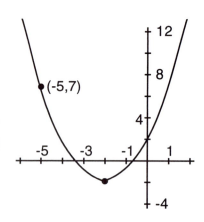

Yes, $7 = 25 - 20 + 2$ is true.

QUADRATIC FUNCTIONS AND RELATIONS   485

8
The graph of the quadratic function $y = x^2 - 6x + 4$ is shown at the right. (5, 4) is not a point on the curve. Is (5, 4) a solution of the quadratic function?

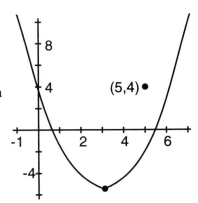

No, $4 = 25 - 30 + 4$ is false.

9
To graph the parabola of a quadratic function, enough of its elements must be plotted to approximate the position of all of its ordered pairs. Two points determine a straight line. Will two points determine the location of a parabola?

No

10
To graph a quadratic function, find at least five of its ordered pair solutions. Plot them and then sketch, from left to right, the _____ that contains them.

parabola

11
Any number of ordered pairs of $y = x^2 + 3x - 2$ can be found using a **table of solutions** such as the one shown at the right. The table shows (2, 8) and (-3, -2) as solutions of the function. Complete the columns for the ordered pairs (-2, ____), (0, ____), and (-4, ____).

x	y
2	8
-3	-2
-2	__
0	__
-4	__

(-2, -4)
(0, -2)
(-4, 2)

12

A table of solutions is an easy way of tabulating ordered pairs of any relation (x, y). The table at the right is for the quadratic function
$$y = x^2 - 2x + 5.$$
Complete the y-column of the table.

x	y
4	13
3	8
1	—
0	—
-1	—

4
5
8

13

Complete the table of solutions shown at the right for the quadratic function
$$y = x^2 + 4x - 3.$$

x	y
1	—
2	—
0	—
-2	—
-5	—

2
9
-3
-7
2

14

The table of solutions found in frame 13 contains the ordered pairs (1, 2), (2, 9), (0, -3), (-2, -7), and (-5, 2). Plot these points on the graph at the right. Notice that the scales on the x and y axes have been adjusted to make it possible to plot these points.

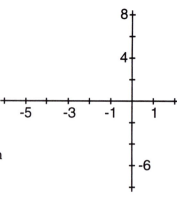

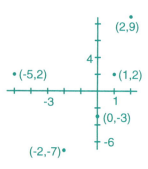

15

The five points plotted in frame 14 are elements of the quadratic function $y = x^2 + 4x - 3$. Draw a smooth curve, from left to right, through the five points. This curve is the graph of the quadratic function.

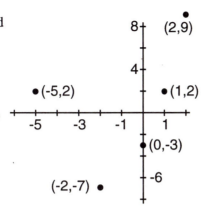

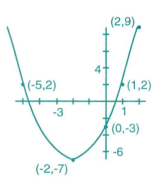

16

Complete the table of solutions shown at the right for the quadratic function
$y = x^2 + 5x + 1$.

x	y
2	___
1	___
-1	___
-3	___
-5	___
-6	___

15
7
-3
-5
1
7

17

The table of solutions found in frame 16 contains the ordered pairs (2, 15), (1, 7), (-1, -3), (-3, -5), (-5, 1), and (-6, 7). Plot these six points on the xy-plane. Notice that the scales on the two axes have been adjusted.

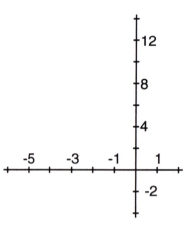

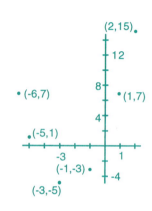

18

The six points plotted in frame 17 are elements of the quadratic function
$$y = x^2 + 5x + 1.$$
Draw the curved line that contains the six points. This curved line is the graph of the quadratic function.

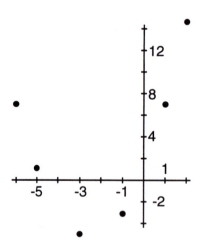

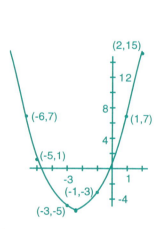

QUADRATIC FUNCTIONS AND RELATIONS   489

19
Two points are all that are necessary for determining a straight line, but tables of solutions for quadratic functions usually contain at least five ordered pairs. To graph a curved line _____ (more, less) than three points are needed.

more

20
Complete all six of the rows in the table of solutions shown at the right for the quadratic function $y = x^2 + x - 6$.

x	y
3	__
1	__
0	__
-1	__
-2	__
-4	__

6
-4
-6
-6
-4
6

21
The xy-plane at the right contains the curve drawn through the six ordered pairs of frame 20. What shape is the curve?

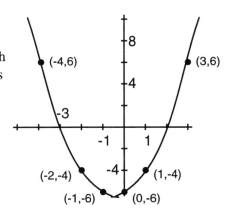

parabola

22
Complete the table of solutions for
$y = x^2 - 2x + 5$.
Place axes where convenient and then sketch the parabola of the quadratic function.

x	y
4	—
2	—
1	—
0	—
-2	—

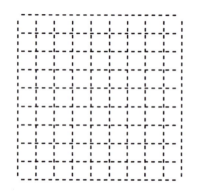

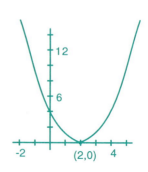

23
Complete the table of solutions for
$y = x^2 - 4x + 4$.
Place axes where convenient and then sketch the parabola of the quadratic function.

x	y
-1	—
0	—
1	—
2	—
3	—

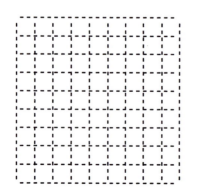

24

Complete the table of solutions for
$$y = x^2 + 3x - 7.$$
Place axes where convenient and then sketch the parabola of the quadratic function.

x	y
-3	__
-2	__
-1	__
0	__
1	__

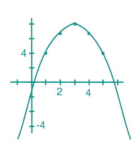

25

Complete the table of solutions for
$$y = -x^2 + 6x - 1.$$
Recall that the negative coefficient for x^2 will cause the parabola to open downwards. Place axes where convenient and then sketch the graph.

x	y
5	__
4	__
3	__
2	__
1	__

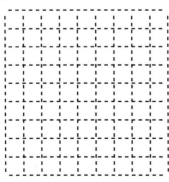

Feedback Unit 3

This quiz reviews the preceding unit. Answers are at the back of the book.

1. Complete the table of solutions for $y = x^2 - 3x - 2$.

x	y
4	__
3	__
2	__
1	__
0	__
-1	__

3. Complete the table of solutions for $y = -x^2 - 4x + 3$.

x	y
-4	__
-3	__
-2	__
-1	__
0	__

2. Graph $y = x^2 - 3x - 2$. Place axes where convenient and then sketch the parabola.

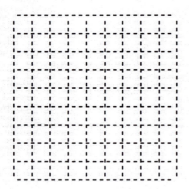

4. Graph $y = -x^2 - 4x + 3$. Place axes where convenient and then sketch the parabola.

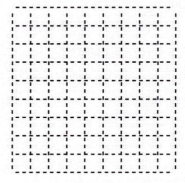

UNIT 4: FUNCTIONS OF THE FORM $f(x) = (x - h)^2 + k$

The following mathematical terms are crucial to an understanding of this unit.

    Vertex   Axis of symmetry
    Completing the square

1
Any equation of the form
    $f(x) = (x - h)^2 + k$
describes a quadratic function which has a parabola like the one shown at the right as its graph.

$f(x) = (x - 3)^2 + 4$ describes a quadratic function. Will the function have a parabola as its graph like the one shown at the right?

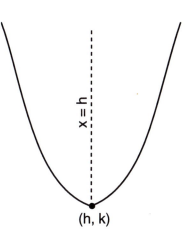

Yes

2

The graph of
$$f(x) = (x + 2)^2 - 7$$
is shown at the right. The point (-2, -7) is the **vertex** of the parabola. The vertical line that bisects the parabola of
$$f(x) = (x + 2)^2 - 7$$
has _____ as its equation.

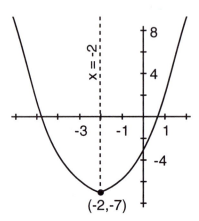

$x = -2$

3

The graph of
$$f(x) = (x - 3)^2 + 4$$
is shown at the right. The vertical line, $x = 3$, is the **axis of symmetry** of the parabola. The vertex (lowest point) of the parabola of
$$f(x) = (x - 3)^2 + 4$$
is _____.

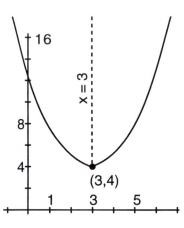

(3, 4)

4

The graph of
$$f(x) = (x - 1)^2 - 3$$
is shown at the right. The vertical line, _____, is the axis of symmetry and the vertex of the parabola is _____.

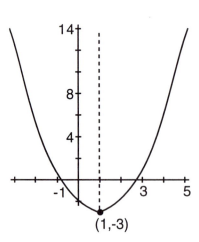

$x = 1$, (1, -3)

5

Any quadratic function of the form $f(x) = -(x - h)^2 + k$ has vertex (h, k) and axis of symmetry $x = h$.

$f(x) = -(x - h)^2 + k$
axis $x = h$ vertex (h, k)

What is the vertex of $f(x) = -(x - 3)^2 + 5$?   (3, 5)

6

(h, k) is the vertex of $f(x) = -(x - h)^2 + k$.
(3, 5) is the vertex of $f(x) = -(x - 3)^2 + 5$.
Find the vertex of $g(x) = -(x - 7)^2 + 9$.   (7, 9)

7

Find the vertex of $j(x) = (x - 6)^2 - 1$.   (6, -1)

8

Find the vertex of $f(x) = -(x - 5)^2 - 7$.   (5, -7)

9

$x + 4$ may be written as the difference $x - (-4)$.
Find the vertex of $m(x) = (x + 4)^2 + 8$.   (-4, 8)

10

Find the vertex of $n(x) = (x + 1)^2 - 4$.   (-1, -4)

11

Find the vertex of $p(x) = (x - 9)^2 + 2$.   (9, 2)

12

Find the vertex of $g(x) = (x + 5)^2 - 4$.   (-5, -4)

13

The parabola of
$f(x) = (x - 2)^2 - 3$
is shown at the right. The vertex of the parabola is the point (2, -3). The equation of its axis of symmetry is _____.

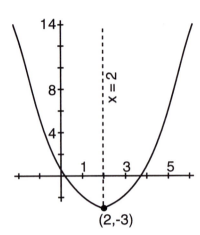

(2,-3)

x = 2

14

Any quadratic function of the form $f(x) = (x - h)^2 + k$ has vertex (h, k) and axis of symmetry x = h.

$$f(x) = (x - h)^2 + k$$

axis x = h   vertex (h, k)

The axis of symmetry of $f(x) = (x + 4)^2 + 1$ is _____.

x = -4

15

The vertical line through (-4, 1) has _____ as its equation.

x = -4

16

The vertex of
$$f(x) = (x + 4)^2 + 1$$
is (-4, 1), and its axis of symmetry is x = -4. Draw the parabola, which opens upwards, using the vertex, axis of symmetry, and a table of solutions.

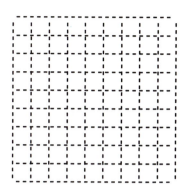

17

Find the vertex and axis of symmetry of $f(x) = -(x + 1)^2 - 4$.

(-1, -4), x = -1

18

Draw the parabola of
$$f(x) = -(x + 1)^2 - 4$$
using the vertex, axis of symmetry, and a table of solutions. Notice that the coefficient of $(x + 1)^2$ is -1 and this will make the parabola open downwards.

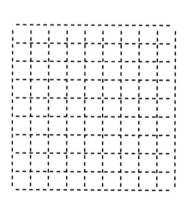

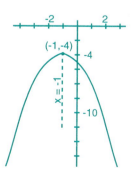

19

Find the vertex and axis of symmetry of $f(x) = (x - 3)^2 - 2$.

(3, -2), x = 3

20

Draw the parabola of
$$f(x) = (x - 3)^2 - 2$$
using the vertex, axis of symmetry, and a table of solutions.

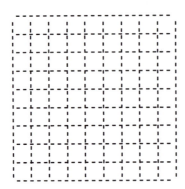

21

The function $y = x^2 + 6x + 7$ is not in the form for easily identifying its vertex and axis. The **completing the square** process learned in Chapter 8 is used below to write the function in the $y = (x - h)^2 + k$ form.

$$\begin{aligned} y &= x^2 + 6x + 7 \\ y &= (x^2 + 6x) + 7 \\ y &= (x^2 + 2 \cdot 3x) + 7 \\ y &= (x^2 + 2 \cdot 3x + 9) + 7 - 9 \\ y &= (x + 3)^2 - 2 \end{aligned}$$

What is the vertex and axis of $y = x^2 + 6x + 7$?

(-3, -2), x = -3

22

To find the vertex and axis of $y = x^2 - 8x + 3$ the completing the square process is used.

$$\begin{aligned} y &= x^2 - 8x + 3 \\ y &= (x^2 - 8x) + 3 \\ y &= (x^2 - 2 \cdot 4x) + 3 \\ y &= (x^2 - 2 \cdot 4x + 16) + 3 - 16 \\ y &= (x - 4)^2 - 13 \end{aligned}$$

What is the vertex and axis of $y = x^2 - 8x + 3$? $(4, -13)$, $x = 4$

23

Use the completing the square process to find the vertex and axis of $y = x^2 - 12x + 33$.

$y = (x^2 - 12x + 36) - 3$
$(6, -3)$, $x = 6$

24

Use the completing the square process to find the vertex and axis of $y = x^2 + 2x - 3$.

$y = (x^2 + 2x + 1) - 4$
$(-1, -4)$, $x = -1$

25

Sketch the graph of
$y = x^2 - 14x + 45$
by finding the vertex, axis, and enough solutions to correctly locate the parabola.

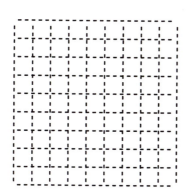

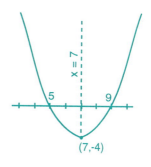

$(7, -4)$

26
Sketch the graph of
  $y = x^2 + 10x + 30$
by finding the vertex, axis, and enough solutions to correctly locate the parabola.

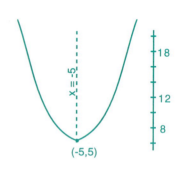
(-5,5)

Feedback Unit 4

This quiz reviews the preceding unit. Answers are at the back of the book.

1. Find the vertex and axis of symmetry of $f(x) = (x - 1)^2 - 3$.

2. Graph. $f(x) = (x - 1)^2 - 3$

   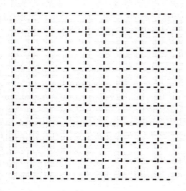

3. Find the vertex and axis of symmetry of $f(x) = -(x - 3)^2 + 5$.

4. Graph. $f(x) = -(x - 3)^2 + 5$

   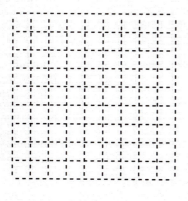

5. Graph. $y = x^2 + 8x + 12$

   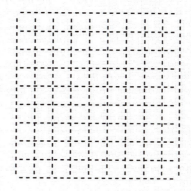

6. Graph. $y = x^2 - 6x + 10$

   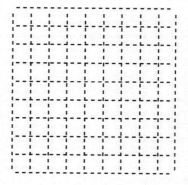

Unit 5: Graphing Circles

The following mathematical terms are crucial to an understanding of this unit.

        Circle   Center
        Radius

1
Both (0, 5) and (0, -5) are solutions of $x^2 + y^2 = 25$.

    $0^2 + 5^2 = 25$ and $0^2 + (-5)^2 = 25$

Is $x^2 + y^2 = 25$ a function?   No

2
Equations like $x^2 + y^2 = 25$ do not describe functions because for some replacements of x there are two correct values for y. Are both (-3, -4) and (-3, 4) solutions of $x^2 + y^2 = 25$?   Yes

3
Plot the solutions for $x^2 + y^2 = 25$ shown in the table below:

x	y
5	0
4	3
4	-3
3	4
3	-4
0	5
0	-5
-3	4
-3	-4
-4	3
-4	-3
-5	0

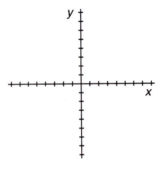

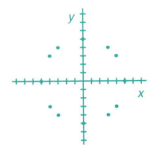

4

On the xy-plane are the 12 solutions of $x^2 + y^2 = 25$ shown in the table of frame 3.

Sketch the graph of $x^2 + y^2 = 25$ by connecting the 12 points by a smooth curve.

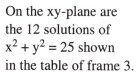

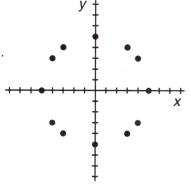

5

The graph of $x^2 + y^2 = 25$ is a **circle** with its **center** at (0, 0) and **radius** 5. Similarly, the graph of $x^2 + y^2 = 9$ is a circle with its center at _____ and radius _____.   (0, 0), 3

6

Equations of the form $x^2 + y^2 = r^2$ have graphs which are circles. The center will be at (0, 0) and the circle will have a radius of r. For example, $x^2 + y^2 = 16$ has a circle as its graph with the center at _____ and radius _____.   (0, 0), 4

504　Chapter 11

7

The graph of
$$x^2 + y^2 = 16$$
is a circle with the center at (0, 0) and radius 4.

Graph $x^2 + y^2 = 16$.

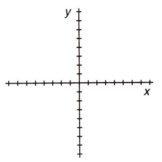

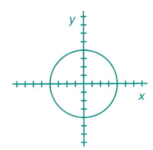

8
Graph $x^2 + y^2 = 4$.

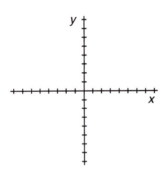

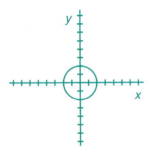

9
Graph $x^2 + y^2 = 36$.

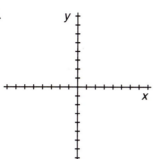

10
The equation of the circle with center at (0, 0) and radius 3 is $x^2 + y^2 = 3^2$ or $x^2 + y^2 = 9$. Similarly, the equation of a circle with center at (0, 0) and radius 2 is _____.

$x^2 + y^2 = 4$

11
The equation of a circle with center at (0, 0) and radius 7 is _____.

$x^2 + y^2 = 49$

12
The equation of a circle with center at (0, 0) and radius 9 is _____.

$x^2 + y^2 = 81$

13
The equation of a circle with center at (0, 0) and radius 5 is _____.

$x^2 + y^2 = 25$

14
The equation of a circle with center at (0, 0) and radius 6 is _____.

$x^2 + y^2 = 36$

15
$x^2 + y^2 = 49$ is the equation of a circle whose center is the origin, (0, 0). Some circles have their center at a point other than the origin. On the xy-plane draw a circle with radius 5 and center at (1, 3).

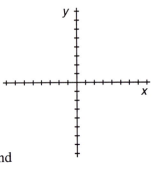

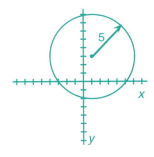

16
The circle drawn for frame 15 has $(x - 1)^2 + (y - 3)^2 = 25$ as its equation.

$$(x - 1)^2 + (y - 3)^2 = 25$$

center (1, 3)   radius 5

From the figure above, what is the center and radius of the circle with equation $(x - 2)^2 + (y + 5)^2 = 16$?

(2, -5), 4

17
Any equation of the form $(x - h)^2 + (y - k)^2 = r^2$ has a circle as its graph.

$$(x - h)^2 + (y - k)^2 = r^2$$

center (h, k)   radius r

For example, the equation $(x - 1)^2 + (y - 3)^2 = 25$ is a circle with the center at _____ and radius ____.

(1, 3), 5

18
Any equation of the form $(x - h)^2 + (y - k)^2 = r^2$ has a circle as its graph.

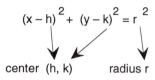

For example, the equation $(x + 4)^2 + (y - 7)^2 = 36$ is a circle with the center at _____ and radius ____.

(-4, 7), 6

19
Graph
$(x - 2)^2 + (y - 1)^2 = 16$
using the facts that the center is (2, 1) and the radius is 4.

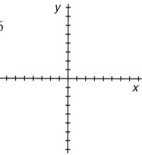

20

Graph
$$(x - 3)^2 + (y + 2)^2 = 49$$
using the facts that the center is (3, -2) and the radius is 7.

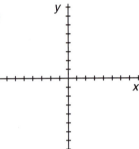

21

Use the form $(x - h)^2 + (y - k)^2 = r^2$ to write an equation for the circle with the center at (2, 5) and radius 3.

$(x - 2)^2 + (y - 5)^2 = 9$

22

Use the form $(x - h)^2 + (y - k)^2 = r^2$ to write an equation for the circle with the center at (-1, 3) and radius 5.

$(x + 1)^2 + (y - 3)^2 = 25$

23

Use the form $(x - h)^2 + (y - k)^2 = r^2$ to write an equation for the circle with the center at (0, 3) and radius 9.

$x^2 + (y - 3)^2 = 81$

24

$x^2 + 10x + y^2 - 2y = 10$ is the equation of a circle, but its center and radius are not easily seen in its present form. The **completing the square** process is used to write the equation in the form $(x - h)^2 + (y - k)^2 = r^2$.

$$\begin{aligned}
x^2 + 10x + y^2 - 2y &= 10 \\
(x^2 + 10x) \quad + \quad (y^2 - 2y) &= 10 \\
(x^2 + 10x + 25) + (y^2 - 2y + 1) &= 10 + 25 + 1 \\
(x + 5)^2 + (y - 1)^2 &= 36
\end{aligned}$$

What is the center and radius of $x^2 + 10x + y^2 - 2y = 10$?

(-5, 1), 6

25

To write $x^2 - 8x + y^2 - 6y = 24$ in a form where the center and radius are more easily seen, the following steps are used.

$$\begin{aligned} x^2 - 8x + y^2 - 6y &= 24 \\ (x^2 - 8x) + (y^2 - 6y) &= 24 \\ (x^2 - 8x + 16) + (y^2 - 6y + 9) &= 24 + 16 + 9 \\ (x - 4)^2 + (y - 3)^2 &= 49 \end{aligned}$$

What is the center and radius of $x^2 - 8x + y^2 - 6y = 24$? (4, 3), 7

26

Graph the circle of $x^2 - 4x + y^2 + 2y = 4$ by first writing the equation in a form where its center and radius are easily seen.

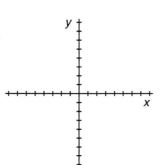

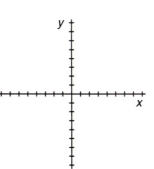

27

Graph the circle of $x^2 + 8x + y^2 + 6y = -9$ by first writing the equation in a form where its center and radius are easily seen.

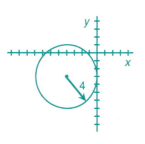

Feedback Unit 5

This quiz reviews the preceding unit. Answers are at the back of the book.

1. Graph. $x^2 + y^2 = 4$

   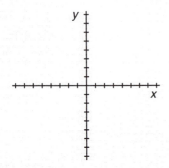

2. Graph. $(x - 1)^2 + (y + 3)^2 = 36$

   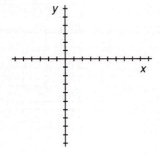

3. Graph. $x^2 + (y - 1)^2 = 49$

   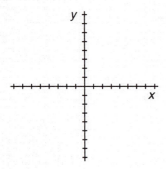

4. Graph. $x^2 + 4x + y^2 - 2y = 11$

   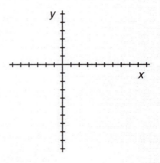

5. Write the equation of a circle with the center at (0, 0) and radius 3.

6. Write the equation of a circle with the center at (3, -2) and radius 4.

QUADRATIC FUNCTIONS AND RELATIONS

UNIT 6: APPLICATIONS

In this Applications Section, the format of the text has been altered. Answers for the problems appear beneath them rather than in the right-hand column. Your studying emphasis should be on learning the best procedures to follow with word problems.

1
If 1 pound of nuts costs $1.20, then 3 pounds cost $1.20 • 3 = $3.60 and n pounds cost _____.

Answer: $1.20n

2
If an alloy contains 34% iron by weight, then 5 pounds of alloy contain .34 • 5 = 1.70 pounds of iron and p pounds contain _____ of iron.

Answer: .34p

3
If 63% of a liquid is water, the amount of water in 17 gallons of the liquid is 17 • 0.63 or 10.71 gallons. What amount of water will there be in g gallons of the liquid?

Answer: .63g

4

The table below summarizes some of the results of the preceding frames. Complete the remaining entries for the table which are for similar situations.

word/phrase/sentence	translation
number of pounds of $1.20 per pound nuts	n
value of n pounds of $1.20 per pound nuts	1.20n
number of pounds of alloy containing 34% iron	p
amount of iron in p pounds of alloy	.34p
gallons of liquid containing 63% water	g
amount of water in g gallons of the liquid	.63g
number of pounds of $1.12 per pound cheese	C
value of C pounds of $1.12 per pound cheese	_____
quarts of radiator fluid containing 45% alcohol	R
amount of alcohol in R quarts of radiator fluid	_____
pints of milk containing 4.7% butterfat	B
amount of butterfat in B pints of milk	_____

Answer:

word/phrase/sentence	translation
value of C pounds of $1.12 per pound cheese	1.12C
amount of alcohol in R quarts of radiator fluid	.45R
amount of butterfat in B pints of milk	.047B

5

A marina needs a 20-gallon mixture of gas and oil as a boat fuel. Gas is $1.14 a gallon and oil is $6.04 a gallon. 18 gallons of gas is mixed with 2 gallons of oil.

Find each of the following:
    a.   The value of the gas in the mixture.
    b.   The value of the oil in the mixture.
    c.   The value of the mixture.

Answer:
    a.   The value of the gas is 18 • 1.14 or $20.52.
    b.   The value of the oil is 2 • 6.04 or $12.08.
    c.   The value of the mixture is the sum of 20.52 and 12.08 or $32.60.

6

A marina needs a 20-gallon mixture of gas and oil as a boat fuel. Gas is $1.14 a gallon and oil is $6.04 a gallon. 18 gallons of gas is mixed with 2 gallons of oil.

Find the value per gallon of the mixture.

Answer:
The value of the mixture is $32.60 and the value per gallon will be found by dividing 32.60 by 20. Hence, the value per gallon is $1.63.

7

The situation described below involves a mixture of 2 kinds of nuts, their separate values, and the value of the mixture.

Two kinds of nuts were put into a single container to make a 30-pound mixture. One kind of nut was worth 65¢ per pound and the other was worth 80¢ per pound. The mixture was worth 70¢ per pound.

Which of the following is stated in the situation?
- a. The number of pounds of 65¢ per pound nuts.
- b. The number of pounds of 80¢ per pound nuts.
- c. The value of the pounds of 65¢ per pound nuts.
- d. The value of the pounds of 80¢ per pound nuts.
- e. The value of the 30 pound mixture.

Answer:
None of these are stated.
The number for (e) can be found by arithmetic. Multiply 30 (the total number of pounds) by 70¢ (the price per pound of the mixture).
30 • .70 = $21.00

8

Two kinds of nuts were put into a single container to make a 30-pound mixture. One kind of nut was worth 65¢ per pound and the other was worth 80¢ per pound. The mixture was worth 70¢ per pound.

To find how many pounds of each kind of nut were used a table of translations is constructed.

word/phrase/sentence	translation
the number of pounds of nuts worth 65¢ per pound	S
the number of pounds of nuts worth 80¢ per pound	E
the total number of pounds of nuts	S + E
Two kinds of nuts were used to make a 30-pound mixture.	_____
the value of the 65¢ per pound nuts	_____
the value of the 80¢ per pound nuts	_____
the total value of the nuts	_____
the total value of the nuts is 30 • .70	_____

Complete the missing entries in the table.

Answer:

word/phrase/sentence	translation
Two kinds of nuts were used to make a 30-pound mixture.	S + E = 30
the value of the 65¢ per pound nuts	.65S
the value of the 80¢ per pound nuts	.80E
the total value of the nuts	.65S + .80E
the total value of the nuts is 30 • .70	.65S + .80E = 21.00

9

Two kinds of nuts were put into a single container to make a 30-pound mixture. One kind of nut was worth 65¢ per pound and the other was worth 80¢ per pound. The mixture was worth 70¢ per pound. How many pounds of each kind of nut were used?

What 2 equations found in the table of translations for frame 8 can be solved?

Answer:
$S + E = 30$ and $.65S + .80E = 21.00$
The first equation states the relationship between the number of pounds. The other equation states the relationship between the values of the 2 kinds of nuts. This equation is equivalent to $65S + 80E = 2100$

10

Solve $S + E = 30$ and $65S + 80E = 2100$ by first solving $S + E = 30$ for S and then substituting in the other equation.

Answer:
$S + E = 30$ is solved for S as $S = 30 - E$
Substituting in $65S + 80E = 2100$ gives $65(30 - E) + 80E = 2100$ which is equivalent to $15E + 1950 = 2100$ or $15E = 150$. When $E = 10$ then $S + E = 30$ can be solved to give $S = 20$.

11

Two kinds of nuts were put into a single container to make a 30-pound mixture. One kind of nut was worth 65¢ per pound and the other was worth 80¢ per pound. The mixture was worth 70¢ per pound.

If 20 pounds of 65¢ per pound nuts are used with 10 pounds of 80¢ per pound nuts, will the situation described above be completely satisfied?

Answer:
Yes. The mixture would contain 30 pounds.
The value of the 65¢ per pound nuts is $20 \cdot .65 = \$13.00$
The value of the 80¢ per pound nuts is $10 \cdot .80 = \$8.00$
The value of the total mixture is $21 which means the value per pound in the mixture is 70¢.

12

The 3-step process learned earlier for solving a word problem is:

    a.   Construct a table of the necessary translations of words, phrases, and sentences.
    b.   Solve any equation(s) obtained in the table.
    c.   Check answers in the original statement of the word problem.

Apply this process to the following problem which is very similar to the problem solved in frames 7-11.

> 50¢ per pound and 90¢ per pound nuts were used to make a 20-pound mixture worth 66¢ per pound. How many pounds of each kind of nut were used in the mixture?

Answer:

a.

word/phrase/sentence	translation
the number of pounds of nuts worth 50¢ per pound	F
the number of pounds of nuts worth 90¢ per pound	N
the total number of pounds of nuts	F + N
Two kinds of nuts were used to make a 20-pound mixture.	F + N = 20
the value of the 50¢ per pound nuts	.50F
the value of the 90¢ per pound nuts	.90N
the total value of the nuts	.50F + .90N
the total value of the nuts is 20 • .66	.50F + .90N = 13.20

    b.   Solve F + N = 20 and 50F + 90N = 1320
           F = 20 − N and 50(20 − N) + 90N = 1320 gives 40N = 320 or N = 8
           When N = 8 then F + N = 20 gives F = 12

    c.   12 pounds of 50¢ per pound nuts and 8 pounds of 90¢ per pound nuts makes a mixture of 20 pounds. The total value of the mixture is 12 • .50 + 8 • .90 = 6 + 7.20 = 13.20 and this makes the value per pound in the mixture 66¢ per pound.

13
Use the 3-step process to solve the following.

A barrel contains a mixture of candy worth 60¢ a pound and candy worth 82¢ a pound. If there are 100 pounds of candy in the barrel and the total cost of the candy is $74.30, how many pounds of each price candy are in the barrel?

Answer:

a.

word/phrase/sentence	translation
the number of pounds of candy worth 60¢ per pound	S
the number of pounds of candy worth 82¢ per pound	E
the total number of pounds of candy	S + E
Two kinds of nuts were used to make a 100-pound mixture.	S + E = 100
the value of the 60¢ per pound candy	.60S
the value of the 82¢ per pound candy	.82E
the total value of the candy	.60S + .82E
the total value of the candy is $74.30	.60S + .82E = 74.30

b. Solve S + E = 100 and 60S + 82E = 7430
S = 35 and E = 65

c. 35 pounds of 60¢ per pound candy and 65 pounds of 82¢ per pound candy makes a mixture of 100 pounds. The total value of the mixture is 35 • .60 + 65 • .82 = 21 + 53.30 = 74.30.

14

Use the 3-step process to solve the following.

A chemist has to make 200 gallons of alcohol that will sell for 65¢ a gallon by mixing two kinds of alcohol. If he mixes one kind worth 80¢ a gallon with another kind worth 60¢ a gallon, how many gallons of each kind will he have to use in the mixture?

Answer:

a.

word/phrase/sentence	translation
the number of gallons of alcohol worth 80¢ per gallon	E
the number of gallons of alcohol worth 60¢ per gallon	S
the total number of gallons of alcohol	E + S
Two kinds of alcohol were used to make a 200-gallon mixture.	E + S = 200
the value of the 80¢ per gallon alcohol	.80E
the value of the 60¢ per gallon alcohol	.60S
the total value of the alcohol	.80E + .60S
the total value of the alcohol is 200 • .65	.80E + .60S = 130.00

b. Solve E + S = 200 and 80E + 60S = 13000
E = 50 and S = 150

c. 50 gallons of 80¢ per gallon alcohol and 150 gallons of 60¢ per gallon alcohol make a mixture of 200 gallons. The total value of the mixture is 50 • .80 + 150 • .60 = 40 + 90 = 130.

FEEDBACK UNIT 6 FOR APPLICATIONS

Use the 3-step process to solve each of the following.

1. An 800-pound mix of cattle feed that is to sell for 14¢ a pound is to be made by mixing a grain worth 8¢ a pound with a grain worth 16¢ a pound. How much of each of the grains is needed in the mixture?

2. A candy dealer has to mix 1000 pounds of candy that will sell for 59¢ a pound from candy that is priced at 55¢ a pound and 65¢ a pound. How much of each should he use?

3. A station operator wanted to mix 70¢ and 95¢ alcohol so that he could have 120 gallons that he could sell for $100.00. How many gallons of each alcohol should he put in the mixture?

4. A pet shop owner wants to mix red gravel worth 50¢ a pound with blue gravel worth 20¢ a pound for use in aquariums. How many pounds of each color does she need to make 100 pounds that she can sell for 38¢ a pound?

SUMMARY FOR CHAPTER 11

The following mathematical terms are crucial to an understanding of this chapter.

Linear function
Quadratic function
Table of solutions
Vertex
Completing the square
Center

f(x) notation
Parabola
Circle
Axis of symmetry
Radius

Functions of the form $f(x) = ax^2 + bx + c$, where a, b, and c are real numbers and $a \neq 0$, were introduced. These quadratic functions were evaluated for different values of x and their graphs were found by constructing a table of solutions, plotting the points, and drawing a smooth curve to connect them.

The graphs of quadratic functions are parabolas, with the parabola opening upward when the coefficient of x^2 is positive and opening downward when the coefficient of x^2 is negative.

Quadratic functions of the form $f(x) = (x - h)^2 + k$ were graphed by determining the vertex and axis of the parabola. The vertex is (h, k), and the axis is $x = h$.

An equation in the form $(x - h)^2 + (y - k)^2 = r^2$ describes a circle with the center at (h, k) with a radius of r. If h and k are both zero, the center of the circle is at the origin, (0, 0).

Chapter 11 Mastery Test

The following questions test the objectives of Chapter 11. Answers are at the back of the book. The number in parentheses which follows each problem indicates the unit in which it can be learned.

1. Which of the following functions is/are linear? (1)
   a. $f(x) = x^2 + 7x - 33$
   b. $g(x) = \frac{5}{7}x + 8$
   c. $k(x) = 16 + 6x$

2. Which of the following functions is/are quadratic? (1)
   a. $k(x) = 2x^2 - 9x - 6$
   b. $m(x) = 6 - x^2$
   c. $f(x) = 14x + 3$

3. Identify the shape of the graph of each function. (2)
   a. $f(x) = 4x + 9$
   b. $g(x) = x^2 - 3x - 25$

4. Complete a table of solutions for $f(x) = x^2 + 2x + 2$, place axes where convenient, and draw its graph. (3)

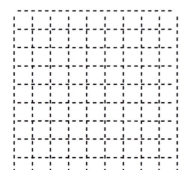

5. Complete a table of solutions for $f(x) = x^2 - 6x + 6$, place axes where convenient, and draw its graph. (3)

6. Find the vertex and axis of $f(x) = (x - 5)^2 - 4$. (4)

7. Find the vertex and axis of $g(x) = (x + 2)^2 + 1$. (4)

8. Graph. $f(x) = (x - 2)^2 - 1$ (4)

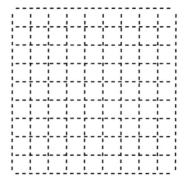

9. Graph. $g(x) = -(x-1)^2 + 5$ (4)

10. Graph. $y = x^2 + 10x + 21$ (4)

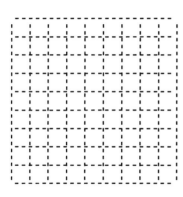

11. Graph. $x^2 + y^2 = 16$ (5)

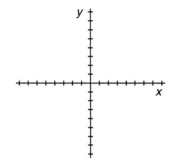

12. Graph. $(x-1)^2 + (y+3)^2 = 9$ (5)

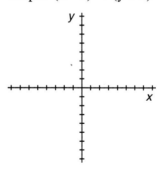

13. Graph. $x^2 + y^2 - 6y = 16$ (5)

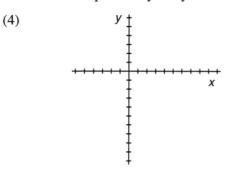

14. Write the equation of a circle with center (-5, 3) and radius 8. (5)

15. Write the equation of a circle with center (-8, 0) and radius 5. (5)

16. Show the 3-step process and solve: A 40-pound mix of coffee that is to sell for $4.20 a pound is to be made by mixing a coffee worth $3.90 a pound with a coffee worth $4.70 a pound. How much of each of the coffees is needed in the mixture? (6)

17. Show the 3-step process and solve: A boat shop owner mixes a 12 gallon container of gas and oil that he can sell for $17.16 when gas is $1.27 a gallon and oil is $2.23 a gallon. How many gallons of each does he use in the mixture? (6)

Chapter 12 Objectives

The following problems illustrate the objectives of this chapter. At this time you are not expected to know how to do these problems. However, if all these problems are thoroughly understood, proceed directly to the Chapter Mastery Test. The number in parentheses which follows each problem indicates the unit in which it can be learned.

1. Write an ordered triple with -3 as its 2nd component, 6 as its 3rd component, and -1 as its 1st component. (1)

2. Find the common solution (x, y, z, w) for the system of equations shown at the right.
   $x = 5$
   $w = -2$
   $z = 4$
   $y = 3$ (1)

For problems 3-14, find the common solution (x, y) or (x, y, z). If no common solution exists or there are an infinite number of them, state whichever is the case.

3. $x + y = 3$
   $x - y = 5$ (2)

4. $x = 4y$
   $2x - 3y = -5$ (2)

5. $3x - 4y = -18$
   $4x - 5y = -23$

6. $2x - 5y = 1$
   $6x - 15y = 0$ (2)

7. $y = x^2$
   $3x - y = 0$ (3)

8. $y = x^2 + 2$
   $x - y = -4$ (3)

9. $y = x^2 - x + 3$
   $3x - y = 1$ (3)

10. $x^2 + y^2 = 16$
    $y = x - 4$ (3)

11. $x - y - z = 0$
    $x + y - z = 4$
    $x - y + z = 2$ (4)

12. $2x + y = 2$
    $x + z = -3$
    $x + y - z = 5$ (4)

13. $2x + 5y - 2z = 3$
    $x + y + z = -4$
    $4x + 10y - 4z = 6$ (4)

14. $x - y + z = 2$
    $x - y - z = -2$
    $x + y + z = 4$ (4)

Chapter 12
Solving Systems of Equations

Unit 1: Solving Pairs of Linear Equations

The following mathematical terms are crucial to an understanding of this unit.

        Ordered pair   Ordered triple
        Ordered n-tuples   Ordered 4-tuple
        Ordered 5-tuple   System of equations
        Common solution

1

    (5, -3) is an **ordered pair**. Its 1st component is 5 and its 2nd component is -3. Is (7, 3, 5) an ordered pair?   No

2

The ordered pair (7, -2) consists of 2 numbers, 7 and -2. An **ordered triple** consists of 3 numbers. For example, (7, 3, 5) is an ordered triple. What is the 2nd component of (7, 3, 5)?   3

3
Ordered n-tuples can have any number of components. An **ordered 4-tuple** like (-3, 6, 2, 9) consists of 4 components. What is the 3rd component of (-3, 6, 2, 9)?

2

4
What is the second component of (-7, 3, 2, 5, 2) which is an **ordered 5-tuple**?

3

5
Write the ordered 4-tuple with 1st component 3, 2nd component -4, 3rd component -1, and 4th component 0.

(3, -4, -1, 0)

6
Write the ordered triple in which all components are -7.

(-7, -7, -7)

7
A solution for $3x + 5y = 13$ is an ordered pair (x, y); each component represents a replacement for its variable. A solution for $x + 2y - z = 6$ is an ordered triple (x, y, z). Any replacement for z is the _____ component of an ordered triple.

3rd

8
To decide if (5, -2, -7) is a solution (x, y, z) of $x + 2y - z = 6$, replace x by 5, y by -2, and z by _____.

-7

9
To check (5, -2, -7) as a solution (x, y, z) for $x + 2y - z = 6$, the following steps are used.

$$x + 2y - z = 6 \quad (5, -2, -7)$$
$$5 + 2(-2) - (-7) = 6$$

Is $5 + 2(-2) - (-7) = 6$ a true statement?

No, $5 - 4 + 7 = 6$ is false.

10

(5, -2) is a solution for $2x + 3y = 4$ because $10 - 6 = 4$.
Is (4, -1) also a solution for $2x + 3y = 4$?

No, $8 - 3 \neq 4$

11

(4, 7, -2) is a solution (x, y, z) for $x + y + z = 9$ because $4 + 7 - 2 = 9$. Is (-3, 8, 4) a solution (x, y, z) for $x + y + z = 9$?

Yes, $-3 + 8 + 4 = 9$

12

Is (2, 3, -6) a solution (x, y, z) for $x^2 - 5y - 2z = 1$?

Yes, $4 - 15 + 12 = 1$

13

Is (6, -2, 3, 1) a solution (x, y, z, w) of $xy + 4wz = 0$?

Yes, $-12 + 12 = 0$

14

The ordered pair (5, 3) is a solution (x, y) of $x = 5$ because it is equivalent to $x + 0y = 5$ and $5 + 0 \cdot 3 = 5$ is true.
Is (7, -3) a solution (x, y) of $5y = -15$?

Yes, $0 \cdot 7 + 5 \cdot -3 = -15$

15

Is (4, -5, 2) a solution (x, y, z) of $2x - y = 3$?

No, $8 + 5 + 0 \cdot 2 \neq 3$

16

Is (5, 7) a solution (x, y) of $2x - y = 3$?

Yes

17

Is (-4, -11) a solution (x, y) of $2x - y = 3$?

Yes

18

Is (0, 3, -2) a solution (x, y, z) of $2x - y = 3$?

No

19

A **system of equations** is a collection of two or more equations. The system shown at the right has 2 equations and _____ variables.

$x + y = 3$
$2x - 8y = 0$

2, x and y

20

A system of equations is a collection of two or more equations. The system shown at the right has 4 equations and ____ variables.

$$5x + y = 7$$
$$3x + y = 2$$
$$z - 2y = 5$$
$$x + z = 4$$

3, x, y and z

21

A **common solution** for a system of equations must be a solution for each of its equations. If the system involves 3 variables, any common solution will be an ordered triple. If the system involves 2 variables, any common solution will be an _____ _____.

ordered pair

22

The ordered pair (1, 3) is a common solution for the system of equations shown at the right.
Is (1, 3) a solution for each equation?

$$2x - y = -1$$
$$3x + y = 6$$

Yes

23

Is (-1, 2) a common solution (x, y) for the system of equations shown at the right.

$$x - y = -3$$
$$x + y = 1$$

Yes

24

Is (3, 5, -2) a common solution (x, y, z) for the system of equations shown at the right.

$$x + y = 8$$
$$y + z = 3$$
$$x + z = 1$$

Yes

25

The system of equations shown at the right has 4 equations with 4 variables. Is (5, 2, -3, 7) a common solution (x, y, z, w) for this system?

$$x = 5$$
$$y = 2$$
$$z = -3$$
$$w = 7$$

Yes

Feedback Unit 1

This quiz reviews the preceding unit. Answers are at the back of the book.

1. Write an ordered triple with 6 as its 2nd component, -4 as its 3rd component, and 5 as its 1st component.

2. Write an ordered 4-tuple with 8 as its 2nd and 3rd components and -5 as its other components.

3. Write the ordered triple (x, y, z) when $z = 7$, $y = 3$, and $x = -1$.

4. Is $(2, 5, -3)$ a solution (x, y, z) for $3x - y + 2y = 5$?

5. Is $(5, -1, 4)$ a solution (x, y, z) for $3y + z = 1$?

6. Are both $(3, 1)$ and $(-3, -5)$ common solutions of the system shown at the right.
$x - y = 2$
$x^2 = 9$

7. Find the common solution (x, y, z, w) for the system of equations shown at the right.
$x = -4$
$w = 6$
$z = -1$
$y = 2$

Unit 2: Systems of 2 Linear Equations

The following mathematical term is crucial to an understanding of this unit.

Substitution

1
At the right is shown a system of 2 equations with 2 variables. The first equation, $3x - y = 2$, is a linear (first degree) equation. Is the second equation also a linear equation?

$3x - y = 2$
$2x + y = 8$

Yes

2

One method for finding the common solution of the system

$$3x - y = 2$$
$$2x + y = 8$$

is by graphing each equation on the xy-plane. The common solution is the ordered pair where the 2 lines intersect. What is the common solution for the system?

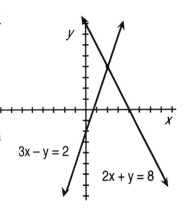

(2, 4)

3

Graph each equation in the system

$$4x - 3y = 12$$
$$x + y = -4$$

and find the common solution using the ordered pair that is the intersection of the 2 lines.

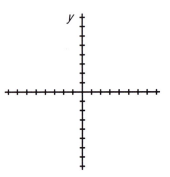

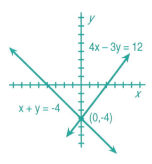

(0, -4)

4

Graph each equation in the system

   2x − y = -3
   x − y = 1

and find the common solution using the ordered pair that is the intersection of the 2 lines.

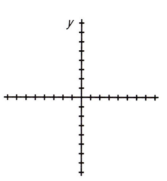

5

Graph each equation in the system

   3x − y = -1
   x − y = 1

and find the common solution using the ordered pair that is the intersection of the 2 lines.

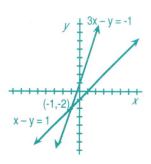

530 CHAPTER 12

6

Another method for finding the common solution for the system shown at the right depends upon "adding" the 2 equations and eliminating the variable y.

$2x + y = 5$
$5x - y = 9$

$$
\begin{aligned}
2x + y &= 5 \\
\underline{5x - y} &= \underline{9} \\
7x + 0 &= 14
\end{aligned}
$$

Use $7x = 14$ or $x = 2$ as a new equation for the system and find the common solution.

If $x = 2$, either $2x + y = 5$ or $5x - y = 9$ gives $y = 1$.
Common solution: (2, 1)

7

Because addition eliminates one of the variables, this process leads to finding a common solution for the system shown at the right.

$-x + 4y = 8$
$x + y = 12$

$$
\begin{aligned}
-x + 4y &= 8 \\
\underline{x + y} &= \underline{12} \\
0 + 5y &= 20
\end{aligned}
$$

Use $5y = 20$ or $y = 4$ to find the common solution.

Solve $-x + 16 = 8$
or $x + 4 = 12$.
(8, 4) is common solution.

8

Find the common solution of the system shown at the right.

$x + 4y = 7$
$3x - 4y = 13$

$\left(5, \frac{1}{2}\right)$

9

Find the common solution of the system shown at the right.

$3x + y = 6$
$-3x - 2y = -10$

$\left(\frac{2}{3}, 4\right)$

10

Find the common solution of the system shown at the right.

$x + y = 2$
$3x - y = 0$

$\left(\frac{1}{2}, \frac{3}{2}\right)$

Solving Systems of Equations

11
Find the common solution of the system shown at the right.

$5x + 2y = 1$
$-5x - 3y = -4$

$(-1, 3)$

12
Find the common solution of the system shown at the right.
Note: $x - 7 = -y$ is equivalent to $x + y = 7$.

$x - 7 = -y$
$2x - y = 8$

$(5, 2)$

13
To find the common solution of the system shown at the right, an equivalent equation for $x + y = 1$ is generated.

$3x - 2y = 8$
$x + y = 1$

$3x - 2y = 8$
$x + y = 1 \rightarrow$ multiply by 2 $\rightarrow$

$3x - 2y = 8$
$2x + 2y = 2$
$\overline{5x + 0 = 10}$

Use $5x = 10$ or $x = 2$ and either of the original equations to find the common solution for the system.

$(2, -1)$

14
To find the common solution of the system shown at the right, an equivalent equation for $2x + y = 3$ is generated.

$5x + y = 12$
$2x + y = 3$

$5x + y = 12$
$2x + y = 3 \rightarrow$ multiply by -1 $\rightarrow$

$5x + y = 12$
$-2x - y = -3$
$\overline{3x = 9}$

Use $3x = 9$ or $x = 3$ and either of the original equations to find the common solution for the system.

$(3, -3)$

15
Find the common solution of the system shown at the right by first generating an equivalent equation for $x + 2y = 3$.

$x + 2y = 3$
$2x + 3y = 5$

$-2x - 4y = -6$
$2x + 3y = 5$
$(1, 1)$

16
Find the common solution of
the system shown at the right.

$5x - y = 11$
$3x - 2y = 8$

$(2, -1)$

17
As a first step in the solution of
the system shown at the right,
generate 2 equivalent equations
by multiplying the terms of
$2x + 5y = 21$ by 2 and the terms
of $3x - 2y = -16$ by 5.

$2x + 5y = 21$
$3x - 2y = -16$

$4x + 10y = 42$
$15x - 10y = -80$

18
The system of equations containing
$2x + 5y = 21$ and $3x - 2y = -16$
can be replaced by the system
shown at the right. Find
the common solution.

$4x + 10y = 42$
$15x - 10y = -80$

$(-2, 5)$

19
As a first step in the solution of
the system shown at the right,
generate 2 equivalent equations
by multiplying the terms of
$5x - 2y = 39$ by 3 and the terms
of $4x + 3y = -1$ by 2.

$5x - 2y = 39$
$4x + 3y = -1$

$15x - 6y = 117$
$8x + 6y = -2$

20
The system of equations containing
$5x - 2y = 39$ and $4x + 3y = -1$
can be replaced by the system
shown at the right. Find
the common solution.

$15x - 6y = 117$
$8x + 6y = -2$

$(5, -7)$

21
Find the common solution of
the system shown at the right.

$9x - 5y = 3$
$2x - y = 1$

$(2, 3)$

22
Find the common solution of the system shown at the right.

$3x - 4y = -5$
$4x - 5y = -5$

(5, 5)

23
Find the common solution of the system shown at the right.

$2x - 5y = 19$
$5x - 2y = 16$

(2, -3)

24
Find the common solution of the system shown at the right.

$7x + 8 = 4y$
$5x + 3y = -35$

(-4, -5)

25
Find the common solution of the system shown at the right.

$9x - 74 = -y$
$3y = 8x - 23$

(7, 11)

26
Some systems of equations are best solved using the **substitution** process. If $y = x + 1$ then $(x + 1)$ may be substituted for y. In $2x - 3y = -5$ substitute $(x + 1)$ for y.

$2x - 3(x + 1) = -5$
$2x - 3x - 3 = -5$
$-x = -2$

27
To find the common solution of the system shown at the right substitution can generate the linear equation $-x = -2$ or $x = 2$. Complete the common solution (2, ___) for the system of equations.

$y = x + 1$
$2x - 3y = -5$

(2, 3)

28
To find the common solution of the system shown at the right substitute $(x - 1)$ for y in the 2nd equation. Solve $3x - 4(x - 1) = 1$.

$y = x - 1$
$3x - 4y = 1$

$x = 3$

29
Complete the substitution process for finding the common solution of the system shown at the right by using the result x = 3.

$y = x - 1$
$3x - 4y = 1$

(3, 2)

30
To find the common solution of the system shown at the right substitute (2y) for x in the 2nd equation. Solve 3(2y) − y = -5.

$x = 2y$
$3x - y = -5$

y = -1

31
Complete the substitution process for finding the common solution of the system shown at the right by using the result y = -1.

$x = 2y$
$3x - y = -5$

(-2, -1)

32
Find the common solution of the system shown at the right using substitution.

$x = 2y - 1$
$3x - y = 12$

3(2y − 1) − y = 12
(5, 3)

33
Find the common solution of the system shown at the right using substitution.

$x - y = 4$
$y = 2x - 7$

x − (2x − 7) = 4
(3, -1)

34
Some systems of equations have no common solution. For example, the system at the right has no common solution. Is there any ordered pair solution for 0x + 0y = 3?

$x + 5y = 7$
$0x + 0y = 3$

No

35

Some systems of equations have no common solution. The lines of $3x - 2y = 6$ and $3x - 2y = -2$ are shown on the xy-plane. Do the lines have a common point?

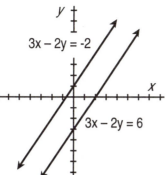

No

36

The system of equations shown at the right contains the same equations as those graphed in frame 35. If the 1st equation is multiplied by -1 and then added to the 2nd equation, what is the result?

$3x - 2y = -2$
$3x - 2y = 6$

$-3x + 2y = 2$
$\underline{3x - 2y = 6}$
$0x + 0y = 8$

37

Whenever 2 distinct equations have parallel lines as their graphs, any attempt to add them and eliminate a variable will give a result like $0x + 0y = 8$. Is it possible for any pair of numbers to make $0x + 0y = 8$ true?

No

38

Some systems of equations have no common solution. Find the common solution for the system shown at the right or state that there is none.

$2x - y = 1$
$4x - 2y = -3$

No solution

39

Some systems of equations have no common solution. Find the common solution for the system shown at the right or state that there is none.

$x - 2y = 8$
$x = 2y$

No solution

40

Some systems of equations have no common solution. Find the common solution for the system shown at the right or state that there is none.

$x = 2y$
$x - 4y = 2$

(-2, -1)

41

The system of equations shown at the right contains 2 equivalent equations because if 2 is multiplied by the 1st equation the result is the 2nd equation. Will every solution of $2x - y = 1$ also be a solution of $4x - 2y = 2$?

$2x - y = 1$
$4x - 2y = 2$

Yes

42

The system of equations shown at the right contains the same equations as in frame 41. If the 1st equation is multiplied by -2 and then added to the 2nd equation, what is the result?

$2x - y = 1$
$4x - 2y = 2$

$-4x + 2y = -2$
$\underline{4x - 2y = 2}$
$0x + 0y = 0$

43

When 2 equations are equivalent, any attempt to add them and eliminate a variable will give a result like $0x + 0y = 0$. Is it possible for any pair of numbers to make $0x + 0y = 0$ true?

Yes, every pair of real numbers will make it true.

44

The common solutions for a pair of equivalent equations will include any solution for each equation. The set of common solutions for the system at the right _____ (is, is not) an infinite set.

$2x - y = 1$
$4x - 2y = 2$

is

45

The equation $0x + 0y = 9$ has no solutions. The equation $0x + 0y = 0$ has _____ (no, one, an infinite number of) solution(s).

an infinite number of

FEEDBACK UNIT 2

This quiz reviews the preceding unit. Answers are at the back of the book.

Find the common solution, if any, for each of the following systems.

1. $2x - y = -7$
   $3x + y = -3$

2. $y = 3x$
   $3x - y = -13$

3. $3x + 5y = 31$
   $4x + 7y = 43$

4. $3x - 2y = -9$
   $6x - y = -3$

5. $3x - 4y = 13$
   $6x - 8y = 10$

6. $x = 2y - 7$
   $2x + y = -29$

7. $x = 3y$
   $8x - 4y = 5$

8. $y = x + 3$
   $3x - y = -9$

9. $3x - 2y = -5$
   $x - 3y = -4$

10. $3x - y = 8$
    $7x - y = 12$

Unit 3: The Common Solution of a Linear and a Quadratic Equation

1
The system shown at the right has 2 equations and 2 variables. Are both equations linear?

$y = x^2$
$x + y = 2$

No, $y = x^2$ is quadratic.

2
The common solutions for the system
$y = x^2$
$x + y = 2$
may be found by graphing each equation on the xy-plane. Is every ordered pair on the parabola shown at the right a solution of $y = x^2$?

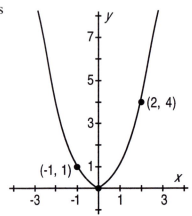

Yes

3
The xy-plane at the right shows the parabola of $y = x^2$ and the straight line of $x + y = 2$. Is every ordered pair on the straight line a solution of $x + y = 2$?

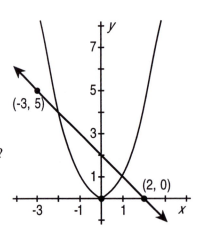

Yes

SOLVING SYSTEMS OF EQUATIONS 539

4
Find 2 common solutions for the system
$$y = x^2$$
$$x + y = 2$$
by naming the ordered pairs for the points common to the parabola and the straight line.

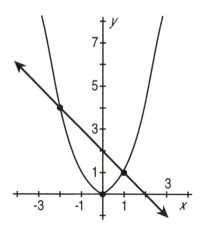

(1, 1) and (-2, 4)

5
The system shown at the right has 2 equations and 2 variables. Which equation is quadratic?

$$y = -(x + 1)^2 + 7$$
$$y = -4x + 7$$

$y = -(x + 1)^2 + 7$

6
The common solution for the system
$$y = -(x + 1)^2 + 7$$
$$y = -4x + 7$$
may be found by graphing each equation on the xy-plane. Is every ordered pair on the parabola shown at the right a solution of $y = -(x + 1)^2 + 7$?

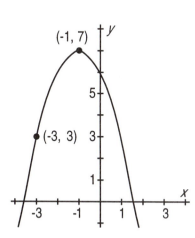

Yes

7

The xy-plane at the right shows the parabola of the quadratic equation $y = -(x + 1)^2 + 7$ and the straight line of $y = -4x + 7$. Is every ordered pair on the straight line a solution of $y = -4x + 7$?

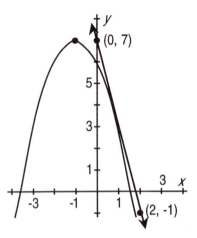

Yes

8

Find the one and only common solution for the system
$$y = -(x + 1)^2 + 7$$
$$y = -4x + 7$$
by naming the ordered pair for the point common to the parabola and the straight line.

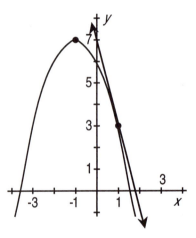

(1, 3)

9

The xy-plane at the right shows the parabola and line graphs for the system
$$y = (x - 3)^2 - 1$$
$$y = -2x + 2$$
Do the equations have any common solutions?

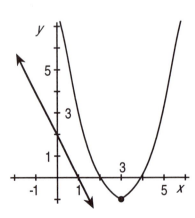

No

SOLVING SYSTEMS OF EQUATIONS   541

10
Systems of equations containing one linear equation
and one quadratic equation may have 2, 1, or 0 common
solutions. Is it possible for a parabola and a straight
line to have 3 or more common points?   No

11
The xy-plane at the
right shows a
circle which is the
graph of the quadratic
equation of the form
$(x - h)^2 + (y - k)^2 = r^2$.
Three straight lines are
shown which intersect
the circle in 0, 1, or 2
points. Is it possible
to draw a straight line
that intersects the circle
in 3 points?   No

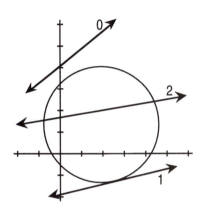

12
The xy-plane at the
right shows an
ellipse which can be the
graph of a quadratic
equation.

Three straight lines are
shown which intersect
the ellipse in 0, 1, or 2
points. Is it possible
to draw a straight line
that intersects the ellipse
in 3 points?   No

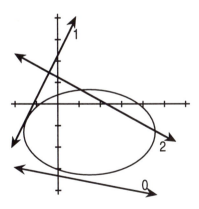

13

The xy-plane at the right shows a hyperbola which can be the graph of a quadratic equation.

Three straight lines are shown which intersect the hyperbola in 0, 1, or 2 points. Is it possible to draw a straight line that intersects the hyperbola in 3 points?

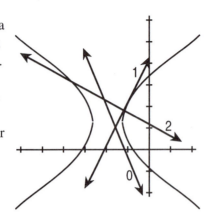

No

14

When the line of a linear equation and the curve of any quadratic equation are graphed on the xy-plane, there can be no more than ____ common points.

2

15

The graphing method for finding common solutions is valuable because it clearly depicts the concept. Is it relatively easy to "see" the common solutions for the system graphed at the right?

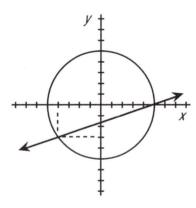

Yes, (-4, -3) and (5, 0)

16
The value of the graphing method for finding common solutions is limited because of difficulties in drawing curves accurately and reading results when the components of ordered pairs are not integers. Can the ordered pair solutions be accurately stated for the system
$$x + y = 4$$
$$y = (x - 2)^2 - 1$$
graphed at the right?

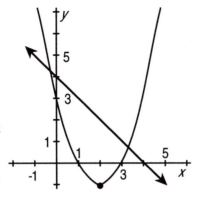

No, components are not integers.

17
To find the common solution(s) for the system shown at the right, substitution is the best approach. To begin this process solve the linear equation for one of its variables.
Solve $x + y = 2$ for y.

$$y = x^2$$
$$x + y = 2$$

$y = -x + 2$

18
To find the common solution(s) for the system shown at the right, substitute $(-x + 2)$ for y in the quadratic equation.
$y = x^2$ becomes $(-x + 2) = x^2$
Solve $x^2 + x - 2 = 0$ by factoring.

$$y = x^2$$
$$x + y = 2$$

$(x + 2)(x - 1) = 0$
$x = -2 \quad x = 1$

19

Using substitution to find the common
solutions for the system at the right,
$\quad\quad y = x^2$
$\quad\quad x + y = 2$

1. Solve $x + y = 2$ for y.
2. Substitute $(-x + 2)$ for y.
3. Solve $x^2 + x - 2 = 0$.
4. Complete the common solutions
   $(-2, \_\_\_), (1, \_\_\_)$. $\quad\quad\quad\quad (-2, 4), (1, 1)$

20

To find the common solution(s) for the
system at the right, substitute $(-4x + 7)$
for y in the quadratic equation.
$\quad y = -(x + 1)^2 + 7$
$\quad y = -4x + 7$

$(-4x + 7) = -(x + 1)^2 + 7$
$(-4x + 7) = -(x^2 + 2x + 1) + 7$
$(-4x + 7) = -x^2 - 2x - 1 + 7$
$x^2 - 2x + 1 = 0$

Solve $x^2 - 2x + 1 = 0$ by factoring. $\quad\quad (x - 1)(x - 1) = 0$
$\quad\quad\quad\quad\quad\quad\quad\quad\quad\quad\quad\quad\quad\quad\quad\quad\quad\quad x = 1$

21

Using substitution to find the common
solution for the system at the right, $y = -(x + 1)^2 + 7$
$\quad\quad\quad\quad\quad\quad\quad\quad\quad\quad\quad\quad\quad y = -4x + 7$

1. Substitute $(-4x + 7)$ for y.
2. Solve $x^2 - 2x + 1 = 0$.
3. Complete the common solution $(1, \_\_\_)$. $\quad\quad (1, 3)$

Solving Systems of Equations

22

To find the common solution(s) for the system at the right, substitute $(y - 1)$ for x in the quadratic equation.

$$(x - 3)^2 + (y - 1)^2 = 9$$
$$x = y - 1$$

$$([y - 1] - 3)^2 + (y - 1)^2 = 9$$
$$(y - 4)^2 + (y - 1)^2 = 9$$
$$y^2 - 8y + 16 + y^2 - 2y + 1 = 9$$
$$2y^2 - 10y + 8 = 0$$
$$y^2 - 5y + 4 = 0$$

Solve $y^2 - 5y + 4 = 0$ by factoring.

$(y - 4)(y - 1) = 0$
$y = 4 \quad y = 1$

23

Using substitution to find the common solution for the system at the right,

$$(x - 3)^2 + (y - 1)^2 = 9$$
$$x = y - 1$$

1. Substitute $(y - 1)$ for x.
2. Solve $y^2 - 5y + 4 = 0$.
3. Complete the common solution (___, 4), (___, 1).

(3, 4), (0, 1)

24

To find the common solution(s) of the system at the right, substitution is the best approach. Begin by solving $4x - y = 4$ for y.

$$y = x^2$$
$$4x - y = 4$$

$y = 4x - 4$

25

Find the common solution(s) of the system at the right by substituting $4x - 4$ for y.

$$y = x^2$$
$$4x - y = 4$$

$4x - 4 = x^2$
$x = 2$
$(2, 4)$

26

To find the common solution(s) of the system at the right, substitution is the best approach. Begin by solving $3x + y = 3$ for y.

$$y = x^2 - x$$
$$3x + y = 3$$

$y = -3x + 3$

27

Find the common solution(s) of the system at the right by substituting -3x + 3 for y.

$y = x^2 - x$
$3x + y = 3$

$x^2 + 2x - 3 = 0$
$x = -3 \quad x = 1$
$(-3, 12) \quad (1, 0)$

28

Earlier in this unit the system
$y = (x - 2)^2 - 1$
$y = -x + 4$
was shown.

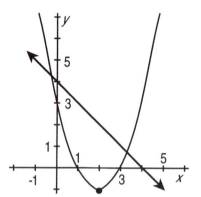

Recall that the location of the common solutions made them difficult to accurately state the ordered pairs.

Is graphing a good method for finding the ordered pairs?

No

29

To find the common solution(s) for the system at the right, substitute (-x + 4) for y in the quadratic equation.

$y = (x - 2)^2 - 1$
$y = -x + 4$

$-x + 4 = (x - 2)^2 - 1$
$-x + 4 = x^2 - 4x + 4 - 1$
$x^2 - 3x - 1 = 0$

Solve $x^2 - 3x - 1 = 0$ by the quadratic formula.

$x = \dfrac{-b \pm \sqrt{b^2 - 4ac}}{2a}$

$x = \dfrac{3 \pm \sqrt{13}}{2}$

30

For $x = \frac{3+\sqrt{13}}{2}$ one solution of the system at the right is found by substituting for x in $y = -x + 4$

$y = (x - 2)^2 - 1$
$y = -x + 4$

$$y = -x + 4 = -\left(\frac{3+\sqrt{13}}{2}\right) + 4$$

$$= \frac{-3-\sqrt{13}}{2} + \frac{8}{2}$$

$$= \frac{5-\sqrt{13}}{2}$$

$\left(\frac{3+\sqrt{13}}{2}, \frac{5-\sqrt{13}}{2}\right)$ is one common solution. Complete the other common solution, $\left(\frac{3-\sqrt{13}}{2}, \underline{}\right)$.

$\left(\frac{3-\sqrt{13}}{2}, \frac{5+\sqrt{13}}{2}\right)$

31

To find the common solution(s) for the system at the right, substitute $2x - 3$ for y in the quadratic equation.

$y = x^2 - 5$
$y = 2x - 3$

$y = x^2 - 5$
$2x - 3 = x^2 - 5$
$x^2 - 2x - 2 = 0$

Solve $x^2 - 2x - 2 = 0$ by the quadratic formula.

$x = \frac{2 \pm \sqrt{12}}{2} = \frac{2 \pm 2\sqrt{3}}{2}$

$x = 1 \pm \sqrt{3}$

32

A common solution of the system at the right can be completed when x is replaced by $(1 + \sqrt{3})$.

$y = x^2 - 5$
$y = 2x - 3$

$y = 2x - 3 = 2(1 + \sqrt{3}) - 3$

$= 2 + 2\sqrt{3} - 3$

$= -1 + 2\sqrt{3}$

$(1 + \sqrt{3}, -1 + 2\sqrt{3})$ is one common solution. Complete the other common solution, $(1 - \sqrt{3}, \underline{})$.

$(1 - \sqrt{3}, -1 - 2\sqrt{3})$

33

Find common solutions of the system at the right. The components of the ordered pairs are irrational numbers.

$y = x^2 + 2$
$y = 3x + 4$

$x^2 - 3x - 2 = 0$
$\left(\dfrac{3+\sqrt{17}}{2}, \dfrac{17+3\sqrt{17}}{2}\right)$ and
$\left(\dfrac{3-\sqrt{17}}{2}, \dfrac{17-3\sqrt{17}}{2}\right)$

34

To find common solutions of the system at the right, substitution is used.

$y = x^2$
$y = x - 5$

$y = x^2$
$x - 5 = x^2$
$x^2 - x + 5 = 0$

The quadratic formula gives: $x = \dfrac{1 \pm \sqrt{-19}}{2}$

Is $\dfrac{1 \pm \sqrt{-19}}{2}$ a real number?

No, complex. $\dfrac{1 \pm i\sqrt{19}}{2}$

35

The system shown at the right has no real numbers in the ordered pairs of its common solution. Do the graphs of the 2 equations have any common points on the xy-plane?

$y = x^2$
$y = x - 5$

No

36

Use substitution to attempt to find the common solution(s) of the system. Find a solution for x whether it be real or complex.

$y = x^2 + 5$
$y = x + 4$

$x = \dfrac{1 \pm i\sqrt{3}}{2}$

37

Using substitution to find any common solution(s) of the system at the right results in $x = \dfrac{1 \pm i\sqrt{3}}{2}$ which means the system (has, has no) _____ real numbers in its common solution.

$y = x^2 + 5$
$y = x + 4$

has no

38
Find the common solution(s) of the system at the right.

$y = x^2 - 2x$
$y = -x - 2$

No real numbers

39
Find the common solution(s) of the system at the right.

$y = x^2 - 2x$
$y = -6x - 4$

$(-2, 8)$

40
Find the common solution(s) of the system at the right.

$y = x^2 - 4$
$y = 4x + 1$

$(-1, -3), (5, 21)$

41
Find the common solution(s) of the system at the right.

$y = x^2$
$x + y = 6$

$(-3, 9), (2, 4)$

42
Find the common solution(s) of the system at the right.

$y = x^2 + x - 3$
$y = 3x - 3$

$(0, -3), (2, 3)$

43
Find the common solution(s) of the system at the right.

$y = x^2 - 4x + 7$
$x - 2y = 0$

No real numbers

44
Find the common solution(s) of the system at the right. The first equation is for a circle; the second for a line.

$x^2 + y^2 = 25$
$x - y = -5$

$x^2 + (x + 5)^2 = 25$
$(0, 5), (-5, 0)$

45
Find the common solution(s) of the system at the right. The first equation is for a circle; the second for a line.

$(x - 3)^2 + y^2 = 25$
$x + y = 8$

$(x - 3)^2 + (-x + 8)^2 = 25$
$(3, 5), (8, 0)$

Feedback Unit 3

This quiz reviews the preceding unit. Answers are at the back of the book.

1. Which of the following are common solutions of the system at the right?

    $y = (x - 3)^2 - 1$
    $y = x + 2$

    (4, 6) (-2, 0) (1, 3) (-4, 2) (6, 8)

2. Which of the following are common solutions of the system at the right?

    $x + 3y = 47$
    $(x - 2)^2 + (y - 5)^2 = 100$

    (-4, 13) (2, 5) (8, 13) (-2, 15) (2, 15)

For problems 3-10 find common solution(s) for the system.

3. $y = x^2$
   $x + y = 2$

4. $x^2 + y^2 = 25$
   $3x - 4y = -25$

5. $y = x^2 + x - 3$
   $y = 3x - 4$

6. $(x + 1)^2 + (y - 1)^2 = 25$
   $3x + y = 3$

7. $x^2 + y^2 = 4$
   $y = -x + 2$

8. $y = x^2 + 5$
   $4x + y = 2$

9. $y = x^2 - x$
   $4x + y = 4$

10. $x^2 + (y - 2)^2 = 9$
    $x - y = 1$

UNIT 4: SYSTEMS OF 3 LINEAR EQUATIONS

1

The system of equations shown at the right contains 3 equations with 3 variables. Is each equation a 1st degree (linear) equation?

$x + y + z = 8$
$x - y + z = 4$
$x + y - z = -2$

Yes

2

A common solution of the system shown at the right will be an ordered triple, (x, y, z), that is a solution of each equation. Is $(1, 2, 5)$ an ordered triple?

$x + y + z = 8$
$x - y + z = 4$
$x + y - z = -2$

Yes

3

To test the ordered triple $(1, 2, 5)$ as a common solution of the system shown at the right, x is replaced by 1 in each equation, y is replaced by 2 in each equation and z is replaced by _____ in each equation.

$x + y + z = 8$
$x - y + z = 4$
$x + y - z = -2$

5

4

$(1, 2, 5)$ is a solution of $x + y + z = 8$ because $1 + 2 + 5 = 8$ is true.
$(1, 2, 5)$ is a solution of $x - y + z = 4$ because $1 - 2 + 5 = 4$ is true.
Is $(1, 2, 5)$ a solution of $x + y - z = -2$?

Yes, $1 + 2 - 5 = -2$ is true.

5

$(1, 2, 5)$ is a solution (x, y, z) for each equation in the system shown at the right. Is $(1, 2, 5)$ a common solution for the system?

$x + y + z = 8$
$x - y + z = 4$
$x + y - z = -2$

Yes

6

An ordered triple (x, y, z) is a solution of $2x + y + z = -3$ if replacement of the 3 variables with their components from (x, y, z) produces a true statement. Is (1, 3, -2) a solution of $2x + y + z = -3$?

No, $2 \cdot 1 + 3 - 2 \ne -3$

7

The ordered triple (-1, 1, 2) is a solution of both $x + y + z = 2$ and $2x + y - z = -3$.
    $-1 + 1 + 2 = 2$ is true. $-2 + 1 - 2 = -3$ is true.
Is (-1, 1, 2) a solution of $x - y + z = 0$?

Yes, $-1 - 1 + 2 = 0$

8

Is (-1, 2, -2) a common solution of the system shown at the right?

$$x + y + z = -1$$
$$2x + y = 0$$
$$x - y + 2z = 4$$

No, it is not a solution of $x - y + 2z = 4$

9

Recall that systems with 2 linear equations and 2 variables are often solved by "adding" the equations and eliminating one of the variables.
Add equations (1) and (2) to eliminate the y's.

(1) $2x + y + z = 8$
(2) $2x - y + z = 2$
(3) $x + y + z = 6$

$2x + y + z = 8$
$\underline{2x - y + z = 2}$
$4x + 2z = 10$

10

The system of frame 9 had 3 equations. By adding (1) and (2) the result may be included as equation (4). Add (2) and (3) and again eliminate the y's.

(1) $2x + y + z = 8$
(2) $2x - y + z = 2$
(3) $x + y + z = 6$
(4) $4x + 2z = 10$

$2x - y + z = 2$
$\underline{x + y + z = 6}$
$3x + 2z = 8$

11

The system of frame 9 had 3 equations, but adding (1) and (2) and then adding (2) and (3) generated (4) and (5).

(1) $2x + y + z = 8$
(2) $2x - y + z = 2$
(3) $x + y + z = 6$
(4) $4x + 2z = 10$
(5) $3x + 2z = 8$

Use addition with (4) and (5) to eliminate the z's.

$4x + 2z = 10$
$\underline{-3x - 2z = -8}$
$x = 2$

12

The common solution of the system at the right is $(2, y, z)$ as was found in frame 11. Use the fact that $x = 2$ with either (4) or (5) to find a value for z.

(1) $2x + y + z = 8$
(2) $2x - y + z = 2$
(3) $x + y + z = 6$
(4) $4x + 2z = 10$
(5) $3x + 2z = 8$

$z = 1$

13

The common solution of the system at the right is $(2, y, 1)$. The value of y remains unknown. Use the fact that $x = 2$ and $z = 1$ with (1), (2), or (3) to find a value for y.

(1) $2x + y + z = 8$
(2) $2x - y + z = 2$
(3) $x + y + z = 6$
(4) $4x + 2z = 10$
(5) $3x + 2z = 8$

$y = 3$

14

In frames 9-13 the solution $(2, 3, 1)$ was found for the system shown at the right. Show that $(2, 3, 1)$ is the common solution of the system.

(1) $2x + y + z = 8$
(2) $2x - y + z = 2$
(3) $x + y + z = 6$

(1) $4 + 3 + 1 = 8$ is true.
(2) $4 - 3 + 1 = 2$ is true.
(3) $2 + 3 + 1 = 6$ is true.

15

Begin the process for finding the common solution of the system shown at the right by adding (1) and (2) and eliminating the x's.

(1) $x - y - z = 6$
(2) $2x - y - z = 5$
(3) $5x + y - z = -2$

$-2x + 2y + 2z = -12$
$\underline{2x - y - z = 5}$
$ y + z = -7$

16

Continue the process for finding the common solution of the system from frame 15 by adding (1) and (3) and again eliminating the x's.

(1) $x - y - z = 6$
(2) $2x - y - z = 5$
(3) $5x + y - z = -2$

$-5x + 5y + 5z = -30$
$\underline{5x + y - z = -2}$
$6y + 4z = -32$

17

In frames 15 and 16 addition was used in eliminating the x's and generating 2 equations. Solve $y + z = -7$ and $6y + 4z = -32$ to find values for y and z.

(1) $x - y - z = 6$
(2) $2x - y - z = 5$
(3) $5x + y - z = -2$

$-4y - 4z = 28$
$\underline{6y + 4z = -32}$
$2y = -4$
$y = -2, \ z = -5$

18

Complete the common solution (___, -2, -5) of the system shown at the right. Any one of the 3 equations can be used with $y = -2$ and $z = -5$.

(1) $x - y - z = 6$
(2) $2x - y - z = 5$
(3) $5x + y - z = -2$

$(-1, -2, -5)$

19

The complete process for finding a common solution for a system of 3 linear equations is:

1. Use a pair of equations to eliminate by addition one of the variables.
2. Use a different pair of equations to eliminate by addition the same variable.
3. Solve the 2 equation-2 variable system generated by steps 1 and 2.
4. Substitute known values into one of the original equations to find the common solution.

The 4-step process listed above might be rephrased as:

To solve a system of 3 linear equations create a system of ___ linear equations.

2

20

Systems of 4 linear equations are solved by repeatedly eliminating, by addition, the same variable until a system of 3 linear equations is generated. Then a system of ___ linear equations is generated.

2

SOLVING SYSTEMS OF EQUATIONS   555

21

Begin the process for finding the common solution of the system shown at the right by adding (1) and (2) and eliminating the x's.

(1) $2x - y + z = -2$
(2) $x + 2y - z = -3$
(3) $4x - y + 2z = -3$

$2x - y + z = -2$
$-2x - 4y + 2z = 6$
$\overline{-5y + 3z = 4}$

22

Continue the process for finding the common solution of the system from frame 21 by adding (1) and (3) and again eliminating the x's.

(1) $2x - y + z = -2$
(2) $x + 2y - z = -3$
(3) $4x - y + 2z = -3$

$-4x + 2y - 2z = 4$
$4x - y + 2z = -3$
$\overline{y = 1}$

23

In frames 21 and 22 addition was used in eliminating the x's, but one result was $y = 1$. Use the fact that $y = 1$ with the equation $-5y + 3z = 4$ to find z.

(1) $2x - y + z = -2$
(2) $x + 2y - z = -3$
(3) $4x - y + 2z = -3$

$z = 3$

24

Complete the common solution (___ , 1, 3) of the system shown at the right. Any one of the 3 equations can be used with $y = 1$ and $z = 3$.

(1) $2x - y + z = -2$
(2) $x + 2y - z = -3$
(3) $4x - y + 2z = -3$

$(-2, 1, 3)$

25

Finding the common solution of the system shown at the right is made easier because (1) involves only x and z. Add (2) and (3) to eliminate the y's.

(1) $3x - z = 13$
(2) $x + y - 2z = 0$
(3) $6x + 2y - 3z = 22$

$-2x - 2y + 4z = 0$
$6x + 2y - 3z = 22$
$\overline{4x + z = 22}$

26

Equation (1) at the right contains x and z. Also the equation generated in frame 25 involves only x and z. Use these equations to find values for x and z.

(1) $3x - z = 13$
(2) $x + y - 2z = 0$
(3) $6x + 2y - 3z = 22$

$x = 5, z = 2$

27

Complete the common solution (5, ___, 2) for the system shown at the right.

(1) $3x - z = 13$
(2) $x + y - 2z = 0$
(3) $6x + 2y - 3z = 22$

(5, -1, 2)

28

Find the common solution (x, y, z) for the system shown at the right.

$2x - 2y - z = 8$
$3x - y - z = 9$
$5x + 2y - 3z = 5$

(3, -2, 2)

29

Find the common solution (x, y, z) for the system shown at the right.

$x + y - z = 2$
$2x + y + z = -9$
$4x + 2y + z = -13$

(-1, -2, -5)

30

Find the common solution (x, y, z) for the system shown at the right.

$2x + 7y + z = 8$
$x - 9y + 3z = 9$
$x - y - 2z = -1$

(3, 0, 2)

31

Find the common solution (x, y, z) for the system shown at the right.

$y + z = -2$
$x + z = -1$
$x + y = 3$

Note: This system is easier to solve than the systems in frames 29 and 30. Generate a two-equation, two-variable system and then solve it.

(2, 1, -3)

32
Find the common solution (x, y, z) for the system shown at the right.

$x + y - z = 10$
$x - y - z = 6$
$y + z = -3$

(3, 2, -5)

33
Find the common solution (x, y, z) for the system shown at the right.

$3x - z = 3$
$2x + 3y + 7z = -1$
$4x - 2y - 3z = 6$

(1, -1, 0)

34
Find the common solution (x, y, z) for the system shown at the right.

$5x - 2y + 7z = 34$
$6x + 2y - 3z = -3$
$4x - 2y - 5z = -27$

(1, 3, 5)

A system of three equations with three variables may have no common solutions. The system shown at the right has no common solution. This is because the terms of the left side of equation (2) are double the terms of the left side of equation (1). Any attempt to eliminate a variable produces a false statement like $0 = -19$. The system has no common solution.

(1) $2x - y + 3z = 23$
(2) $4x - 2y + 6z = 27$
(3) $x - 3y + 7z = 1$

35
Begin the process for finding a common solution for the system shown at the right. Multiply the terms of equation (2) by -2 and add to equation (3). What statement is obtained?

(1) $x - y + 3z = 1$
(2) $3x - y + z = 2$
(3) $6x - 2y + 2z = 1$

$0 = -3$

36

In the system of frame 35, attempts to find the common solution resulted in the false statement 0 = -3. This means the system has _____ (how many) common solutions.

0

37

Begin the process for finding a common solution for the system shown at the right. Multiply the terms of equation (3) by -1 and add to equation (1). What statement is obtained?

(1) $x + y - z = 3$
(2) $x + y + z = 1$
(3) $x + y - z = -1$

$0 = 4$

38

In the system of frame 37, attempts to find the common solution resulted in the false statement 0 = 4. What does this mean?

There is no common solution

39

A system of three equations with three variables may have infinitely many common solutions. The system shown at the right illustrates this possibility. Equation (3) is equivalent to equation (1). If each term of equation (1) is multiplied by 4, what equivalent equation is generated?

(1) $x - 3y + 2z = 3$
(2) $3x - 3y - z = 1$
(3) $4x - 12y + 8z = 12$

equation (3)

40

When 2 equations in a system are equivalent, attempts to find a common solution will result in true statements like 0 = 0. Are 2 equations of the system at the right equivalent?

(1) $x - 2y + 3z = 1$
(2) $3x - 6y + 9z = 3$
(3) $x + y - z = 6$

Yes, (1) and (2)

41

When 2 equations in a system are equivalent, the system has infinitely many common solutions. This means the system shown at the right has _____ (how many) common solutions.

(1) $x - 2y + 3z = 1$
(2) $3x - 6y + 9z = 3$
(3) $x + y - z = 6$

infinitely many

42

Will the system shown at the right have infinitely many common solutions?

(1) $3x - 3y + z = 1$
(2) $-3x + 3y - z = -1$
(3) $2x - y + 7z = 8$

Yes, equations (1) and (2) are equivalent

FEEDBACK UNIT 4

This quiz reviews the preceding unit. Answers are at the back of the book.

Indicate whether the following systems have no, one, or an infinite number of common solutions. If a system has one common solution, find it.

1. $x - y + z = 2$
   $x + y + 3z = 12$
   $2x + y + z = 7$

2. $x + y + z = 8$
   $x - 2y - z = 2$
   $3x + 3y + 3z = 24$

3. $2x + y + z = -1$
   $x + y + z = 0$
   $x - y + z = 0$

4. $2x + 3y - z = 2$
   $2x + 5y + z = -1$
   $4x + 10y + 2z = 1$

5. $2x + z = -5$
   $3x + y = -1$
   $x - y = -3$

6. $2x - 3y + z = -16$
   $3x - y - z = -14$
   $x + 2y - z = 5$

7. $3x + 2y = -9$
   $3y - z = -11$
   $5x + 2z = -1$

Unit 5: Applications

In this Applications Section, the format of the text has been altered. Answers for the problems appear beneath them rather than in the right-hand column. Your studying emphasis should be on learning the best procedures to follow with word problems.

1

The formula D = RT relates distance (D) to the speed (R) and to the time (T). The distance a car can travel in 3 hours at 40 miles per hour is D = 40 • 3 = 120 miles. If a car travels 7 hours and its speed is represented by R, use the formula to write an open expression to represent its distance.

Answer: 7R

2

The formula D = RT gives a method for showing distance as a product of rate and time. If a horse travels 8 miles per hour and its time is represented by T, use the formula to represent its distance.

Answer: 8T

3

5R represents the distance traveled in 5 hours at R miles per hour. Use the formula D = RT to represent the distance traveled in 5 hours at (R + 3) miles per hour.

Answer: 5(R + 3)

4

Use the formula D = RT to represent the distance traveled in (T + 6) hours at 73 miles per hour.

Answer: 73(T + 6)

5

Use the formula D = RT to represent the distance traveled if a boat goes downstream 26 miles per hour for (T – 5) hours.

Answer: 26(T – 5)

6

The situation described below involves 2 cars, their speeds, their times traveled, and their distances.

> Car A and Car B run the same course with Car B traveling 15 miles per hour slower than Car A. Car A finishes the course in 5 hours and Car B in 8 hours.

Make a drawing. Use arrows to represent the distance of each car.

Answer:
The arrows must go in the same direction and must have the same length because the 2 cars make the same trip (travel the same distance). The drawing below would fit the situation.

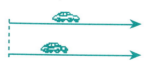

7

Car A and Car B run the same course with Car B traveling 15 miles per hour slower than Car A. Car A finishes the course in 5 hours and Car B in 8 hours.

Which of the following is stated in the situation?
- a. The rate of Car A.
- b. The rate of Car B.
- c. The time traveled by Car A.
- d. The time traveled by Car B.
- e. The distance traveled by Car A.
- f. The distance traveled by Car B.

Answer: Only (c) and (d) are stated. Car A travels 5 hours. Car B travels 8 hours.

8

To solve the problem below, the 3-step process learned earlier is used.

> Car A and Car B run the same course with Car B traveling 15 miles per hour slower than Car A. Car A finishes the course in 5 hours and Car B in 8 hours. Find the speed of each car and the length of the course.

The 3-step process begins with a table of translations like the one shown below.

word/phrase/sentence	translation
rate of Car A	A
rate of Car B	B
15 miles per hour slower than Car A	_____
Car B travels 15 miles per hour slower than Car A	_____
time of Car A	5
time of Car B	8
distance (D = RT) of Car A	_____
distance (D = RT) of Car B	_____
Car A and Car B travel the same distance	_____

Complete the missing entries in the table of translations.

Answer:

word/phrase/sentence	translation
15 miles per hour slower than Car A	A − 15
Car B travels 15 miles per hour slower than Car A	B = A − 15
distance (D = RT) of Car A	5A
distance (D = RT) of Car B	8B
Car A and Car B travel the same distance	5A = 8B

9

To solve the problem below, the 3-step process learned earlier is used.

> Car A and Car B run the same course with Car B traveling 15 miles per hour slower than Car A. Car A finishes the course in 5 hours and Car B in 8 hours. Find the speed of each car and the length of the course.

Use the table of translations constructed in frame 8. Find 2 equations and solve them.

Answer:
$B = A - 15$ and $5A = 8B$
$A = 40$ and $B = 25$

10

Car A and Car B run the same course with Car B traveling 15 miles per hour slower than Car A. Car A finishes the course in 5 hours and Car B in 8 hours. Find the speed of each car and the length of the course.
Find the distance traveled by each car using the results from frame 9. Then confirm that the answers make the original problem statement true.

Answer:
When Car A travels 40 mph for 5 hours, it goes 200 miles.
When Car B travels 25 mph for 8 hours, it goes 200 miles.
Car B travels 15 mph slower than Car A: $40 - 25 = 15$

11

The situation described below involves a boat making 2 trips at different speeds and times.

> A boat goes downstream and then returns to its original starting point. Going downstream it has a speed of 12 miles per hour and going upstream it has a speed of 8 miles per hour. The round trip will take 10 hours.

Make a drawing. Use arrows to represent the distance of each part of the trip.

> Answer:
> The arrows must go in opposite directions, but must have the same length because the 2 trips are the same length (distance). The drawing below would fit the situation.

12

> A boat goes downstream and then returns to its original starting point. Going downstream it has a speed of 12 miles per hour and going upstream it has a speed of 8 miles per hour. The round trip will take 10 hours.

Which of the following is stated in the situation?
    a. The rate of the boat going downstream.
    b. The rate of the boat going upstream.
    c. The time of the boat going downstream.
    d. The time of the boat going upstream.
    e. The distance traveled downstream.
    f. The distance traveled upstream.

> Answer: Only (a) and (b) are stated. The downstream rate is 12 mph; upstream rate is 8 mph.

13
To solve the problem below, the 3-step process learned earlier is used.

A boat goes downstream and then returns to its original starting point. Going downstream it has a speed of 12 miles per hour and going upstream it has a speed of 8 miles per hour. The round trip will take 10 hours. Find the time it will spend going in each direction and the total distance it will travel.

The 3-step process begins with a table of translations like the one shown below.

word/phrase/sentence	translation
rate of downstream trip	12
rate of upstream trip	8
time of downstream trip	D
time of upstream trip	U
The round trip will take 10 hours.	_____
distance (D = RT) of downstream trip	_____
distance (D = RT) of upstream trip	_____
Both trips travel the same distance	_____

Complete the missing entries in the table of translations.

Answer:

word/phrase/sentence	translation
The round trip will take 10 hours.	$D + U = 10$
distance (D = RT) of downstream trip	$12D$
distance (D = RT) of upstream trip	$8U$
Both trips travel the same distance	$12D = 8U$

14

To solve the problem below, the 3-step process learned earlier is used.

> A boat goes downstream and then returns to its original starting point. Going downstream it has a speed of 12 miles per hour and going upstream it has a speed of 8 miles per hour. The round trip will take 10 hours. Find the time it will spend going in each direction and the total distance it will travel.

Use the table of translations constructed in frame 13. Find 2 equations and solve them.

Answer:
$D + U = 10$ and $12D = 8U$
$D = 4$ and $U = 6$

15

> A boat goes downstream and then returns to its original starting point. Going downstream it has a speed of 12 miles per hour and going upstream it has a speed of 8 miles per hour. The round trip will take 10 hours. Find the time it will spend going in each direction and the total distance it will travel.

Use the results from frame 14 to find the time traveled going downstream/upstream and the total distance traveled. Then confirm that the answers make the original problem statement true.

Answer:
Going downstream for 4 hours at 12 mph the boat goes 48 miles.
Going upstream for 6 hours at 8 mph the boat goes 48 miles.
The total distance traveled is 96 miles.
The round trip will take 10 hours because $4 + 6 = 10$.

16

A man took a two-day auto trip. The first day he traveled 10 hours, but he increased his speed the second day by 10 miles per hour and traveled only 8 hours. He was 620 miles from his starting point after the second day.

Make a drawing. Use arrows to represent the distance of each day of the trip.

Answer:
The drawing might look like the one below.

17

A man took a two-day auto trip. The first day he traveled 10 hours, but he increased his speed the second day by 10 miles per hour and traveled only 8 hours. He was 620 miles from his starting point after the second day. Find his average speed each day.

A table of translations for the problem shown above will have 8 lines: 2 lines will be for the rate each day, 2 lines will be for the time each day, 2 lines will be for the distance each day, and 2 will be equations relating rate/time and distance. Construct a table of translations for the problem.

Answer:

word/phrase/sentence	translation
rate of 1st day	F
rate of 2nd day	S
he increased his speed the 2nd day by 10 mph	$F + 10 = S$
time of 1st day	10
time of 2nd day	8
distance ($D = RT$) of 1st day	$10F$
distance ($D = RT$) of 2nd day	$8S$
620 miles from his starting point after the 2nd day	$10F + 8S = 620$

18

A man took a two-day auto trip. The first day he traveled 10 hours, but he increased his speed the second day by 10 miles per hour and traveled only 8 hours. He was 620 miles from his starting point after the second day. Find his average speed each day.

Complete the 3-step process for solving this problem. The table of translations was completed in frame 17.

Answer:
b. $F + 10 = S$ and $10F + 8S = 620$ give $F = 30$ and $S = 40$

c. The 1st day he averaged 30 mph and the 2nd day 40 mph. The 1st day he traveled 300 miles; the 2nd day 320 miles.

19

Make a drawing for the following situation.

Mr. George drove part of a trip at 30 miles per hour and the rest of the trip at 45 miles per hour. He drove the same length of time at each rate of speed and traveled a total of 150 miles.

Answer:
The drawing might look like this.

20

Use the 3-step process and solve the following problem.

Mr. George drove part of a trip at 30 miles per hour and the rest of the trip at 45 miles per hour. He drove the same length of time at each rate of speed and traveled a total of 150 miles. How much time was required for the entire trip?

Answer:

a.

word/phrase/sentence	translation
rate of 1st part	30
rate of 2nd part	45
time of 1st part	F
time of 2nd part	S
He drove the same time at each rate of speed	F = S
distance (D = RT) of 1st part	30F
distance (D = RT) of 2nd part	45S
traveled a total of 150 miles	30F + 45S = 150

b. F = S and 30F + 45S = 150 gives F = 2 and S = 2.

c. He drove 2 hours each part of the trip for a total time traveled of 4 hours. The 1st part of the trip he traveled 60 miles; the 2nd part 90 miles.

21

Make a drawing for the following situation.

An airplane flew for 9 hours at a fixed rate of speed. It then increased its speed by 20 miles per hour for 2 more hours. The total distance traveled was 3890 miles.

Answer:
The drawing might look like this.

22

Use the 3-step process and solve the following problem.

An airplane flew for 9 hours at a fixed rate of speed. It then increased its speed by 20 miles per hour for 2 more hours. The total distance traveled was 3890 miles. Find the speed for each part of the trip.

Answer:

a.

word/phrase/sentence	translation
rate of 1st part	F
rate of 2nd part	S
increased its speed by 20 mph	$F + 20 = S$
time of 1st part	9
time of 2nd part	2
distance (D = RT) of 1st part	9F
distance (D = RT) of 2nd part	2S
total distance traveled was 3890 miles	$9F + 2S = 3890$

b. $F + 20 = S$ and $9F + 2S = 3890$ gives $F = 350$ and $S = 370$.

c. The plane flew 350 mph for the 1st part of the trip; 370 mph for 2nd part. The plane traveled 3150 miles in 1st part and 740 miles in 2nd part for a total of 3890 miles.

23

Make a drawing for the following situation.

Two cars left the same town traveling in opposite directions. One car left 2 hours before the other and traveled at 48 miles per hour. The other car traveled at 63 miles per hour and by noon they were 540 miles apart.

Answer:
The drawing might look like this.

24

Use the 3-step process and solve the following problem.

Two cars left the same town traveling in opposite directions. One car left 2 hours before the other and traveled at 48 miles per hour. The other car traveled at 63 miles per hour and by noon they were 540 miles apart. What time did the first car start?

Answer:

a.

word/phrase/sentence	translation
rate of 1st car	48
rate of 2nd car	63
time of 1st car	F
time of 2nd car	S
One car left 2 hours before the other	F − 2 = S
distance (D = RT) of 1st car	48F
distance (D = RT) of 2nd car	63S
by noon they were 540 miles apart	48F + 63S = 540

b. F − 2 = S and 48F + 63S = 540 gives F = 6 and S = 4.

c. The 1st car traveled 6 hours; the 2nd car traveled 4 hours.
The 1st car traveled 288 miles and the 2nd car 252 miles.
At noon the 1st car had traveled 6 hours, so it started at 6 a.m.

25

Make a drawing for the following situation.

A horseback rider left home one morning and traveled 5 hours to the nearest town. The next day she started back home but traveled 3 miles per hour slower. After 4 hours she was still 20 miles from home.

Answer:
The drawing might look like this.

26

Use the 3-step process and solve the following problem.

A horseback rider left home one morning and traveled 5 hours to the nearest town. The next day she started back home but traveled 3 miles per hour slower. After 4 hours she was still 20 miles from home. How far was it to town from her home?

Answer:

a.

word/phrase/sentence	translation
rate of trip to town	T
rate of trip heading home	H
heading home (she) traveled 3 mph slower	$H = T - 3$
time of trip to town	5
time of trip heading home	4
distance (D = RT) to town	5T
distance (D = RT) heading home	4H
heading home was still 20 miles from home	$4H + 20 = 5T$

b. $H = T - 3$ and $4H + 20 = 5T$ gives $T = 8$ and $H = 5$.

c. Going to town took 5 hours at 8 mph so the distance was 40 miles. Heading home for 4 hours at 5 mph gives distance of 20 miles. Heading home 20 miles means that there were 20 miles more to go.

Feedback Unit 5 for Applications

Use the 3-step process to solve each of the following.

1. On a 262-mile trip, Mr. Morgan drove for 3 hours at a constant rate. For the next 4 hours he drove 20 miles per hour faster. How fast was he driving the first 3 hours?

2. Two cars left the same town and followed the same route. If one car was traveling at 60 miles per hour and the other at 52 miles per hour, how long would it take for the cars to be 44 miles apart?

3. A moving van traveled 55 miles per hour on the first leg of a trip and only 33 miles per hour on the return trip. If it took the van 16 hours to make the round trip, how far were the towns from each other?

4. Two jets started toward each other from airports that were 12,000 miles apart. One jet flew 20 miles per hour slower than the other and they met in 10 hours. What was the speed of the slower jet?

Summary for Chapter 12

The following mathematical terms are crucial to an understanding of this chapter.

  Ordered pair   Ordered triple
  Ordered n-tuples   Ordered 4-tuple
  Ordered 5-tuple   System of equations
  Common solution   Substitution

The systems of equations that were studied in this chapter were as follows:

1. Two linear equations with two variables—Both the elimination method by addition and the substitution method were used. It was shown that it is possible to have a system with no solution, a system with one solution, and a system with an infinite number of solutions.

2. The common solution(s) of a system that included one linear equation and one second-degree or quadratic equation—It was shown that it is possible to have a system with no solution, a system with one solution, and a system with two solutions.

3. The common solution of a system of three equations with three variables was solved by an elimination method to produce an ordered triple. It was shown that it is possible to have a system with no solution, a system with one solution, and a system with an infinite number of solutions.

Chapter 12 Mastery Test

The following questions test the objectives of Chapter 12. Answers are at the back of the book. The number in parentheses which follows each problem indicates the unit in which it can be learned.

1. Write an ordered triple with 2 as its 2nd component, -9 as its 3rd component, and 3 as its 1st component. (1)

2. Find the common solution (x, y, z, w) for the system of equations shown at the right.
   $x = 6$
   $w = 4$
   $z = -8$
   $y = 0$ (1)

For problems 3-14, find the common solution (x, y) or (x, y, z). If no common solution exists or there are an infinite number of them, state whichever is the case.

3. $x + y = 7$
   $-x + y = -3$ (2)

4. $y = 2x$
   $x - y = 1$ (2)

5. $y = x^2$
   $2x + y = 15$ (3)

6. $2x + 3y = 11$
   $3x + y = -1$ (2)

7. $2x + 3y - z = -7$
   $5x - 3y - z = 8$
   $x - y + z = 6$ (4)

8. $y + 2z = -2$
   $3x + y = -1$
   $5x - 2z = -1$ (4)

9. $x - 3y = -1$
   $3x - 9y = 2$ (2)

10. $y = x^2$
    $y = 2x$ (3)

11. $y = x^2 + x + 3$
    $y = 4x + 1$ (3)

12. $4x + 7y + z = 0$
    $3x - y + 2z = 1$
    $8x + 14y + 2z = -1$ (4)

13. $y = x^2 - 4$
    $y = 2x - 5$ (3)

14. $x + y - z = 0$
    $x - y - z = -4$
    $x + y + z = 0$ (4)

15. Use the 3-step process and solve: A boat left its dock and went downstream to Big City at 12 mph. On its return trip upstream to the dock the boat could only travel 8 mph. If total traveling time was 10 hours, what is the distance from the dock to Big City. (5)

16. Use the 3-step process and solve: Two bicyclers left the same point traveling the same route, but one traveled 8 mph and the other 5 mph. How long would it take for the bicyclers to be 14 miles apart? (5)

Square Root Table

Number	Square Root	Number	Square Root	Number	Square Root
1	1.000	36	6.000	71	8.426
2	1.414	37	6.083	72	8.485
3	1.732	38	6.164	73	8.544
4	2.000	39	6.245	74	8.602
5	2.236	40	6.325	75	8.660
6	2.449	41	6.403	76	8.718
7	2.646	42	6.481	77	8.775
8	2.828	43	6.557	78	8.832
9	3.000	44	6.633	79	8.888
10	3.162	45	6.708	80	8.944
11	3.317	46	6.782	81	9.000
12	3.464	47	6.856	82	9.055
13	3.606	48	6.928	83	9.110
14	3.742	49	7.000	84	9.165
15	3.873	50	7.071	85	9.220
16	4.000	51	7.141	86	9.274
17	4.123	52	7.211	87	9.327
18	4.243	53	7.280	88	9.381
19	4.359	54	7.348	89	9.434
20	4.472	55	7.416	90	9.487
21	4.583	56	7.483	91	9.539
22	4.690	57	7.550	92	9.592
23	4.796	58	7.616	93	9.644
24	4.899	59	7.681	94	9.695
25	5.000	60	7.746	95	9.747
26	5.099	61	7.810	96	9.798
27	5.196	62	7.874	97	9.849
28	5.292	63	7.937	98	9.899
29	5.385	64	8.000	99	9.950
30	5.477	65	8.062	100	10.000
31	5.568	66	8.124		
32	5.657	67	8.185		
33	5.745	68	8.246		
34	5.831	69	8.307		
35	5.916	70	8.367		

Answers for All Tests and Feedback Exercises

Chapter 1
Objectives

1. 413
2. {49, 51}
3. {20, 21, 22}
4. 14
5. 84
6. Yes
7. 17
8. a. 7x + 5 = 31
9. a. x + y = y + x
10. integers
11. No
12. $\frac{1}{2}$
13. $\frac{12}{5}$
14. Yes
15. (7x + 5y) + 4
16. 4(5x) or (x • 5)4
17. 7x − 9
18. $\frac{1}{12}x - \frac{5}{4}$
19. $\frac{7}{2}$
20. $\frac{-9}{2}$

Chapter 1
Feedback Unit 1

1. 814
2. {34, 36}
3. Yes
4. {1, 2, 3, 5, 6, 10, 15, 30}
5. {23, 29}
6. Yes
7. No
8. 1
9. 2
10. 60
11. 18
12. 36
13. 50
14. Yes
14. statement
15. conditional equation

Chapter 1
Feedback Unit 2

1. No
2. -7
3. No
4. rational
5. irrational
6. rational
7. No
8. Yes
9. $\frac{-2}{3}$
10. $\frac{9}{7}$

Chapter 1
Feedback Unit 3

1. commutative
2. cannot
3. No
4. No
5. associative
6. Yes
7. 0
8. Yes, 0
9. No
10. 1

Chapter 1
Feedback Unit 4

1. x + 15
2. 35x
3. 16y + 6
4. 21 + 14x
5. 14x + 5
6. 13x + 34
7. 3x + 13
8. -x − 17
9. -6x + 13
10. 5x − 20
11. -7x − 1
12. -8x − 7
13. $x - \frac{13}{20}$
14. $\frac{1}{4}x + \frac{1}{6}$
15. -5x
16. -2x
17. $\frac{1}{8}x - \frac{14}{5}$
18. $\frac{23}{10}x + 1$
19. -6
20. $-2x + \frac{1}{2}$

ANSWERS 581

CHAPTER 1
FEEDBACK UNIT 5

1. $\frac{7}{3}$
2. $\frac{-13}{33}$
3. $\frac{3}{40}$
4. $\frac{24}{7}$
5. $\frac{14}{5}$
6. $\frac{-11}{3}$
7. $\frac{16}{5}$
8. $\frac{11}{2}$
9. 4
10. 10
11. 7
12. 3
13. 5
14. 3
15. -6
16. $\frac{13}{3}$
17. $\frac{1}{3}; \frac{1}{6} + \frac{2}{3} = \frac{5}{6}$ is true
18. $\frac{-9}{2}; \frac{-27}{8} - \frac{7}{8} = \frac{-9}{4} - 2$ is true
19. $\frac{7}{2}; \frac{7}{3} - \frac{1}{6} = \frac{7}{4} + \frac{5}{12}$ is true

CHAPTER 1
FEEDBACK UNIT 6

1. $27.97
2. $19.12
3. $1,625
4. $109.69
5. 43%

CHAPTER 1
MASTERY TEST

1. 352
2. {62, 64}
3. {2, 3, 5, 7, 11, 13}
4. 13
5. 84
6. Yes
7. 22
8. b. $2x - 7 = 5$
9. a. $x(yz) = (xy)z$
10. counting numbers
11. Yes
12. $\frac{-3}{4}$
13. -2 or $\frac{-2}{1}$
14. Yes
15. 3(xy)
16. 5x + 4
17. -3x + 20
18. $\frac{1}{24}x - \frac{2}{3}$
19. $\frac{9}{4}$
20. $\frac{-18}{5}$
21. $55.93
22. $59.57

CHAPTER 2
OBJECTIVES

1. 7^{13}
2. 5^{28}
3. 3^6
4. Yes
5. 7
6. 13
7. -4
8. 9^7
9. z^4
10. $\sqrt{35}$
11. $4\sqrt{5}$
12. $2\sqrt[4]{42}$
13. $4\sqrt[5]{4x^3}$
14. $6\sqrt{10} - 5$
15. $12 - 12\sqrt{6} - 2\sqrt{10} + 4\sqrt{15}$
16. $2\sqrt{21}$
17. $3\sqrt{10}$
18. $4\sqrt{6}$
19. 3
20. $4\sqrt[3]{2}$
21. $5\sqrt[8]{7}$
22. $x\sqrt[5]{x^4}$
23. $8\sqrt{6} - 6\sqrt{5}$
24. 2
25. -25
26. $-16 - 3i$

Chapter 2
Feedback Unit 1

1. 216
2. 81
3. Yes
4. No
5. Yes
6. x^{15}
7. x^8
8. x^{11}
9. x^{14}
10. x^{20}
11. y^{12}
12. x^6
13. $\dfrac{1}{x^4}$
14. x

Chapter 2
Feedback Unit 2

1. -15
2. 3
3. 312
4. 21
5. 927
6. -2
7. 2
8. -5
9. x^3
10. a^5
11. z^3
12. 57^2
13. 9^5
14. 13^3
15. a^3
16. 21^{12}

Chapter 2
Feedback Unit 3

1. $42\sqrt{15}$
2. $2\sqrt{70}$
3. $-15\sqrt{30}$
4. $3\sqrt[3]{42}$
5. $2\sqrt[5]{90}$
6. $11\sqrt{3}$
7. $-5\sqrt{3} - 4\sqrt{17}$
8. $-2\sqrt{5} - \sqrt{6}$
9. $\sqrt[3]{7} - 2\sqrt[3]{3}$
10. $3\sqrt{5} - 3\sqrt{3}$
11. $2\sqrt{21} + 2\sqrt{42}$
12. $6\sqrt{15} - 5$
13. $8\sqrt[6]{3} - 6\sqrt[6]{15}$
14. $-2 - 5\sqrt{5}$
15. $\sqrt{30} + 7\sqrt{6} - 2\sqrt{5} - 14$
16. $2\sqrt{21} + 2\sqrt{30} - 5\sqrt{7} - 5\sqrt{10}$
17. $3\sqrt[7]{40} + \sqrt[7]{5} + 18\sqrt[7]{8} + 6$
18. $10\sqrt[8]{49} + 37\sqrt[8]{7} - 36$ or $10\sqrt[4]{7} + 37\sqrt[8]{7} - 36$

Chapter 2
Feedback Unit 4

1. 15
2. $7\sqrt{2}$
3. $-2\sqrt{10}$
4. $8\sqrt{2}$
5. $2\sqrt{14}$

6. 8
7. 6
8. $2\sqrt{2}$
9. $2\sqrt{7}$
10. $3\sqrt{7}$
11. $15\sqrt{2}$
12. $-2\sqrt{5}$
13. $2\sqrt{10}$
14. $4\sqrt{5}$
15. $5\sqrt[3]{3}$
16. $3\sqrt[4]{2}$
17. $2\sqrt[5]{3}$
18. $2\sqrt[5]{3}$
19. $3\sqrt[4]{3^3 \cdot 5}$
20. $7\sqrt[9]{3}$
21. $x\sqrt[5]{x^4}$
22. $x^3\sqrt[6]{x}$
23. $x^5\sqrt[7]{x^3}$
24. $10\sqrt{2}$
25. $-\sqrt{2} - \sqrt{3}$

CHAPTER 2
FEEDBACK UNIT 5

1. negative
2. $\sqrt{-2}$, i, 3i
3. -8
4. 3
5. -1
6. -3
7. 4

8. $-17 - 3i$
9. $-12i$
10. $-2 - 17i$
11. 15
12. 0
13. $18 - 19i$
14. $-14i$
15. $2 - i$

CHAPTER 2
FEEDBACK UNIT 6

1. [diagram]
2. [diagram]
3. [diagram]

4.
	Rate (R)	Time (T)	Distance (D)
airboat	18 mph	2 hours	36 miles

5.
	Rate (R)	Time (T)	Distance (D)
car	R	8 hours	8R miles

6.
	Rate (R)	Time (T)	Distance (D)
1st train	45 mph	3	135 miles
2nd train	62 mph	3	186 miles

distance apart is 186 – 135 or 51 miles

CHAPTER 2
MASTERY TEST

1. 6^{13}
2. 7^{15}
3. 5^4
4. Yes
5. 5
6. -19
7. -6

584   Answers

8. 7^8
9. y^6
10. $\sqrt{15}$
11. $-\sqrt{3}$
12. $-10\sqrt[3]{12}$
13. $-3\sqrt[6]{5x^2}$
14. $-8\sqrt{3} + 14\sqrt{2}$
15. $-3 + 4\sqrt{15}$
16. $3\sqrt{5}$
17. $9\sqrt{11}$
18. $-8\sqrt{3}$
19. 3
20. $4\sqrt[3]{3}$
21. $7\sqrt[5]{9}$
22. $x^3\sqrt[7]{x^3}$
23. $4\sqrt{7} + 38\sqrt{3}$
24. -6
25. 4
26. $1 + i$
27.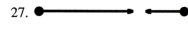
28.

Chapter 3
Objectives

1. $\dfrac{3}{5}$
2. $\dfrac{\sqrt{5}}{5}$
3. $2\sqrt{21}$
4. $\dfrac{\sqrt{55}}{10}$
5. $\dfrac{7\sqrt{2}}{2}$
6. $\dfrac{\sqrt{33}}{3}$
7. $\dfrac{\sqrt{6}}{9}$
8. $\sqrt[7]{8}$
9. $2\sqrt[3]{6}$
10. $\dfrac{\sqrt[9]{5^2}}{7}$
11. $\dfrac{4\sqrt[6]{3^4}}{9}$ or $\dfrac{4\sqrt[3]{3^2}}{9}$
12. $\dfrac{\sqrt[4]{7}}{5}$
13. $-2\sqrt[3]{3}$
14. $4\sqrt[3]{18}$
15. $\dfrac{20 + 4\sqrt{6}}{19}$
16. $-\sqrt{6} - 2\sqrt{3}$
17. $27 - 16\sqrt{3}$
18. $-8 - 3\sqrt{7}$
19. $\dfrac{-7 - \sqrt{10}}{3}$
20. $\dfrac{8 - \sqrt{15}}{7}$

Chapter 3
Feedback Unit 1

1. $\dfrac{5\sqrt{14}}{7}$
2. $\dfrac{2\sqrt{42}}{7}$
3. $\sqrt{5}$
4. $\dfrac{2\sqrt{2}}{3}$
5. $\dfrac{3\sqrt{7}}{28}$
6. $\dfrac{3\sqrt{5}}{10}$
7. $\dfrac{-2\sqrt{6}}{7}$
8. $\dfrac{\sqrt{35}}{5}$
9. $\dfrac{\sqrt{15}}{4}$
10. $\dfrac{5\sqrt{3}}{3}$

Chapter 3
Feedback Unit 2

1. $\dfrac{\sqrt[7]{3}}{2\sqrt[7]{5}}$
2. $\dfrac{2\sqrt[3]{5}}{\sqrt[3]{3}}$
3. $\dfrac{6\sqrt[5]{2}}{11y\sqrt[5]{3^2 y^2}}$
4. $\dfrac{2x\sqrt[7]{3x^3}}{3\sqrt[7]{10}}$

ANSWERS 585

5. $\dfrac{5x^2}{6}$

6. $\dfrac{\sqrt[5]{4}}{y\sqrt[5]{3x^2}}$

7. $\dfrac{1}{3}$

8. $\dfrac{x\sqrt[4]{x}}{\sqrt[4]{3}}$

3. $-\sqrt{21} + 2\sqrt{7}$

4. $\dfrac{21 - 8\sqrt{5}}{11}$

5. $-11 + 8\sqrt{2}$

6. $\dfrac{10 - 3\sqrt{6}}{23}$

7. $\dfrac{-19 + 3\sqrt{35}}{2}$

8. $\dfrac{23 - \sqrt{21}}{127}$

CHAPTER 3
FEEDBACK UNIT 3

1. $\dfrac{\sqrt[6]{10 \cdot 3^4}}{3}$

2. $\dfrac{\sqrt[3]{51}}{3}$

3. $\dfrac{\sqrt[5]{2^3 \cdot 3}}{4}$

4. $\dfrac{4\sqrt[7]{2 \cdot 5^6}}{35}$

5. 2

6. $\dfrac{2\sqrt[5]{2^4 \cdot 3}}{9y^2}$

7. $\dfrac{3\sqrt[4]{5^3}}{25}$

8. $\dfrac{-\sqrt[7]{3 \cdot 7^6}}{3}$

CHAPTER 3
FEEDBACK UNIT 4

1. $-5 - 5\sqrt{2}$

2. $\dfrac{13 + 3\sqrt{5}}{31}$

CHAPTER 3
FEEDBACK UNIT 5

1.
word/phrase/sentence	translation
a number	N
a number increased by 14	N + 14
A number increased by 14 is 34.	N + 14 = 34

2.
word/phrase/sentence	translation
a number	N
15 less than a number	N − 15
15 less than a number is 53.	N − 15 = 53

3.
word/phrase/sentence	translation
a number	N
7 multiplied by a number	7N
7 multiplied by a number is 56.	7N = 56

4.
word/phrase/sentence	translation
a number	N
3 times a number	3N
19 more than 3 times a number	3N + 19
19 more than 3 times a number is 52.	3N + 19 = 52

5.
word/phrase/sentence	translation
a number	N
a number times 11	11N
17 decreased by a number times 11	17 − 11N
17 decreased by a number times 11 is -16.	17 − 11N = −16

Chapter 3 Mastery Test

1. $\dfrac{7}{4}$
2. $\dfrac{\sqrt{5}}{5}$
3. $4\sqrt{26}$
4. $\dfrac{\sqrt{42}}{21}$
5. 5
6. $\dfrac{\sqrt{70}}{5}$
7. $\dfrac{\sqrt{15}}{7}$
8. $\sqrt[7]{7}$
9. $\dfrac{7\sqrt[3]{17}}{3}$
10. $\dfrac{\sqrt[9]{3^2}}{5}$
11. $\dfrac{\sqrt[6]{5^5}}{15}$
12. $\dfrac{2\sqrt[4]{3}}{5}$
13. $-4\sqrt[3]{3}$
14. $\dfrac{7\sqrt[3]{3}}{2}$
15. $\dfrac{8\sqrt{6}+20}{19}$
16. $\dfrac{10\sqrt{2}+3\sqrt{15}}{13}$
17. $\dfrac{31-17\sqrt{5}}{44}$
18. $\dfrac{-43+30\sqrt{2}}{7}$
19. $\dfrac{-73-7\sqrt{21}}{172}$
20. $\dfrac{-31-11\sqrt{3}}{46}$

21.

word/phrase/sentence	translation
a number	N
4 times a number	4N
17 increased by 4 times a number	4N + 17
17 increased by 4 times a number is 77.	4N + 17 = 77

22.

word/phrase/sentence	translation
a number	N
twice a number	2N
twice a number decreased by 12	2N – 12
Twice a number decreased by 12 is 46.	2N – 12 = 46

Chapter 4 Objectives

1.
    a. $5^{\frac{4}{3}}$
    b. $7^{\frac{1}{6}}$
    c. $2^{\frac{7}{4}}$
    d. $11^{\frac{5}{2}}$
2.
    a. $\sqrt{17}$
    b. $\sqrt[4]{5^3}$
    c. $\sqrt[5]{(-2)^3}$
    d. $\sqrt[3]{6^7}$
3.
    a. $11^{\frac{3}{4}}$
    b. No
    c. $3^{\frac{5}{4}}$
    d. $(-3)^{\frac{1}{3}}$
4. $6^{\frac{17}{15}}$
5. $y^{\frac{29}{24}}$
6. $\sqrt[15]{x^{29}}$
7. $\sqrt[35]{x^{17}}$
8. $x^{\frac{20}{21}}$
9. $x^{\frac{3}{10}}$
10. $\sqrt[21]{x^8}$
11. $\sqrt[8]{x^{11}}$
12. $x^{\frac{1}{4}}$
13. $x^{\frac{3}{14}}$
14. x^{10}
15. $\sqrt[3]{x^{20}}$

Answers

Chapter 4
Feedback Unit 1

1. a. $5^{\frac{2}{3}}$
   b. $11^{\frac{5}{7}}$
   c. $7^{\frac{3}{2}}$
   d. $3^{\frac{4}{5}}$
2. a. $\sqrt[7]{2^5}$
   b. $\sqrt[4]{6^7}$
   c. $\sqrt[3]{-7}$
   d. $\sqrt{3^5}$
3. a. $3^{\frac{7}{9}}$
   b. No
   c. $11^{\frac{2}{3}}$
   d. $(-5)^{\frac{2}{3}}$

Chapter 4
Feedback Unit 2

1. 3^{11}
2. 5^{-9}
3. x^{-3}
4. y^4
5. $2^{\frac{4}{5}}$
6. $7^{\frac{3}{13}}$
7. $x^{\frac{5}{6}}$
8. $y^{\frac{11}{14}}$
9. $x^{\frac{3}{2}}$
10. $y^{\frac{8}{5}}$
11. $\sqrt[15]{x^{11}}$
12. $\sqrt[14]{x^{13}}$
13. $z\sqrt[8]{z^3}$
14. $x^{18}\sqrt{x^5}$

Chapter 4
Feedback Unit 3

1. $x^{\frac{5}{12}}$
2. $x^{\frac{1}{12}}$
3. $x^{\frac{3}{20}}$
4. $x^{\frac{7}{4}}$
5. $x^{\frac{33}{20}}$
6. $x^{\frac{-11}{24}}$
7. $x\sqrt[35]{x^{11}}$
8. $\sqrt[9]{x^5}$
9. $\sqrt[40]{x^7}$
10. $x^2\sqrt[6]{x}$

Chapter 4
Feedback Unit 4

1. $x^{\frac{15}{28}}$
2. $x^{\frac{1}{5}}$
3. $x^{\frac{1}{12}}$
4. $x^{\frac{9}{2}}$
5. $x^{\frac{3}{2}}$
6. x
7. $y^{\frac{8}{3}}$
8. $z^{\frac{3}{2}}$
9. x^6
10. $y^6\sqrt[3]{y^2}$
11. x^5
12. w^3

Chapter 4
Feedback Unit 5

1.
word/phrase/sentence	translation
a number	N
a number increased by 11	N + 11
A number increased by 11 equals 48.	N + 11 = 48

    N = 37 and it is true that: 37 increased by 11 equals 48.

2.
word/phrase/sentence	translation
a number	N
15 less than a number	N – 15
15 less than a number is 17.	N – 15 = 17

    N = 32 and it is true that: 15 less than 32 is 17.

3.
word/phrase/sentence	translation
a number	N
7 multiplied by a number	7N
7 multiplied by a number is 56.	7N = 56

    N = 8 and it is true that: 7 multiplied by 8 is 56.

4.
word/phrase/sentence	translation
a number	N
twice a number	2N
19 more than twice a number	2N + 19
19 more than twice a number equals 27.	2N + 19 = 27

    N = 4 and it is true that: 19 more than twice 4 equals 27.

5.
word/phrase/sentence	translation
a number	N
the product of a number and 9	9N
4 less than the product of a number and 9	9N – 4
4 less than the product of a number and 9 is 59.	9N – 4 = 59

    N = 7 and it is true that: 4 less than the product of 7 and 9 is 59.

Chapter 4
Mastery Test

1. a. $11^{\frac{1}{6}}$
   b. $7^{\frac{5}{2}}$
   c. $6^{\frac{3}{7}}$
   d. $(-7)^{\frac{4}{9}}$

2. a. $\sqrt[3]{13^7} = 13^2 \sqrt[3]{13}$
   b. $\sqrt{2}$
   c. $\sqrt[6]{(-3)^2}$
   d. $\sqrt[7]{5^6}$

3. a. $17^{\frac{2}{3}}$
   b. $(-7)^{\frac{2}{5}}$
   c. $2^{\frac{4}{3}}$
   d. No

4. $3^{\frac{59}{40}}$
5. y
6. $\sqrt[10]{x^7}$
7. $x\sqrt[12]{x^5}$
8. $x^{\frac{11}{20}}$
9. $x^{\frac{11}{24}}$
10. $x\sqrt[8]{x}$
11. $x\sqrt[10]{x}$
12. $x^{\frac{2}{5}}$
13. $x^{\frac{3}{2}}$
14. x^8
15. $x^6\sqrt[3]{x^2}$

16.

word/phrase/sentence	translation
a number	N
3 times a number	3N
13 more than 3 times a number	3N + 13
13 more than 3 times a number equals 34.	3N + 13 = 34

N = 7 and it is true that: 13 more than 3 times 7 equals 34.

17.

word/phrase/sentence	translation
a number	N
the product of a number and 8	8N
7 less than the product of a number and 8	8N – 7
7 less than the product of a number and 8 is 113.	8N – 7 = 113

N = 15 and it is true that: 7 less than the product of 15 and 8 is 113.

CHAPTER 5
OBJECTIVES

1. $-5x^3 - x^2 - 2x + 7$
2. $-12x^2 + 12x$
3. $-3x^3 - 5x^2 + 8x - 15$
4. $-7x^3 - x^2 - 4x - 4$
5. $10x^2 - x - 1$
6. $4x^2 + 2x - 1$
7. $10x^3 + x^2 + 10x - 16$
8. $5x^3 - 2x^2 + 6x - 1$
9. $15x^3 - 5x^2 + 5x$
10. $4x^3 - 25x$
11. $x^2 + 5x - 14$
12. $kr - kv - sr + sv$
13. $6x^2 - x - 1$
14. $25x^2 - 70x + 49$
15. $9x^2 - 4$
16. $18x^2 + 3x - 28$
17. $2x^3 - 13x^2 + 22x - 3$
18. $27x^3 - 8$
19. $6x^3 - 23x^2 + 10x + 25$
20. $x + 5$ R 1
21. $3x - 1$
22. $7x + 4$ R 2
23. $2x^2 + x - 2$
24. $3x - 4$ R 12

CHAPTER 5
FEEDBACK UNIT 1

1. $-3 - x + 7x^2 - 9x^4$
2. $6x^2 - 12x$
3. $x - 4$
4. $-2x^3 + 3x^2 - 2x + 1$
5. $2x^3 + 2x^2 + 8x - 18$
6. $-6x^3 - 7x^2 - 2x - 2$
7. $4x - 12$
8. $2x^3 - 2x^2 - x - 7$
9. $4x^2 + x$
10. $-11x^2 + 4x$
11. $-5x^2 + 2x - 6$

CHAPTER 5
FEEDBACK UNIT 2

1. $4x^3 + 2x^2 - 10x$
2. $-x^3 + 9x$
3. $-21x^3 - 18x^2 + 15x$
4. $-8x^2 + 2x$
5. $x^2 + 5x - 36$
6. $6x^2 - x - 1$
7. $4x^2 + 5x - 6$
8. $16x^2 - 9$
9. $3x^2 + 14x - 24$
10. $10x^2 - 21x + 9$
11. $4x^3 - 2x^2 + 7x$
12. $28x^2 + 25x + 3$
13. $x^3 - 3x^2 - 16x + 6$
14. $2x^3 - 5x^2 - 28x + 15$
15. $36x^2 + 84x + 49$
16. $10x^3 - 15x^2 + 2x - 3$

CHAPTER 5
FEEDBACK UNIT 3

1. $x + 5$
2. $4x - 3$ R -20
3. $5x + 2$ R 1
4. $2x + 3$ R 20
5. $2x^2 - x - 1$
6. $2x + 5$ R -2
7. $3x^2 + x - 1$
8. $4x^2 - 20x + 25$

Chapter 5
Feedback Unit 4

1.

word/phrase/sentence	translation
integer	N
next consecutive integer	N + 1
the sum of the consecutive integers	N + (N + 1)
The sum of the consecutive integers is 35.	N + (N + 1) = 35

N = 17 and N + 1 = 18
It is true that: 17 and 18 are consecutive integers which have a sum of 35.

2.

word/phrase/sentence	translation
integer	N
next consecutive integer	N + 1
twice the smaller integer	2N
twice the smaller integer decreased by the larger	2N − (N + 1)
Twice the smaller integer decreased by the larger is 10.	2N − (N + 1) = 10

N = 11 and N + 1 = 12
It is true that: 11 and 12 are consecutive integers and twice 11 decreased by 12 is 10.

3.

word/phrase/sentence	translation
integer	N
next consecutive even integer	N + 2
twice the smaller integer	2N
twice the smaller integer increased by the larger	2N + (N + 2)
Twice the smaller integer increased by the larger is 44.	2N + (N + 2) = 44

N = 14 and N + 2 = 16
It is true that: 14 and 16 are consecutive even integers and twice 14 increased by 16 is 44.

4.

word/phrase/sentence	translation
integer	N
next consecutive odd integer	N + 2
three times the smaller integer	3N
twice the larger integer	2(N + 2)
Three times the smaller increased by twice the larger is 39.	3N + 2(N + 2) = 39

N = 7 and N + 2 = 9
It is true that: 7 and 9 are consecutive odd integers and 3 times 7 increased by twice 9 is 39.

5.

word/phrase/sentence	translation
integer	N
next consecutive integer	N + 1
twice the smaller	2N
twice the smaller decreased by the larger	2N − (N + 1)
Twice the smaller decreased by the larger is 42.	2N − (N + 1) = 42

N = 43 and N + 1 = 44
It is true that: 43 and 44 are consecutive integers and twice 43 decreased by 44 is 42.

6.

word/phrase/sentence	translation
integer	N
next consecutive integer	N + 1
the square of the smaller	N^2
3 times the larger	3(N + 1)
the square of the smaller decreased by 3 times the larger	$N^2 - 3(N + 1)$
The square of the smaller decreased by 3 times the larger is 25.	$N^2 - 3(N + 1) = 25$

N = 7 and N + 1 = 8 or N = -4 and N + 1 = -3
It is true that: 7 and 8 are consecutive integers and the square of 7 decreased by 3 times 8 is 25. It is also true that: -4 and -3 are consecutive integers and the square of -4 decreased by 3 times -3 is 25.

CHAPTER 5
MASTERY TEST

1. $2x^2 + 8x$
2. $-3x^2 - 4x$
3. $8x^3 + 4x - 25$
4. $2x^3 + 4x^2 + 4x - 13$
5. $-x^3 - 3x^2 + 12x - 3$
6. $-x^3 - x - 2$
7. $7x^5 - 4x^4 - x^3 - 8x^2 - 4$
8. $7x^4 - 7x^3 + 2x^2 + 8x - 2$
9. $6x^2 + 16x$
10. $10x^4 - 5x^3 + 15x^2$
11. $x^2 + 5x + 6$
12. $dm - dn + km - kn$
13. $10x^2 - 21x + 9$
14. $9x^2 - 48x + 64$
15. $4x^2 - 9$
16. $6x^2 + 13x - 5$
17. $x^3 + x^2 - x - 10$
18. $8x^3 + 2x^2 - 13x + 14$
19. $8x^3 - 27$
20. $x + 4$
21. $x - 7$
22. $2x + 5$ R -1
23. $2x^2 + x - 1$
24. $4x^2 + 10x + 25$

25.

word/phrase/sentence	translation
integer	N
next consecutive integer	N + 1
the sum of the two integers	N + (N + 1)
The sum of two consecutive integers is 85.	N + (N + 1) = 85

N = 42 and N + 1 = 43
It is true that: 42 and 43 are consecutive integers and the sum of 42 and 43 is 85.

26.

word/phrase/sentence	translation
integer	N
next consecutive odd integer	N + 2
twice the larger	2(N + 2)
twice the larger increased by the smaller	2(N + 2) + N
Twice the larger increased by the smaller is 145.	2(N + 2) + N = 145

N = 47 and N + 2 = 49
It is true that: 47 and 49 are consecutive odd integers and twice 49 increased by 47 is 145.

CHAPTER 6
OBJECTIVES

1. $-(8x - 9)$
2. $7(x - 2)$
3. $3(2x^2 + x - 5)$
4. $2(2x^2 + x - 7)$
5. $(3x - 7)(x - 1)$
6. $(x - 3)(6x^2 - 1)$
7. $(3x - 1)(2x^2 - 5)$
8. $(x - 8)(x + 7)$
9. $(x + 8y)(x + 6y)$
10. $(x - 6y)^2$
11. $(x - 20y)(x + 2y)$
12. $(x + 21y)(x - 3y)$
13. $(x^3 + 3y)(x^3 - 3y)$
14. $(x - y)(7x + 10y)$
15. $(3x - 4y)(3x + 4y)$
16. $(x + 5y)(6x - y)$
17. $(2x + 3y)^2$
18. $(9x^2 + 5y^4)(9x^2 - 5y^4)$
19. $(2x^2 + y^3)(2x^2 - y^3)$
20. $(x + 6)(x^2 - 6x + 36)$
21. $(x - 5)(x^2 + 5x + 25)$
22. $(3x + 4y^4)(9x^2 - 12xy^4 + 16y^8)$
23. $(x^3 - 4y)(x^6 + 4x^3y + 16y^2)$
24. $(a - b - 13)(a - b + 2)$
25. $(x - y - w)(x - 8y - 8w)$
26. $(x + y + 7)(x - y - 7)$
27. $(3x + y + z)(3x + y - z)$
28. $(x + y + 10z)(x - y - 10z)$
29. $2(x - y)(x - 10y)$
30. $2(x - 5y)(2x - y)$
31. $(k - r)(k^2 + kr + r^2)(k + r)(k^2 - kr + r^2)$
32. $(m^6 + m^3 + 1)(m - 1)(m^2 + m + 1)$
33. $5(2x + 3y)(2x - 3y)$
34. $7(x - 4y)(x + y)$
35. $10(1 + 2x - 2y)(1 - 2x + 2y)$

Chapter 6
Feedback Unit 1

1. $2(3x - 5)$
2. $5(3x + 4)$
3. $7x(x - 3)$
4. $6xy(2x + 3y)$
5. $-9(2x + 1)$
6. $7x^2yz(5xz - 3y)$
7. $-xw^3(x^2w - 5)$
8. $3x(x^2 - 3x + 4)$
9. $x^2(x^2 - 3x - 7)$
10. $3(x^2 - 5x - 4)$
11. $7x(2x^2 - x + 5)$
12. $3(3x^2 + 2x + 7)$
13. $4xy(2x + 7 - y)$
14. $(2x + 1)(5x - 3)$
15. $(x + 3)(7x + 4)$
16. $(3x - 2)(4x - 1)$
17. $(x - 2)(3x^2 + 4)$
18. $(x - 3)(5x^2 - 1)$
19. $(r - t)(s + x)$
20. $(x + b)(t - w)$

Chapter 6
Feedback Unit 2

1. $(x - 3)(x + 1)$
2. $(x + 6)(x - 3)$
3. $(x - 10)(x - 3)$
4. $(x - 9)(x + 8)$
5. $(x + 7)(x + 4)$
6. $(x + 7)(x - 3)$
7. $(x - 2y)(x - y)$
8. $(x + 8y)(x + 3y)$
9. $(x - 7y)(x + 6y)$
10. $(x + 4y)(x + y)$
11. $(x - 7y)(x + 2y)$
12. $(x + 5y)(x - 5y)$
13. $(x - 6y^2)(x + 6y^2)$
14. $(x^2 + 9y^3)(x^2 - 9y^3)$
15. $(x^4 - 2y)(x^4 + 2y)$
16. $2(x - 7)(x + 1)$
17. $5(x + 5)(x - 2)$
18. $3x(x - 4)(x - 2)$
19. $2xy(x - 4)(x + 1)$
20. $2y(x - 9)(x + 4)$

Chapter 6
Feedback Unit 3

1. $(3x + 2)(5x - 1)$
2. prime
3. $(x + 3)(4x - 3)$
4. $(b - c)^2$
5. prime
6. $(x + 3)(3x + 1)$
7. $(x + 5)(7x - 2)$
8. prime
9. $(10x + 3z)(10x - 3z)$
10. $(2x + 3)(x - 1)$
11. prime
12. $(3x + 7)(2x - 3)$
13. $(5x - 2)(x + 4)$
14. $(5x + 6z)^2$
15. $(9 - d)(9 + d)$
16. $(4x + 5)(3x + 4)$

Chapter 6
Feedback Unit 4

1. $(x + 1)(x^2 - x + 1)$
2. $(a^2 - 2)(a^4 + 2a^2 + 4)$
3. $(3z - 1)(9z^2 + 3z + 1)$
4. $(2a + 3b)(4a^2 - 6ab + 9b^2)$
5. $(4x^2 - 3y)(16x^4 + 12x^2y + 9y^2)$
6. $(b + a)(b^2 - ab + a^2)$
7. $(10z - 3x^2)(100z^2 + 30zx^2 + 9x^4)$
8. $(xy + z)(x^2y^2 - xyz + z^2)$

9. $(2ab^2 - 1)(4a^2b^4 + 2ab^2 + 1)$
10. $(m^2 - 4rs^2)(m^4 + 4m^2rs^2 + 16r^2s^4)$
11. $(x + y)(x^2 - xy + y^2)$
12. $(x - y)(x^2 + xy + y^2)$

Chapter 6
Feedback Unit 5

1. $5xy^2(2xy - 3)$
2. $(x + 7)(x - 3)$
3. $(x + 5)(3x - 2)$
4. $(x - 5)(x + 4)$
5. $(3x + 2)(2x - 1)$
6. $7a^2b^2(7a^2b + 4)$
7. $(2x - 3y^2)(2x + 3y^2)$
8. $(x + y - 5)(x + y + 3)$
9. $(1 - x + y)(3 + x - y)$
10. $(2x + 7)(5x - 1)$
11. $(x + 9)(x - 8)$
12. $(x + y)^2$
13. $(2a^2 - 5b)(2a^2 + 5b)$
14. $(a - b)^2$
15. $3x^2w^3(3w^2 + 2)$
16. $(x + 17)(x - 1)$
17. $(a - b + 6)(a - b - 1)$
18. $(y + 2w)(4y - 3w)$
19. $(2x + 1 + z)(2x + 1 - z)$
20. $(1 + a - 5b)(1 - a + 5b)$

Chapter 6
Feedback Unit 6

1. $(x + 5)(x - 3)$
2. $3(x - 5)(x + 1)$
3. $5(a + b)(a - b)$
4. $(2x - y)(4x^2 + 2xy + y^2)$
5. $2(3x - 2y^2)(9x^2 + 6xy^2 + 4y^4)$
6. $5x^2y^2(x + 3)$
7. $a(x^2 - 3y^2)$

8. $(a^2 + b^2)(a^4 - a^2b^2 + b^4)$
9. $4x(r^2 + s)(r^4 - r^2s + s^2)$
10. $5(x - 5)(x + 1)$
11. $(a + 5b)(a - 5b)$
12. $13a^2b^2c^2(a^2c + 4)$
13. $(a^2 + b^2)(a + b)(a - b)$
14. $3(a + b)(a^2 - ab + b^2)(a - b)(a^2 + ab + b^2)$
15. $3(3x + 4)(2x - 1)$
16. $(x + y - 3)(x + y - 1)$
17. $(x + 3)(9x + 2)$
18. $5(x - 2y)(x^2 + 2xy + 4y^2)$
19. $(x - 2)(x + 2)(x - 1)(x + 1)$
20. $(x^2 + 4)(x - 2)(x + 2)$

CHAPTER 6
FEEDBACK UNIT 7

1.

word/phrase/sentence	translation
width (of the rectangle)	W
length	W + 4
perimeter (P = 2L + 2W)	2(W + 4) + 2W
The perimeter of a rectangle is 28.	2(W + 4) + 2W = 28

W = 5 and L = 9
It is true that: The length, 9, is 4 more than the width, 5, and the perimeter of the rectangle is 28.

2.

word/phrase/sentence	translation
length (of the rectangle)	L
width	L − 2
perimeter (P = 2L + 2W)	2L + 2(L − 2)
The perimeter of a rectangle is 60.	2L + 2(L − 2) = 60

L = 16 and W = 14
It is true that: The width, 14, is 2 less than the length, 16, and the perimeter of the rectangle is 60.

3.

word/phrase/sentence	translation
width (of the rectangle)	W
length	2W
perimeter (P = 2L + 2W)	2(2W) + 2W
The perimeter of a rectangle is 42.	2(2W) + 2W = 42

W = 7 and L = 14
It is true that: The length, 14, is twice the width, 7, and the perimeter of the rectangle is 42.

4.

word/phrase/sentence	translation
width (of the rectangle)	W
length	2W + 1
perimeter (P = 2L + 2W)	2(2W + 1) + 2W
The perimeter of a rectangle is 44.	2(2W + 1) + 2W = 44

W = 7 and L = 15
It is true that: The length, 15, is 1 more than twice the width, 7, and the perimeter of the rectangle is 44.

5.

word/phrase/sentence	translation
width (of the rectangle)	W
length	3W − 1
perimeter (P = 2L + 2W)	2(3W − 1) + 2W
The perimeter of a rectangle is 22.	2(3W − 1) + 2W = 22

W = 3 and L = 8
It is true that: The length, 8, is 1 less than 3 times the width, 3, and the perimeter of the rectangle is 22.

6.

word/phrase/sentence	translation
width (of the rectangle)	W
length	W + 2
area (A = LW)	W(W + 2)
The area of a rectangle is 63.	W(W + 2) = 63

W = 7 and L = 9
It is true that: The length, 9, is 2 more than the width, 7, and the area of the rectangle is 63.

Chapter 6
Mastery Test

1. $-5x(7x^2 + 4)$
2. $-9(3x - 1)$
3. $x^2y^2(19x^2 - 3x + 7)$
4. $(2x - 1)(6x - 5)$
5. $(5a + 2)(12a - 1)$
6. $(y - z)(x + b)$
7. $(x - 5)(x + 7)$
8. $(a - 10)(a + 2)$
9. $(x - 4)(x - 1)$
10. $(b - 17c)(b - 2c)$
11. $(x + 7y)(x - 3y)$
12. $(c + 5d)(c - 5d)$
13. $(x + 9y)^2$
14. $(5x - 3)(x + 2)$
15. $(x + 3)(4x + 3)$
16. $(4x + 3y)(3x - 2y)$
17. $(2a - 3b)(5a + 2b)$
18. $(10x + 7)(10x - 7)$
19. $(8x - 3)^2$
20. $(y + 1)(y^2 - y + 1)$
21. $(x - 2)(x^2 + 2x + 4)$
22. $(x - 3y^2)(x^2 + 3xy^2 + 9y^4)$
23. $(5x + 4)(25x^2 - 20x + 16)$
24. $(x - 12)(x - 4)$
25. $(x - 4a - 4b)(x + 3a + 3b)$
26. $(5x^3 + r - s)(5x^3 - r + s)$
27. $(r + s + 3)(r - s - 3)$
28. $(b + c + w)(b + c - w)$
29. $7(x - 5)(x + 2)$
30. $4(x - 2y)(x^2 + 2xy + 4y^2)$
31. $13w^5(x + y)(x - y)$
32. $(x^2 + 9)(x - 3)(x + 3)$
33. $2(3x^2 + 11x + 2)$
34. $(x - 3)(x + 3)(x - 1)(x + 1)$
35. $5(x + 3)(x + 4)$

36.

word/phrase/sentence	translation
width (of the rectangle)	W
length	W + 6
perimeter (P = 2L + 2W)	2(W + 6) + 2W
The perimeter of a rectangle is 88.	2(W + 6) + 2W = 88

W = 19 and L = 25
It is true that: The length, 25, is 6 more than the width, 19, and the perimeter of the rectangle is 88.

37.

word/phrase/sentence	translation
width (of the rectangle)	W
length	3W + 1
perimeter (P = 2L + 2W)	2(3W + 1) + 2W
The perimeter of a rectangle is 90.	2(3W + 1) + 2W = 90

W = 11 and L = 34
It is true that: The length, 34, is 1 more than 3 times the width, 11, and the perimeter of the rectangle is 90.

Answers

Chapter 7
Objectives

1. $\dfrac{6y^5}{7z}$
2. $\dfrac{-z}{2x^3 y}$
3. -6
4. $\dfrac{-x}{x-6}$
5. $\dfrac{7x+y}{6x-y}$
6. $\dfrac{2x+5y}{2x-5y}$
7. $3x + 7y$
8. $x + 2y$
9. $\dfrac{1}{3x(5x-y)}$
10. $\dfrac{-(x-4y)}{x-3y}$
11. $\dfrac{6x-11}{2x-17}$
12. $\dfrac{2x^2 - 5x - 1}{2x - 1}$
13. $\dfrac{2x^2 + 7x - 19}{(x-3)(x-2)(x+2)}$
14. $\dfrac{3x^2 + 14x + 55}{(2x+1)(2x-1)(x+7)}$
15. $\dfrac{x^2 - 2x - 14}{(x^2 + 4x + 16)(x - 4)}$
16. $\dfrac{2x^2 - 14x - 25}{(4x^2 - 6x + 9)(2x + 3)}$
17. $\dfrac{2x(x+1)}{x+5}$
18. $\dfrac{-xy}{x+y}$

Chapter 7
Feedback Unit 1

1. $2xy$
2. $\dfrac{-2x^2 y^4}{z}$
3. $\dfrac{2z^2}{5x^2 y^4}$
4. $\dfrac{-z^2}{2y^3}$
5. $\dfrac{-6z^3}{y^3}$
6. $\dfrac{1}{y}$
7. $\dfrac{-1}{2x}$
8. $\dfrac{7x-3}{7x}$
9. $\dfrac{x-3}{x+6}$
10. $\dfrac{3x-1}{2x}$
11. $\dfrac{3+z}{x^2 + 3xy + 9y^2}$
12. $\dfrac{4x^2 - 2x + 1}{2x + 1}$
13. $\dfrac{-(3+x)}{x-7}$
14. $\dfrac{x+2}{5x+7}$

Chapter 7
Feedback Unit 2

1. $\dfrac{5(x+9)}{x(2x+3)}$
2. $\dfrac{x-1}{5}$
3. $\dfrac{1}{4}$
4. $4x + 9$
5. $\dfrac{x-2}{x^2 + x + 1}$
6. $\dfrac{10x + y}{5}$
7. $\dfrac{2x-3}{x+10}$
8. -1
9. $\dfrac{x+7}{3}$
10. $\dfrac{x(x+2)}{x-9}$

Chapter 7
Feedback Unit 3

1. $\dfrac{19}{x}$
2. $\dfrac{9x-11}{2x+3}$
3. $\dfrac{-x^2 + 8x - 1}{x^2 - 17}$
4. $\dfrac{2x^2 + x + 26}{(3x+4)(x-7)}$
5. $\dfrac{2}{(x-5)(x-3)}$
6. $24x^3$
7. $(x+7)(x+9)(x-4)$
8. $(3x+2)(2x+1)(2x-1)$
9. $(5x+1)^2(5x-1)$
10. $(x-4)(x^2 + 4x + 16)$

Chapter 7
Feedback Unit 4

1. $\dfrac{25xy + 24x}{80y^3}$

2. $\dfrac{9 - 28x}{8xy}$

3. $\dfrac{2x^2 + 2x + 8}{(x-1)(x-7)(x+3)}$

4. $\dfrac{4x^2 - 37x - 15}{(x+3)(x+8)(x-8)}$

5. $\dfrac{4x^2 - 20x + 9}{(x-5)(x^2+5x+25)}$

6. $\dfrac{x^2 - 16x + 18}{(2x-3)(x+3)(x-3)}$

7. $\dfrac{2}{3x-8}$

8. $\dfrac{2x}{x-2}$

Chapter 7
Feedback Unit 5

1. $\dfrac{1}{7}$

2. $\dfrac{n}{7}$

3. $\dfrac{1}{10}, \dfrac{1}{6}$

4. $\dfrac{1}{10} + \dfrac{1}{6}$

5. $\dfrac{n}{10}, \dfrac{n}{6}$

6. $\dfrac{n}{10} + \dfrac{n}{6}$

7. $\dfrac{15}{4}$ days or $3\dfrac{3}{4}$ days

Chapter 7
Mastery Test

1. $\dfrac{2y^2}{3x^2z^3}$

2. $\dfrac{3x^3z}{2y}$

3. -6

4. $\dfrac{-x}{x-5}$

5. $\dfrac{x^2+4x+16}{x-9}$

6. $\dfrac{-(5+x)}{2x+11}$

7. $\dfrac{x-1}{x-6}$

8. -1

9. $\dfrac{3x-5}{18x^2+13x+2}$

10. $\dfrac{5x-3}{7y}$

11. $\dfrac{6x-30}{x^2+x-2}$

12. $\dfrac{3x^2-8x+12}{x^3-7}$

13. $\dfrac{3x^2-3x-28}{(x-3)(x-7)(x+7)}$

14. $\dfrac{6x^2-26x+69}{(x^2-3x+9)(x+3)(x-3)}$

15. $\dfrac{17x^2-20x-7}{(x+1)(3x+2)(3x-2)}$

16. $\dfrac{-x^2+7x+11}{(2x+3)^2(x-2)}$

17. $\dfrac{3x(x-1)}{x-8}$

18. xy

19. $\dfrac{1}{10}, \dfrac{n}{10}$

20. $\dfrac{1}{10} + \dfrac{1}{12}, \dfrac{n}{10} + \dfrac{n}{12}$

Chapter 8
Objectives

1. 8

2. $\sqrt{3}$

3. $\dfrac{21}{8}$

4. $\dfrac{-16}{5}$

5. $\dfrac{8\sqrt{3}}{3}$

6. $\dfrac{11\sqrt{7}-\sqrt{35}}{7}$

7. $\dfrac{7-i}{10}$

8. 1

9. $\dfrac{28}{45}$

10. $\dfrac{-24}{7}$

11. 46

12. -11, 2

13. 9, -9

14. 24, -3

15. 2, 14

16. $\dfrac{8}{3}, \dfrac{-8}{3}$

17. $\dfrac{11}{2}$

18. 3

19. 5

20. $\dfrac{-9 \pm \sqrt{73}}{2}$

21. $5 \pm \sqrt{29}$

ANSWERS   599

22. $-2 \pm 3\sqrt{2}$
23. $-2 \pm i\sqrt{6}$
24. $4, -2 \pm 2i\sqrt{3}$
25. -13
26. $\frac{-10}{3}$
27. 8
28. -10
29. $\frac{\sqrt{5}}{3}, \frac{-\sqrt{5}}{3}$
30. $8, -8$

Chapter 8
Feedback Unit 1

1. $\frac{17}{6}$
2. $\frac{-7}{3}$
3. $\frac{6}{5}$
4. 2
5. $\frac{5}{6}$
6. $\frac{-21}{5}$
7. $\frac{-8}{3}$
8. $\frac{-12}{5}$
9. $\frac{4\sqrt{5}}{5}$
10. $\frac{8\sqrt{2} - \sqrt{14}}{2}$
11. $\frac{11\sqrt{3} + 6\sqrt{2}}{3}$
12. $\frac{20 + 37i}{29}$

Chapter 8
Feedback Unit 2

1. $\frac{13}{8}$
2. $\frac{-9}{4}$
3. 3
4. $\frac{37}{7}$
5. $\frac{19}{18}$
6. $\frac{57}{40}$
7. $\frac{-40}{49}$
8. no solution
9. $\frac{-18}{7}$
10. $\frac{22}{5}$
11. 71
12. -15

Chapter 8
Feedback Unit 3

1. $6, -1$
2. $-4, 2$
3. $0, 3, \frac{-7}{3}$
4. $-5, \frac{11}{2}$
5. $-11, 2$
6. $9, -9$
7. $2, \frac{3}{4}$
8. $0, \frac{3}{2}, \frac{-4}{3}$

9. $\frac{5}{2}$
10. $\frac{-1}{2}$
11. $5, -2$
12. -2

Chapter 8
Feedback Unit 4

1. $x^2 + 12x + 36$
2. $x^2 - 10x + 25$
3. $x^2 + 7x + \frac{49}{4}$
4. $x^2 - 8x + 16$
5. $x^2 - 9x + \frac{81}{4}$
6. $x^2 + 15x + \frac{225}{4}$
7. $x^2 + \frac{8}{5}x + \frac{16}{25}$
8. $x^2 - \frac{7}{2}x + \frac{49}{16}$

Chapter 8
Feedback Unit 5

1. $\frac{-7 \pm \sqrt{57}}{2}$
2. $\pm\sqrt{43}$
3. $\frac{5 \pm i\sqrt{23}}{4}$
4. $\frac{2 \pm \sqrt{19}}{3}$
5. $\frac{-5 \pm \sqrt{13}}{2}$
6. $\frac{4 \pm i\sqrt{2}}{2}$

600 ANSWERS

7. $-5, \dfrac{5 \pm 5i\sqrt{3}}{2}$

8. $3, \dfrac{-3 \pm 3i\sqrt{3}}{2}$

Chapter 8
Feedback Unit 6

1. -3
2. $\dfrac{9}{2}$
3. 1
4. $\dfrac{21}{2}$
5. $\dfrac{7}{5}$
6. 1
7. -9
8. no solution
9. $\pm 3\sqrt{5}$
10. $3, -3$

Chapter 8
Feedback Unit 7

1.
word/phrase/sentence	translation
1st number	F
2nd number	S
twice the 1st number	2F
the sum of twice the 1st number and the 2nd number is 4	2F + S = 4
3 times the 1st number	3F
3 times the 1st number diminished by 2nd number	3F – S
3 times the 1st number diminished by 2nd is 1.	3F – S = 1

   $F = 1$ and $S = 2$
   It is true that: The sum of twice 1 and the 2nd number is 4 and 3 times 1 diminished by 2 is 1.

2.
word/phrase/sentence	translation
1st number	F
2nd number	S
twice the 1st number	2F
3 times the 2nd number	3S
the sum of twice the 1st number and 3 times the 2nd	2F + 3S
the sum of twice the 1st number and 3 times the 2nd is 13	2F + 3S = 13
3 times the 1st number	3F
3 times the 1st number increased by the 2nd	3F + S
3 times the 1st number increased by the 2nd is 9.	3F + S = 9

   $F = 2$ and $S = 3$
   It is true that: The sum of twice 2 and 3 times 3 is 13 and 3 times 2 increased by 3 is 9.

3.
word/phrase/sentence	translation
1st number	F
2nd number	S
twice the 1st number	2F
the sum of twice the 1st number and the 2nd number	2F + S
the sum of twice the 1st number and the 2nd number is 16	2F + S = 16
3 times the 1st number	3F
the 2nd subtracted from 3 times the 1st number	3F – S
The 2nd subtracted from 3 times the 1st is 14.	3F – S = 14

   $F = 6$ and $S = 4$
   It is true that: The sum of twice 6 and 4 is 16 and 4 subtracted from 3 times 6 is 14.

4.

word/phrase/sentence	translation
1st number	F
2nd number	S
5 times the 1st number	5F
the 2nd number is 5 times the 1st number.	S = 5F
3 times the 1st number	3F
2 times the 2nd number	2S
3 times the 1st subtracted from 2 times the 2nd number	2S – 3F
3 times the 1st subtracted from 2 times the 2nd number is 14.	2S – 3F = 14

F = 2 and S = 10
It is true that: 10 is 5 times 2 and 3 times 2 subtracted from 2 times 10 is 14.

5.

word/phrase/sentence	translation
1st number	F
2nd number	S
3 less than the 2nd number	S – 3
the 1st number is 3 less than the 2nd number	F = S – 3
the sum of the numbers	F + S
The sum of the numbers is 17.	F + S = 17

F = 7 and S = 10
It is true that: 7 is 3 less than 10 and the sum of 7 and 10 is 17.

6.

word/phrase/sentence	translation
1st number	F
2nd number	S
1st number increased by the 2nd number	F + S
the 1st number increased by the 2nd number is 12	F + S = 12
1st number decreased by the 2nd number	F – S
The 1st number decreased by the 2nd number is 2.	F – S = 2

F = 7 and S = 5
It is true that: 7 increased by 5 is 12 and 7 decreased by 5 is 2.

CHAPTER 8
MASTERY TEST

1. $\frac{-19}{4}$
2. $\sqrt{3}$
3. $\frac{6}{25}$
4. $\frac{-11}{4}$
5. $2\sqrt{5}$
6. $\frac{2\sqrt{3}-\sqrt{21}}{3}$
7. $\frac{7+4i}{5}$
8. $\frac{13}{7}$
9. $\frac{21}{16}$
10. $\frac{27}{64}$
11. 85
12. 9, -7
13. 8, -8
14. 18, -4
15. 4, 7
16. $\frac{7}{5}, \frac{-7}{5}$
17. $\frac{9}{2}$
18. $2, \frac{-1}{2}$
19. 12
20. $\frac{-7 \pm \sqrt{37}}{2}$
21. $\frac{9 \pm \sqrt{93}}{2}$
22. $-2 \pm 2\sqrt{3}$
23. $\frac{-1 \pm 2i\sqrt{6}}{2}$
24. $3, \frac{-3 \pm 3i\sqrt{3}}{2}$
25. -2
26. 0
27. 9
28. -11
29. $\frac{\sqrt{3}}{2}, \frac{-\sqrt{3}}{2}$
30. $\sqrt{39}, -\sqrt{39}$

31.
word/phrase/sentence	translation
1st number	F
2nd number	S
twice the 1st number	2F
the sum of twice the 1st number and the 2nd number	2F + S
the sum of twice the 1st and the 2nd is 24	2F + S = 24
3 times the 1st number	3F
3 times the 1st number diminished by the 2nd number	3F – S
3 times the 1st diminished by the 2nd is 11.	3F – S = 11

F = 7 and S = 10
It is true that: The sum of twice 7 and 10 is 24 and 3 times 7 diminished by 10 is 11.

32.
word/phrase/sentence	translation
1st number	F
2nd number	S
5 times the 1st number	5F
the 2nd number is 5 times the 1st number	S = 5F
9 times the 1st number	9F
9 times the 1st number diminished by the 2nd number	9F – S
9 times the 1st diminished by the 2nd is 8.	9F – S = 8

F = 2 and S = 10
It is true that: 10 is 5 times 2 and 9 times 2 diminished by 10 is 8.

Answers 603

Chapter 9
Objectives

1. [number line: open circle at -4, shaded right]
2. [number line: open circle at 4, shaded right]
3. [number line: closed circle at 2, shaded right]
4. $x > \frac{19}{7}$
5. $x < -3$
6. $x \leq \frac{5}{4}$
7. $x > \frac{13}{5}$
8. $>, \geq$
9. $<, \leq$
10. $>, \geq$
11. $11, -11$
12. $4, \frac{-4}{3}$
13. $\frac{2}{5}, \frac{-6}{5}$
14. no solution
15. [number line: closed circles at -2 and 6]
16. [number line: open circles at -6 and -2]
17. [number line: closed circles at -2 and 4]
18. $x < 1$ and $x > \frac{-3}{5}$
19. all real numbers
20. $x > \frac{11}{3}$ or $x < -1$

Chapter 9
Feedback Unit 1

1. [number line: open circle at -2, shaded right]
2. [number line: closed circle at 4, shaded left]
3. [number line: closed circle at 2, shaded right]
4. [number line: open circle at 0, shaded left]
5. [number line: open circle at 2, closed circle at 4]
6. [number line: closed circles at 0 and 2]
7. [number line: closed circle at 2, open circle at 4]
8. [number line: open circle at 0, closed circle at 6]
9. [number line: closed circle at 2, open circle at 4]
10. [number line: no points]

Chapter 9
Feedback Unit 2

1. $x \leq 11$
2. $x > \frac{-7}{2}$
3. $x \leq -8$
4. $x \leq \frac{5}{7}$
5. $x > \frac{-14}{9}$
6. $x \geq 4$
7. $x \leq \frac{3}{2}$
8. $x > 0$

Chapter 9
Feedback Unit 3

1. $>, \geq$
2. $\geq, \leq$
3. $<, \leq$
4. $>, \geq$
5. $>, \geq$
6. $<, \leq$
7. $\geq, \leq$
8. $<, \leq$
9. $<, \leq$
10. $>, \geq$

Chapter 9
Feedback Unit 4

1. $5, -5$
2. $4, -8$
3. $6, 4$
4. $5, -9$
5. no solution
6. 12
7. no solution
8. $14, 0$
9. no solution
10. $3, -11$
11. $\frac{10}{3}, -2$
12. $1, -6$

Chapter 9
Feedback Unit 5

1. [number line: open circles at -2 and 4]
2. [number line: open circles at -2 and 6]
3. [number line: closed circles at 2 and 4]

4. [number line from -8 to 8, no points]
5. [number line from -8 to 8, closed points at -8 and 8]
6. [number line from -8 to 8, open points at -4 and 0, shaded between]
7. [number line from -8 to 8, closed points at -4 and 4, shaded outside]
8. [number line from -8 to 8, open points at -8 and -4, shaded between]

Chapter 9
Feedback Unit 6

1. $x > 12$ or $x < -2$
2. no solution
3. $x \geq -5$ or $x \leq -13$
4. $x \leq 7$ and $x \geq 5$
5. $x > 3$ or $x < \frac{-5}{3}$
6. $x \leq \frac{1}{2}$ and $x \geq \frac{-7}{2}$
7. all real numbers
8. $x \geq -1$ or $x \leq \frac{-5}{2}$
9. $x < \frac{4}{3}$ and $x > \frac{-4}{9}$
10. $x \leq -2$ and $x \geq \frac{-14}{3}$

Chapter 9
Feedback Unit 7

1.
word/phrase/sentence	translation
units (ones) digit	U
tens digit	T
sum of the digits	U + T
the sum of the digits is 10	U + T = 10
4 less than the tens digit	T – 4
The ones digit is 4 less than the tens digit.	U = T – 4

U = 3 and T = 7 and the number is 73.
It is true that: The sum of the digits of 73 is 10 and the ones digit, 3, is 4 less than the tens digit, 7.

2.
word/phrase/sentence	translation
units (ones) digit	U
tens digit	T
twice the ones digit	2U
tens digit is twice the ones digit	T = 2U
sum of the digits	U + T
The sum of the digits is 12.	U + T = 12

U = 4 and T = 8 and the number is 84.
It is true that: The tens digit, 8, is twice the ones digit, 4, and the sum of the digits of 84 is 12.

3.
word/phrase/sentence	translation
units (ones) digit	U
tens digit	T
sum of the digits	U + T
3 is the sum of the digits	3 = U + T
the number with ones digit U and tens digit T	10T + U
6 times the ones digit	6U
The number is 6 times the ones digit.	10T + U = 6U

U = 2 and T = 1 and the number is 12.
It is true that: The sum of the digits of 12 is 3 and the number 12 is 6 times the ones digit, 2.

ANSWERS 605

4.

word/phrase/sentence	translation
units (ones) digit	U
tens digit	T
3 more than the tens digit	T + 3
ones digit is 3 more than the tens digit	U = T + 3
the number with ones digit U and tens digit T	10T + U
sum of the digits	U + T
4 times the sum of the digits	4(U + T)
The number is 4 times the sum of the digits.	10T + U = 4(U + T)

U = 6 and T = 3 and the number is 36
It is true that: The ones digit, 6, is 3 more than the tens digit, 3, and the number, 36, is 4 times the sum of the digits.

5.

word/phrase/sentence	translation
units (ones) digit	U
tens digit	T
twice the tens digit	2T
ones digit is twice the tens digit	U = 2T
the number with ones digit U and tens digit T	10T + U
new number with the digits reversed	10U + T
36 more than the original number	(10T + U) + 36
The new number is 36 more than the original number.	10U + T = (10T + U) + 36

U = 8 and T = 4 and the number is 48
It is true that: The ones digit, 8, is twice the tens digit, 4, and the new number with digits reversed, 84, is 36 more than the original number, 48.

6.

word/phrase/sentence	translation
units (ones) digit	U
tens digit	T
3 times the tens digit	3T
ones digit is 3 times the tens digit	U = 3T
the number with ones digit U and tens digit T	10T + U
new number with the digits reversed	10U + T
54 more than the original number	(10T + U) + 54
The new number is 54 more than the original number.	10U + T = (10T + U) + 54

U = 9 and T = 3 and the number is 39
It is true that: The ones digit, 9, is 3 times the tens digit, 3, and the new number with digits reversed, 93, is 54 more than the original number, 39.

Chapter 9 Mastery Test

1. [number line: open circle at 4, arrow left through -8]
2. [number line: open circle at 4, closed arrow right from just past 4]
3. [number line: open circles at -4 and 4, arrow left from -4 and right from 4]
4. $x > 2$
5. $x < 4$
6. $x \geq \frac{3}{4}$
7. $x < \frac{5}{2}$
8. $>, \geq$
9. $<, \leq$
10. $>, \geq$
11. $8, -8$
12. $\frac{9}{4}, \frac{-3}{4}$
13. no solution
14. $6, -1$
15. [number line: closed dots at -4 and 4, arrows extending outward]
16. [number line: open circles at -8 and -4, segment between]
17. [number line: closed dots at -4 and 0, segment between]
18. $x < 7$ and $x > -2$
19. $x \geq 3$ or $x \leq -5$
20. all real numbers

21.

word/phrase/sentence	translation
units (ones) digit	U
tens digit	T
twice the ones digit	2U
1 less than twice the ones digit	2U − 1
tens digit is 1 less than twice the ones digit	T = 2U − 1
sum of the digits	T + U
sum of the digits is 11	T + U = 11

$U = 4$ and $T = 7$ and the number is 74

It is true that: The tens digit, 7, is 1 less than twice the ones digit, 4, and the sum of the digits is 11.

22.

word/phrase/sentence	translation
units (ones) digit	U
tens digit	T
twice the tens digit	2T
1 more than twice the tens digit	2T + 1
ones digit is 1 more than twice the tens digit	U = 2T + 1
the number with ones digit U and tens digit T	10T + U
new number with the digits reversed	10U + T
27 more than the original number	(10T + U) + 27
new number is 27 more than the original number	10U + T = (10T + U) + 27

$U = 5$ and $T = 2$ and the number is 25

It is true that: The ones digit, 5, is 1 more than twice the tens digit, 2, and the new number with digits reversed, 52, is 27 more than the original number, 25.

ANSWERS   607

CHAPTER 10
OBJECTIVES

1. 2, 3, 4, 5
2. Yes
3. $\frac{2}{3}$
4. Yes
5.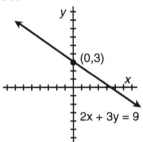
6.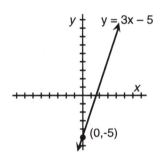
7. $\frac{-8}{7}$
8. 0
9. (0, -1)
10. $\left(0, \frac{-10}{7}\right)$
11. $\frac{-2}{5}$
12. $\frac{1}{2}$
13. $y = \frac{4}{5}x - 5$
14. $y = -3x - 9$
15. $y = \frac{5}{7}x + \frac{3}{7}$

CHAPTER 10
FEEDBACK UNIT 1

1. False
2. False
3. No
4. 0
5. 3, -3, 5
6. No
7. 0
8. Yes
9. Yes
10. Yes
11. Yes
12. No

CHAPTER 10
FEEDBACK UNIT 2

1. | x | y |
    |----|-----|
    | 6 | 16 |
    | 2 | 4 |
    | 0 | -2 |
    | -3 | -11 |
    | -5 | -17 |

2. | x | y |
    |---|---------------|
    | 8 | -1 |
    | 6 | 0 |
    | 3 | $\frac{3}{2}$ |
    | 0 | 3 |

3.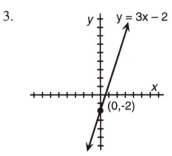

4.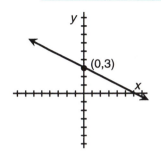

CHAPTER 10
FEEDBACK UNIT 3

1. undefined
2. 3
3. 0
4. $\frac{1}{2}$
5. 10
6. -1

CHAPTER 10
FEEDBACK UNIT 4

1. (0, 2), -3
2. (0, -1), 1
3. $\left(0, \frac{-7}{8}\right), \frac{-2}{5}$
4. (0, b), m
5. (0, -5), $\frac{-2}{3}$
6. (0, 3), 2
7. (0, -1), $\frac{-5}{3}$
8. (0, 2), -1
9. (0, -2), $\frac{2}{3}$
10. (0, -2), $\frac{3}{5}$

Chapter 10
Feedback Unit 5

1. $y = 2x + 3$
2. $y = -x + 7$
3. $y = \frac{-5}{2}x$
4. $y = x - 1$
5. $y = \frac{3}{4}x - \frac{9}{2}$
6. $y = -3$
7. $y = 6x + 7$
8. $y = -x - 1$
9. $y = \frac{-7}{9}x + \frac{1}{3}$
10. $y = -x + 1$

Chapter 10
Feedback Unit 6

1.

word/phrase/sentence	translation
number of dimes	D
number of quarters	Q
51 coins in all	$D + Q = 51$
value of dimes	.10D
value of quarters	.25Q
total value of the coins	.10D + .25Q
total value of $7.20	.10D + .25Q = 7.20

$D = 37$ and $Q = 14$
It is true that: With 37 dimes and 14 quarters there are 51 coins in all, and the total value of the coins is $7.20.

2.

word/phrase/sentence	translation
number of $1 bills	B
number of $5 bills	F
23 bills	$B + F = 23$
value of $1 bills	1B
value of $5 bills	5F
value of the bills is $51	$1B + 5F = 51$

$B = 16$ and $F = 7$
It is true that: With 16 $1 bills and 7 $5 bills there are 23 bills in all, and the total value of the bills is $51.

3.

word/phrase/sentence	translation
number of less expensive cards	C
number of deluxe cards	D
3 times as many less expensive cards as deluxe	$3D = C$
value of less expensive cards	.75C
value of deluxe cards	1.25D
value of the cards is $38,500	.75C + 1.25D = 38,500

$C = 33,000$ and $D = 11,000$
It is true that: With 33,000 less expensive cards and 11,000 deluxe cards that there are 3 times as many less expensive cards, and the total value of the cards is $38,500.

4.

word/phrase/sentence	translation
number of pennies	P
number of nickels	N
7 times as many pennies as nickels	7N = P
value of pennies	.01P
value of nickels	.05N
total value of the coins	.01P + .05N
total value of $1.44	.01P + .05N = 1.44

P = 84 and N = 12

It is true that: With 84 pennies and 12 nickels there are 7 times as many pennies as nickels, and the total value of the coins is $1.44.

Chapter 10
Mastery Test

1. 4, -6, 3, -1
2. Yes
3. $\frac{-4}{5}$
4. No
5.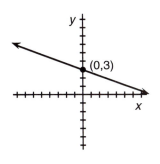
6.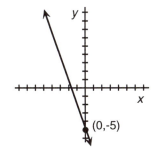
7. -3
8. $\frac{-1}{5}$
9. (0, 13)
10. $\left(0, \frac{-9}{4}\right)$
11. 13
12. $\frac{-2}{5}$
13. $y = \frac{-3}{5}x + 6$
14. $y = -4x + 22$
15. $y = -8x - 11$

610   ANSWERS

16.

word/phrase/sentence	translation
number of $1 bills	B
number of $5 bills	F
19 bills	B + F = 19
value of $1 bills	1B
value of $5 bills	5F
value of the bills is $43	1B + 5F = 43

B = 13 and F = 6
It is true that: With 13 $1 bills and 6 $5 bills there are 19 bills in all, and the total value of the bills is $43.

17.

word/phrase/sentence	translation
number of dimes	D
number of quarters	Q
9 times as many dimes as quarters	D = 9Q
value of dimes	.10D
value of quarters	.25Q
total value of the coins	.10D + .25Q
total value of $12.65	.10D + .25Q = 12.65

D = 99 and Q = 11
It is true that: With 99 dimes and 11 quarters there are 9 times as many dimes as quarters, and the total value of the coins is $12.65.

CHAPTER 11
OBJECTIVES

1. a
2. c
3. a. parabola
   b. straight line
4.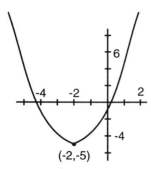

5.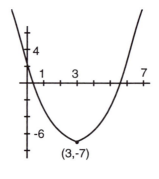

6. (4, -1), x = 4
7. (-1, 3), x = -1

8.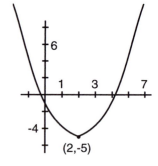

ANSWERS 611

9.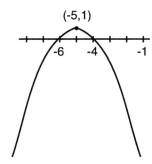

10.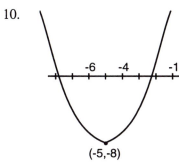

11.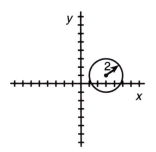

12.

13.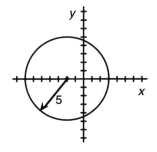

14. $(x - 2)^2 + (y + 4)^2 = 16$
15. $x^2 + (y - 2)^2 = 1$

Chapter 11
Feedback Unit 1

1. a, c, d
2. a, b, d
3. a. 13
   b. -7
   c. -15
   d. -9
4. a. 22
   b. -2
   c. -8
   d. $\dfrac{-74}{25}$

Chapter 11
Feedback Unit 2

1. parabola
2. straight line
3. parabola
4. parabola
5. straight line
6. parabola

Chapter 11
Feedback Unit 3

1.
x	y
4	2
3	-2
2	-4
1	-4
0	-2
-1	2

2.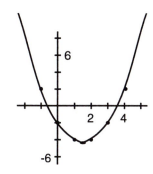

3.
x	y
-4	3
-3	6
-2	7
-1	6
0	3

4.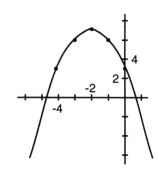

612 ANSWERS

CHAPTER 11
FEEDBACK UNIT 4

1. $(1, -3)$, $x = 1$
2.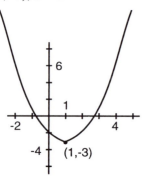

3. $(3, 5)$, $x = 3$
4.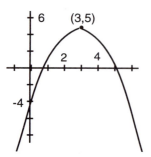

5.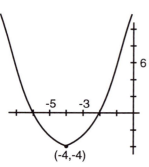

6.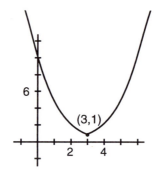

CHAPTER 11
FEEDBACK UNIT 5

1.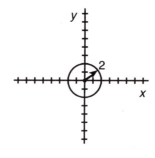

2.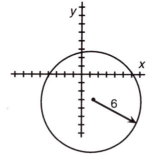

3.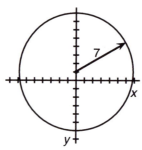

4.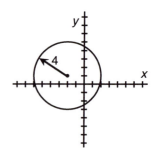

5. $x^2 + y^2 = 9$
6. $(x - 3)^2 + (y + 2)^2 = 16$

CHAPTER 11
FEEDBACK UNIT 6

1.

word/phrase/sentence	translation
the number of pounds of grain worth 8¢ per pound	C
the number of pounds of grain worth 16¢ per pound	E
the total number of pounds of grain	C + E
an 800 pound mix of grain	C + E = 800
the value of the 8¢ per pound grain	.08C
the value of the 16¢ per pound grain	.16E
the total value of the grain (800 lbs at 14¢ each)	800 • .14 = 112
the total value of the grain is $112	.08C + .16E = 112

C = 200 and E = 600
It is true that: With 200 pounds of 8¢ grain and 600 pounds of 16¢ grain, there is a mix of 800 pounds. Also, the value of the mixture is $112 which is 14¢ per pound.

2.

word/phrase/sentence	translation
the number of pounds of candy worth 55¢ per pound	C
the number of pounds of candy worth 65¢ per pound	E
the total number of pounds of candy	C + E
a 1000 pound mix of candy	C + E = 1000
the value of the 55¢ per pound candy	.55C
the value of the 65¢ per pound candy	.65E
the total value of the candy (1000 lbs at 59¢ each)	1000 • .59 = 590
the total value of the candy is $590	.55C + .65E = 590

C = 600 and E = 400
It is true that: With 600 pounds of 55¢ candy and 400 pounds of 65¢ candy, there is a mix of 1000 pounds. Also, the value of the mixture is $590 which is 59¢ per pound.

3.

word/phrase/sentence	translation
the number of gallons of alcohol worth 70¢ per gallon	C
the number of gallons of alcohol worth 95¢ per gallon	E
the total number of gallons of alcohol	C + E
120 gallons of alcohol	C + E = 120
the value of the 70¢ per gallon alcohol	.70C
the value of the 95¢ per gallon alcohol	.95E
the total value of the alcohol	$100
the total value of the candy is $100	.70C + .95E = 100

C = 56 and E = 64
It is true that: With 56 gallons of 70¢ alcohol and 64 gallons of 95¢ alcohol, there is a mix of 120 gallons. Also, the value of the mixture is $100.

614 ANSWERS

4.

word/phrase/sentence	translation
the number of pounds of red gravel	R
the number of pounds of blue gravel	B
the total number of pounds of gravel	R + B
a 100 pound mix of gravel	R + B = 100
the value of the red gravel	.50R
the value of the blue gravel	.20B
the total value of the gravel (100 lbs at 38¢ each)	100 • .38 = 38
the total value of the gravel is $38	.50R + .20B = 38

R = 60 and B = 40
It is true that: With 60 pounds of red gravel and 40 pounds of blue gravel, there is a mix of 100 pounds. Also, the value of the mixture is $38 which is 38¢ per pound.

CHAPTER 11
MASTERY TEST

1. b, c
2. a, b
3. a. straight line
   b. parabola
4.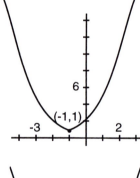
5.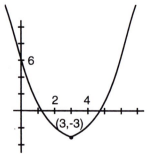
6. (5, -4), x = 5
7. (-2, 1), x = -2
8.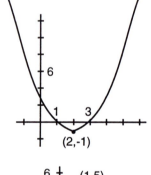
9.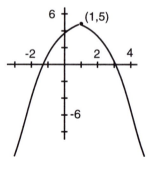
10.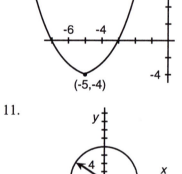
11.
12.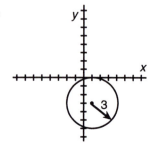

13.

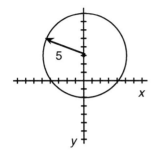

14. $(x+5)^2 + (y-3)^2 = 64$
15. $(x+8)^2 + y^2 = 25$
16.

word/phrase/sentence	translation
the number of pounds of coffee worth $3.90 per pound	C
the number of pounds of coffee worth $4.70 per pound	E
the total number of pounds of coffee	C + E
a 40 pound mix of coffee	C + E = 40
the value of the $3.90 per pound coffee	3.90C
the value of the $4.70 per pound coffee	4.70E
the total value of the coffee (40 lbs at $4.20 each)	40 • 4.20 = 168
the total value of the coffee is $168	$3.90C + $4.70E = 168

C = 25 and E = 15
It is true that: With 25 pounds of $3.90 coffee and 15 pounds of $4.70 coffee, there is a mix of 40 pounds. Also, the value of the mixture is $168 which is $4.20 per pound.

17.

word/phrase/sentence	translation
the number of gallons of gas worth $1.27 per gallon	G
the number of gallons of oil worth $2.23 per gallon	L
the total number of gallons of gas and oil	G + L
a 12 gallons mix of gas and oil	G + L = 12
the value of the $1.27 per gallon gas	1.27G
the value of the $2.23 per gallon oil	2.23L
the total value of the mixture	$17.16
the total value of the mixture is $17.16	1.27G + 2.23L = 17.16

G = 10 and L = 2
It is true that: With 10 gallons of $1.27 gas and 2 gallons of $2.23 oil, there is a mix of 12 gallons. Also, the value of the mixture is $17.16.

Chapter 12
Objectives

1. (-1, -3, 6)
2. (5, 3, 4, -2)
3. (4, -1)
4. (-4, -1)
5. (-2, 3)
6. no solution
7. (0, 0), (3, 9)
8. (2, 6), (-1, 3)
9. (2, 5)
10. (0, -4), (4, 0)
11. (3, 2, 1)
12. infinite solutions
13. infinite solutions
14. (1, 1, 2)

Chapter 12
Feedback Unit 1

1. (5, 6, -4)
2. (-5, 8, 8, -5)
3. (-1, 3, 7)
4. No
5. Yes
6. Yes
7. (-4, 2, -1, 6)

Chapter 12
Feedback Unit 2

1. (-2, 3)
2. no solution
3. (2, 5)
4. $\left(\frac{1}{3}, 5\right)$
5. no solution
6. (-13, -3)
7. $\left(\frac{3}{4}, \frac{1}{4}\right)$
8. (-3, 0)
9. (-1, 1)
10. (1, -5)

Chapter 12
Feedback Unit 3

1. (1, 3), (6, 8)
2. (8, 13), (2, 15)
3. (-2, 4), (1, 1)
4. (-3, 4)
5. (1, -1)
6. (2, -3), (-1, 6)
7. (0, 2), (2, 0)
8. (-3, 14), (-1, 6)
9. (-4, 20), (1, 0)
10. (3, 2), (0, -1)

Chapter 12
Feedback Unit 4

1. (1, 2, 3)
2. infinite solutions
3. (-1, 0, 1)
4. no solution
5. (-1, 2, -3)
6. (-2, 5, 3)
7. (-1, -3, 2)

CHAPTER 12
FEEDBACK UNIT 5

1.

word/phrase/sentence	translation
rate for 1st part of trip	F
rate for 2nd part of trip	S
he drove 10 mph faster for the 2nd part	S = F + 20
time for 1st part of trip	3
time for 2nd part of trip	4
distance (D = RT) for 1st part of trip	3F
distance (D = RT) for 2nd part of trip	4S
a 262-mile (total) trip	3F + 4S = 262

F = 26 and S = 46. His rate the first 3 hours was 26 mph.

It is true that: Mr. Morgan's rate for the 2nd part of the trip, 46 mph, was 20 mph faster than his rate the 1st part of the trip, 26. On the 1st part of his trip the distance was 3 • 26 or 78 miles and on the 2nd part of his trip the distance was 4 • 46 or 184 miles. The total distance was 262 miles.

2.

word/phrase/sentence	translation
time of trip for 1st car	F
time of trip for 2nd car	S
traveling time was equal	F = S
rate of 1st car	60
rate of 2nd car	52
distance of 1st car	60F
distance of 2nd car	52S
the difference of the distances is 44	60F − 52S = 44

$F = \frac{11}{2}$ and $S = \frac{11}{2}$. $\frac{11}{2}$ hours is $5\frac{1}{2}$ hours or 5 hours, 30 minutes.

It is true that: Going 60 mph for $5\frac{1}{2}$ hours, means the distance traveled by the 1st car is 330 miles. Going 52 mph for $5\frac{1}{2}$ hours, means the distance traveled by the 2nd car is 286 miles. The cars are 44 miles apart in $5\frac{1}{2}$ hours.

3.

word/phrase/sentence	translation
time for 1st part of trip	F
time for 2nd part of trip	S
total traveling time was 16 hours	F + S = 16
rate for 1st part of trip	55
rate for 2nd part of trip	33
distance for 1st part of trip	55F
distance for 2nd part of trip	33S
the distances were equal (round trip)	55F = 33S

F = 6 and S = 10.   Distance between the 2 towns was 330 miles.

It is true that: The time for the 1st part of trip was 6 hours and for the 2nd part 10 hours for a total of 16 hours. The distance traveled in the 1st part of trip is 6 • 55 or 330 miles and the distance traveled in the 2nd part of trip is 10 • 33 or 330 miles.

4.

word/phrase/sentence	translation
rate for 1st plane	F
rate for 2nd plane	S
1st plane flew 20 mph slower than the 2nd	F = S − 20
time for 1st plane	10
time for 2nd plane	10
distance (D = RT) for 1st plane	10F
distance (D = RT) for 2nd plane	10S
a 12,000-mile distance before they meet	10F + 10S = 12,000

F = 590 and S = 610.   The speed of the slower plane was 590 mph.

It is true that: With speeds of 590 mph and 610 mph, the 1st plane flew 20 mph slower than the 2nd. The 1st plane's distance was 10 • 590 or 5900 miles and the 2nd plane's distance was 10 • 610 or 6100 miles. The total distance was 12,000 miles.

CHAPTER 12
MASTERY TEST

1. (3, 2, -9)
2. (6, 0, -8, 4)
3. (5, 2)
4. (-1, -2)
5. (-5, 25), (3, 9)
6. (-2, 5)
7. (1, -2, 3)
8. (-1, 2, -2)
9. no solution
10. (0, 0), (2, 4)
11. (1, 5), (2, 9)
12. no solution
13. (1, -3)
14. (-2, 2, 0)

ANSWERS 619

15.

word/phrase/sentence	translation
time of trip downstream	D
time of trip upstream	U
total traveling time was 10 hours	D + U = 10
rate of trip downstream	12
rate of trip upstream	8
distance downstream	12D
distance upstream	8U
roundtrip makes the distances equal	12D = 8U

D = 4 and U = 6.

It is true that: Going downstream at 12 mph for 4 hours, means the distance from the dock to Big City is 48 miles. Going upstream at 8 mph for 6 hours, means the distance traveled was also 48 miles.

16.

word/phrase/sentence	translation
time of trip for 1st bicycler	F
time of trip for 2nd bicycler	S
traveling time was equal	F = S
rate of 1st bicycler	8
rate of 2nd bicycler	5
distance of 1st bicycler	8F
distance of 2nd bicycler	5S
the difference of the distances is 14	8F − 5S = 14

$F = \frac{14}{3}$ and $S = \frac{14}{3}$. $\frac{14}{3}$ hours is $4\frac{2}{3}$ hours or 4 hours, 40 minutes.

It is true that: Going 8 mph for $4\frac{2}{3}$ hours, means the distance traveled by the 1st bicycler is $37\frac{1}{3}$ miles. Going 5 mph for $4\frac{2}{3}$ hours, means the distance traveled by the 2nd bicycler is $23\frac{1}{3}$ miles. The bicyclers are 14 miles apart in $4\frac{2}{3}$ hours.

Index

A

Absolute value	378
Addition	
expression	167
imaginary numbers	92
nth root	70
polynomial fractions	275
Additive Property for Equivalent Equations	292
Additive Property of Inequality	362
"And" connective	358
Ascending order	169
Associative Law of Addition	17
Associative Law of Multiplication	18
Axis of symmetry	494

B

Base	49, 147
Between	3
Binary operation	15
Binomial	168

C

Cancellation	261
Center (of a circle)	503, 506
Checking solutions	37
Circle	338, 375, 503, 539
Closed dot	355
Coefficient	7, 68, 147
Common denominator	268
Common factor	4, 205, 216
Common factor method	233
Common solution	526
Commutative Law of Addition	17
Commutative Law of Multiplication	18
Completely simplified (Radical fraction)	121
Completeness Property	19
Completing the square	321-323, 498, 508
Complex fraction	266, 278
Complex numbers	290
Composite number	4
Conditional equation	8
Conjugate	128, 295
Consecutive counting numbers	2
Consecutive integers	10
Constant	168
Counting numbers	2
Cube	61
Cube root	61
Cube-root symbol	61
Cubed	48
Cubic equation	305
Cyclic process	186

D

Dense	12
Descending order	170
Difference	436
Difference of two cubes	229, 235
Difference of two squares	184, 215, 226, 234
Distance between two numbers	386
Distributive Law of Addition Over Multiplication	19

Dividend	188
Dividing	
polynomial fractions	264
power expressions	54, 152
Division sign	16
Divisor	188
Domain	414

E

Element	2, 412
Ellipse	339, 376, 541
Empty set	3
Equation	7
Equivalent	
equations	291, 536
inequalities	361
open expression	7
Evaluate	5
Even numbers	2
Exponent	48, 147

F

f(x) notation	474
Factor	3
Factored completely	207
Factored form	206
Factoring polynomials	202-252
Factors	47
Fifth root	56
First component	414
First degree equation	304
FOIL	
multiplication	180, 211
polynomial	180, 212, 218, 234
Formula for the slope	438
Fourth degree equation	305
Fourth power	48
Function	416

G

Graph	
of an equation	336
of number line inequality	355
of a linear function	421
Greater than	352
Greater than or equal to	352

H

Highest common factor (HCF)	4, 205
Hyperbola	339, 377, 542

I

Identity	8, 298
Identity Element for Addition	17
Identity Element for Multiplication	18
Imaginary number i	86
Index of the radical	64
Inequality (one variable)	355
Inequality (two variables)	368
Infinite	12
Integer factors	203
Inverse Law of Addition	17
Inverse Law of Multiplication	19
Irrational number	12

L

Least common multiple (LCM)	5, 299
Less than	352
Less than or equal to	352

Like terms	22, 70, 170
Line segment	390
Linear equation	28, 290
Linear function	419, 473

M

Member	2
Minuend	173
Minus sign	15
Monomial	168
Multiples	5
Multiplication (of power expressions)	51, 148
Multiplicative Property for Equivalent Equations	292
Multiplicative Property of Inequality	364
Multiplicity	307
Multiply	
imaginary numbers	90
nth roots	68
polynomial fractions	262

N

Negative integers	10
Next consecutive integer	10
Nth root	65
Numerical expression	5

O

Odd numbers	3
Open dot	355
Open expression	6
Opposite	13, 293
Opposite of a polynomial	257
"Or" connective	357
Order of operations	6
Ordered 4-tuple	524
Ordered 5-tuple	524
Ordered n-tuples	524
Ordered pair (x, y)	332, 412, 523
Ordered pair solution	333, 336
Ordered triple	523
Origin	421

P

Parabola	339, 376, 478, 538
Parentheses	6
Perfect cubes	50, 62, 229
Perfect fifth powers	50
Perfect fourth powers	50
Perfect square integer	59
Perfect squares	50
Polynomial	169
Polynomial equation	305
Polynomial over the integers	169
Positive integer factors	204
Power expression	51, 147
Prime (trinomial)	212
Prime number	3
Prime polynomial	54, 223

Q

Quadratic	
equation	291, 305
formula	327
function	475
Quotient	188

R

Radical sign, $\sqrt{}$	56
Radicand	56
Radius	503, 506
Raising a power to a power	53, 156
Rationalizing the denominator	106, 124
Rays	390
Real number line	9
Reciprocal	13, 293
Reduce (a fraction)	102, 114, 255
Relation	414
Remainder	188
Remove the parentheses	24
Repeating decimal	11
Replacement set	290
Reverse orientation	362
Root	291

S

Second degree equation	305
Set	412
of counting numbers	2
of integers	10
of rational numbers	11
of real numbers	9
Set selector method	413
Simplest form (of a radical)	60
Simplify	
a cube root	80
a square root	77
an inequality	363
an open expression	21
polynomial fraction	256
power multiplications	51

Slope	
of a linear function	442
through two points	437
undefined	441
Slope formula	438
Slope-intercept form	448
Solution	291
inequality with 2 variables	368
of a linear equation	31
Solve	
absolute value equation	382
absolute value inequality	392-398
Solve an equation	290, 296
Solved for y	445
Square brackets	6
Square root	56
Squared	48
Statement	7, 289
Sub-one	438
Sub-two	438
Subset	3
Substitution	237, 533
Subtraction (Polynomial Fractions)	275
Subtrahend	173
Sum and difference	184
Sum of two cubes	231, 235
Symbol for absolute value	379
Symbols of inequality	352
System of equations	525

T

Table of solutions	485
Table of square roots	58
Table of values	424
Terminating decimal	11

Terms	21
Third degree equation	305
Trinomial	168
Trinomial perfect square	183, 214, 224, 234
Truth set	37, 290

V

Variable	7
Variable factors	204
Vertex	494

X

x-axis	421
xy-plane	335, 370, 421

Y

y-axis	421
y-intercept	444

Z

Zero as a slope	440